The Legacy of Cell Fusion

The Legacy of Cell Fusion

Edited by

Siamon Gordon

Glaxo Professor of Cellular Pathology
Sir William Dunn School of Pathology,
University of Oxford

Oxford   New York   Tokyo
OXFORD UNIVERSITY PRESS
1994

Oxford University Press, Walton Street, Oxford OX2 6DP
Oxford   New York   Toronto
Delhi   Bombay   Calcutta   Madras   Karachi
Kuala Lumpur   Singapore   Hong Kong   Tokyo
Nairobi   Dar es Salaam   Cape Town
Melbourne   Auckland   Madrid
and associated companies in
Berlin   Ibadan

Oxford is a trade mark of Oxford University Press

Published in the United States
by Oxford University Press Inc., New York

© Oxford University Press, 1994

All rights reserved. No part of this publication may be
reproduced, stored in a retrieval system, or transmitted, in any form by or any
means, without the prior permission in writing of Oxford
University Press. Within the UK, exceptions are allowed in respect of any
fair dealing for the purpose of research or private study, or criticism or
review, as permitted under the Copyright, Designs and Patents Act, 1988, or
in the case of reprographic reproduction in accordance with the terms of
licences issued by the Copyright Licensing Agency. Enquiries concerning
reproduction outside those terms and in other countries should be sent to
the Rights Department, Oxford University Press, at the address above.

This book is sold subject to the condition that it shall not,
by way of trade or otherwise, be lent, re-sold, hired out, or otherwise
circulated without the publisher's prior consent in any form of binding
or cover other than that in which it is published and without a similar
condition including this condition being imposed
on the subsequent purchaser.

A catalogue record for this book is available from the British Library

Library of Congress Cataloging-in-Publication Data
The Legacy of cell fusion / edited by S. Gordon. — 1st ed.
Based on articles from a symposium held at Oxford University,
24–25 Sept. 1992.
Includes bibliographical references.
1. Cell hybridization—Congresses.   I. Gordon, Siamon.
QH451.L44   1994   574.87'612—dc20   94-4189
ISBN 0 19 854772 2

Typeset by EXPO Holdings, Malaysia
Printed in Great Britain by
Bookcraft Ltd, Midsomer Norton, Avon

Preface

The technique of Sendai virus-induced cell fusion introduced by Henry Harris, John Watkins, and their colleagues 30 years ago gave rise to a remarkable range of experimental analyses in the field of somatic cell genetics. A symposium at the Sir William Dunn School of Pathology, Oxford, where the method was developed, was held on 24 and 25 September 1992 to evaluate the legacy of cell fusion. A panel of international speakers reviewed the development and current status of nuclear, cytoplasmic, and membrane functions of heterokaryons, origins of the monoclonal antibody technique, and genetic analysis through proliferating hybrid cells. Special emphasis was placed on gene mapping, cell differentiation, and suppression of malignancy, now recognized to be an important defect in many natural human tumours. The occasion also marked the pending retirement of Henry Harris, who delivered a valedictory appraisal of 'unfinished business'. Authors were asked to place their topics in perspective so that this resultant volume would look forward, as well as to the past achievements of cell hybridization.

The symposium was made possible by generous grants from Squibb (now Bristol-Myers Squibb) and the Cancer Research Campaign, UK. I would like to thank Valerie Boasten, Pam Woodward, Dennis McMiken, Dr Mike Bramwell, and all other members of the School of Pathology who helped to organize the conference, and the staff of Oxford University Press for arranging publication. This volume is dedicated to Henry Harris, pioneer of cell fusion.

Oxford S. G.
March 1994

Contents

List of contributors	xi

I Differentiation and gene regulation

1.	Heterokaryons reveal that differentiation requires continuous regulation *Helen M. Blau*	3
2.	The instability of differentiation in hepatomas *S. J. Goss*	17
3.	Nuclear protein sorting in heterokaryons and homokaryons *Nils R. Ringertz*	31
4.	Some insights into the replication of damaged DNA in mammalian cells *R. T. Johnson, C. S. Downes, D. B. Godfrey, D. H. Hatton, A. J. Ryan, and S. Squires*	50
5.	The role of the nucleolus in the transfer of information from nucleus to cytoplasm *Masakazu Hatanaka*	68
6.	Transcription by immobile RNA polymerases *P. R. Cook*	84
7.	Lateral mobility of membrane proteins—a journey from heterokaryons to laser tweezers *Michael Edidin*	101

8. Intracellular transport of secretory proteins within intact transient heterokaryons and homokaryons — 115
Conny Valtersson, Masahiro Mizuno, Anne H. Dutton, and S. J. Singer

II Gene mapping

9. Radiation hybrid mapping: an idea whose time has finally arrived — 133
David R. Cox

10. The cloning of tumour suppressor genes — 140
David E. Housman

III Monoclonal antibodies

11. The road to monoclonal antibodies — 153
R. G. H. Cotton

IV Genetic suppression of tumour formation

12. The genetic analysis of human cancer — 169
Eric J. Stanbridge

13. Genes in control of cell proliferation and tumorigenesis in *Drosophila* — 183
Bernard M. Mechler

14. Genetics and molecular biology of tumour formation in Xiphophorus — 199
Joachim Altschmied and Manfred Schartl

15. Loss-of-function mutations in human cancer — 215
Webster K. Cavenee

16. Cancer suppression by the retinoblastoma gene — 227
Wen-Hwa Lee, David Goodrich, and Eva Lee

17. The p53 pathway, past and future *David P. Lane*	241
18. Somatic cell genetics and the search for colon cancer genes *H. J. W. Thomas, A.-M. Frischauf, and E. Solomon*	249
19. Roles of the *myc* gene in cell proliferation and differentiation, as deduced from its role in tumorigenesis *George Klein*	261
20. Genes that suppress the action of mutated *ras* genes *Makoto Noda*	270
21. Unfinished business *Henry Harris*	286
Index	297

Contributors

Joachim Altschmied
Biocenter, Department of Physiological
                Chemistry I,
University of Würzburg,
Am Hubland, D-97074 Würzburg,
Germany

Helen M. Blau
Department of Molecular
            Pharmacology,
Stanford University School of
Medicine,
Stanford,
CA 94305-5332, USA

Webster K. Cavenee
Ludwig Institute for Cancer Research
        and Department of Medicine,
Center for Molecular Genetics,
University of California at San Diego,
9500 Gilman Drive, La Jolla, CA
92093-0660, USA

P. R. Cook
Sir William Dunn School of Pathology,
University of Oxford,
South Parks Road,
Oxford OX1 3RE, UK

R. G. H. Cotton
The Murdoch Institute,
Royal Children's Hospital,
Flemington Road,
Parkville, Victoria 3052,
Australia

David R. Cox
Department of Genetics, M326,
School of Medicine,
Stanford University,
CA, USA

C. R. Downes
CRC Mammalian Cell DNA Repair
                Research Group,
Department of Zoology,
University of Cambridge,
Downing Street, Cambridge
CB2 3EJ, UK

Anne H. Dutton
Department of Biology,
University of California at San Diego,
La Jolla, CA 92093-0322,
USA

Michael Edidin
Department of Biology,
The Johns Hopkins University,
Baltimore,
MD 21218, USA

A.-M. Frischauf
Imperial Cancer Research Fund,
PO Box 123,
Lincoln's Inn Fields,
London WC2A 3PX, UK

D. B. Godfrey
CRC Mammalian Cell DNA Repair
  Research Group,
Department of Zoology,
University of Cambridge,
Downing Street, Cambridge
CB2 3EJ, UK

David Goodrich
Center for Molecular Medicine/
  Institute of Biotechnology,
University of Texas Health Science
Center at San Antonio,
15355 Lambda Drive,
San Antonio, TX 78245, USA

S. J. Goss
Sir William Dunn School of Pathology,
University of Oxford,
South Parks Road,
Oxford OX1 3RE, UK

Henry Harris
Sir William Dunn School of Pathology,
University of Oxford,
South Parks Road,
Oxford OX1 3RE, UK

Masakazu Hatanaka
Department of Molecular Virology,
Institute for Virus Research,
Kyoto University,
Sakyo-ku,
Kyoto 606, Japan

D. H. Hatton
CRC Mammalian Cell DNA Repair
  Research Group,
Department of Zoology,
University of Cambridge,
Downing Street, Cambridge
CB2 3EJ, UK

David E. Housman
Center for Cancer Research,
Massachusetts Institute of Technology,
Cambridge,
MA 02139, USA

R. T. Johnson
CRC Mammalian Cell DNA Repair
  Research Group,
Department of Zoology,
University of Cambridge,
Downing Street, Cambridge
CB2 3EJ, UK

George Klein
Department of Tumor Biology,
Karolinska Institutet, Box 60400,
S-10401 Stockholm,
Sweden

David P. Lane
CRC Cell Transformation Research
  Group,
Department of Biochemistry,
University of Dundee,
Dundee DD1 4HN, UK

Eva Lee
Center for Molecular Medicine/
  Institute of Biotechnology,
University of Texas Health Science
Center at San Antonio,
15355 Lambda Drive,
San Antonio, TX 78245, USA

Wen-Hwa Lee
Center for Molecular Medicine/
  Institute of Biotechnology,
University of Texas Health Science
Center at San Antonio,
15355 Lambda Drive,
San Antonio, TX 78245, USA

Bernard M. Mechler
Department of Developmental
                      Genetics,
Deutsches Krebsforschungszentrum,
Im Neuenheimer Feld 280,
D-69120 Heidelberg 1,
Germany

Masahiro Mizuno
Department of Biology,
University of California at San Diego,
La Jolla, CA 92093-0322, USA.
Present address: Department of
                      Geriatric Medicine,
Faculty of Medicine,
Kyoto University,
Kyoto 606, Japan

Makoto Noda
Department of Viral Oncology,
Cancer Institute,
Kami-Ikebukuro 1-37-1,
Toshima-ku, Tokyo 170,
Japan
Present address: Department of
                      Molecular Oncology,
Kyoto University Medical School,
Yoshido-Konoe-Cho,
Sakyo-ku,
Kyoto 606, Japan

Nils R. Ringertz
Department of Medical Cell Genetics,
Medical Nobel Institute,
Karolinska Institutet,
Box 60400, Stockholm 10401,
Sweden

A. J. Ryan
CRC Mammalian Cell DNA Repair
                      Research Group,
Department of Zoology,
University of Cambridge,
Downing Street, Cambridge
CB2 3EJ, UK

Manfred Schartl
Biocenter,
Department of Physiological
Chemistry I,
University of Würzburg,
Am Hubland,
D-97074 Würzburg,
Germany

S. J. Singer
Department of Biology,
University of California at San Diego,
B-022 La Jolla,
CA 92093, USA

E. Solomon,
Imperial Cancer Research Fund,
PO Box 123,
Lincoln's Inn Fields,
London WC2A 3PX, UK

S. Squires
CRC Mammalian Cell DNA Repair
                      Research Group,
Department of Zoology,
University of Cambridge,
Downing Street, Cambridge
CB2 3EJ, UK

Eric J. Stanbridge
Department of Microbiology and
                      Molecular Genetics,
California College of Medicine,
University of California, Irvine,
CA 92717, USA

H. J. W. Thomas
Imperial Cancer Research Fund,
PO Box 123,
Lincoln's Inn Fields,
London WC2A 3PX, UK.
Present address: Department of
Medicine,
St Mary's Hospital, London, UK

Conny Valtersson
Department of Biology,
University of California at San Diego,
La Jolla,
CA 92093-0322, USA.
Present address: Department of Cellular
and Neuropathology,
Karolinska Institutet, Haddinge
Sjukhus,
141 86 Huddinge,
Sweden

I

Differentiation and gene regulation

1. Heterokaryons reveal that differentiation requires continuous regulation

Helen M. Blau

Summary

As vertebrates develop, totipotent cells in the early embryo give rise to cell types specialized for function in tissues. In Waddington's epigenetic landscape, differentiated cells enter grooves, or valleys (Waddington 1940), a depiction which suggests that their fate is not easily changed. Although this appears to be the case, the mechanism by which this stability is maintained and propagated to differentiated cell progeny remains unknown. In this review I present evidence which suggests that the differentiated state requires continuous regulation. The implication of this type of control mechanism is that at any given time the balance of positive and negative regulators present in a cell is critical. To sustain threshold concentrations of key regulators and ensure stability, feedback mechanisms seem essential. Yet, to date, relatively few feedback loops with a role in differentiation have been elucidated. A genetic complementation approach appears to hold promise of identifying novel regulators capable of tipping the balance, acting in a feedback loop, and reinforcing the decision either to grow or to differentiate.

Nuclear transplantation reveals that genes are not lost during differentiation

Gurdon's experiments (Gurdon 1962) showed that the function of a highly differentiated cell nucleus could be altered when placed in another environment, such as that of an enucleated egg. Indeed, the genetic information present in intestinal nuclei sufficed to produce entire feeding tadpoles. This dramatic change in nuclear function was questioned at first, because the relatively low frequency of occurrence could have been characteristic of a subpopulation of cells, possibly residual stem cells. However, later experiments by Gurdon *et al.* (1975) and DiBerardino and Hoffner (1983) showed a marked increase in the frequency of tadpoles if nuclei were first 'conditioned' by injection into oocytes and then transplanted into enucleated eggs (DiBerardino *et al.*, 1986; for review see DiBerardino (1988)). From these experiments it was clear that genetic material is not lost or permanently inactivated during vertebrate differentiation. However, the possibility remained that the reactivation of dormant genes relied on cues only expressed during the progression from undifferentiated zygote to specialized tissue.

Heterokaryon experiments show that silent genes can be activated in differentiated cells

Somatic cell hybrid experiments showed that the differentiated state of a cell could be altered without recourse to the regulatory hierarchy characteristic of development (Blau *et al.* 1983, 1985). Critical to this discovery were heterokaryons, stable cell fusion products. Henry Harris and Nils Ringertz independently pioneered the use of interspecific heterokaryons in the study of regulation of gene expression. A key feature of heterokaryons that distinguishes them from most other hybrids is that they are stable. There is no nuclear fusion or cell division after cell fusion. Consequently, all genetic material remains intact within its own nucleus. The advantage of such short-term non-dividing fusion products is that they make possible an assessment of the influence of two or more sets of cytoplasmic and nuclear components on gene expression with minimal disruption. Thus, heterokaryons permit an analysis of changes in gene activity in the context of the whole cell (see cover photograph). Moreover, since growth and genetic selection are not required to obtain the fusion product of interest, changes in gene expression can be monitored immediately after the fusion event and at well-defined time intervals thereafter. It was with heterokaryons that Harris and coworkers made the remarkable discovery that cells of different species could be fused leading to changes in RNA and DNA synthesis and nuclear protein distribution (Harris and Watkins 1965; Harris *et al.* 1969). Ringertz observed similar changes in cross-species heterokaryons formed with chick erythrocytes; despite their 'terminal differentiation' chick erythrocyte nuclei in heterokaryons swelled, resumed RNA synthesis, and contained human nuclear proteins (Ringertz *et al.* 1971; reviewed in Ringertz and Savage (1976)). Other stable fusion experiments with intraspecific hybrids by Harris, Stanbridge, and colleagues (Harris and Klein 1969; Peehl and Stanbridge 1982; Harris 1988) revealed the prescient finding that the malignant state is often recessive to the differentiated state, a finding amply borne out by the recent discovery of more than a dozen tumour suppressor genes.

Until the report of Blau *et al.* (1983) of activation of silent muscle genes in non-muscle cells, only coexpression or extinction of genes had been reported in heterokaryons (Carlsson *et al.* 1974; Konieczny *et al.* 1983; Wright and Aronoff 1983; Lawrence and Coleman 1984). Upon fusion of muscle cells with primary cell types representing all three embryonic lineages (endoderm, ectoderm, and mesoderm), genes were activated that encoded a wide range of products including enzymes, membrane components, and contractile proteins (Fig. 1.1; Blau *et al.* 1985). Gene dosage, or the balance of regulators contributed by the two fused cell types, was critical (Pavlath and Blau 1986; Miller *et al.* 1988; Schäfer *et al.* 1990). When the number of muscle cell nuclei exceeded that of liver nuclei, extinction of liver genes and activation of muscle genes was observed; conversely, when the number of liver cell nuclei exceeded that of muscle nuclei, muscle gene expression was repressed. The success of these experiments was probably due to a combination of three features:

1. The choice of cell type: primary diploid cells, rather than aneuploid transformed cells were used.

2. Culture conditions: heterokaryons were maintained in media that promoted differentiation, not proliferation. This was important for two reasons. First, conditions that stimulate proliferation are antagonistic to muscle differentiation. Second, the heterokaryons did not divide for up to two weeks, permitting an analysis of gene expression over much longer periods of time than had previously been possible.

3. The differentiated state: the muscle cells used were multinucleated, well-differentiated myotubes and gene dosage in the resulting fusion product usually favoured muscle.

These findings were soon corroborated by others: silent myogenic genes (Wright 1984 *a,b*), haematopoietic genes (Baron and Maniatis 1986), hepatic genes (Spear and Tilghman 1990), and pancreatic genes (Wu *et al.* 1991) were activated in fibroblasts upon fusion with cell types derived from muscle, blood, liver, and pancreas, respectively.

Differentiation requires continuous regulation

Several aspects of these heterokaryon experiments suggest that gene expression in the differentiated cells of eukaryotes, like the cells of prokaryotes (Jacob and Monod 1961), is dynamic and subject to continuous regulation (for reviews see Blau and Baltimore (1991) and Blau (1992)). First, they suggest that genes are available for expression in cells that normally never express them. For instance, genes typical of mesoderm are readily activated in ectodermal cell types. Second, genes are activated in the absence of DNA replication (Chiu and Blau 1984; Blau *et al.* 1985). The frequency of gene activation did not differ when cells were continuously exposed to an inhibitor of DNA synthesis prior to and after fusion (Fig. 1.1). Thus, if activation of silent genes requires changes in chromatin structure (Holtzer *et al.* 1983; Brown 1984; Weintraub 1985), these changes are mediated by mechanisms independent of DNA synthesis. Third, cells in which differentiation is well under way are as capable of inducing the expression of previously silent tissue-specific genes as cells initiating differentiation. From this finding, it appears that the activity of *trans*-acting regulators is required not transiently at the onset of differentiation, but continuously to maintain it. Fourth, gene dosage, or the balance of regulators contributed by the two fused cell types, is a critical determinant of whether genes are repressed or activated (Fig. 1.2). Fifth, differences among cell types are observed in the kinetics, frequency, and effects of gene dosage on gene

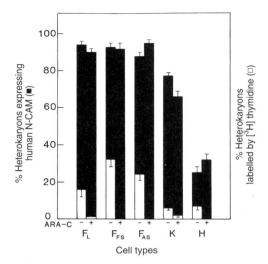

Fig. 1.1. Human muscle genes are activated in the presence or absence of DNA replication. Heterokaryons containing nuclei from muscle and either MRC5 fetal lung fibroblasts (F_L), fetal skin fibroblasts (F_{FS}), adult skin fibroblast (F_{AS}), keratinocytes (K), or HepG2 hepatocytes (H) were cultured in the presence (+) or absence (−) of cytosine arabinoside (ARA-C) and assayed for human muscle N-CAM expression by immunofluorescence (black bars), and for DNA replication by the incorporation of [^{3}H$^+$]thymidine (open bars). Reproduced with permission from Blau (1989).

activation or gene repression in heterokaryons (Fig. 1.3). This result is likely to reflect differences in the combination of proteins that interact in each type of heterokaryon. Taken together, these observations provide strong support for regulation of differentiation by continuous active control.

If differentiation is continuously regulated, the stoichiometry, or relative concentration, of positive and negative regulators present at any given time must be critical to the expression of the differentiated state. The effective concentration of a regulator is altered not only by changing its rate of synthesis or degradation, but also by altering the concentration of the proteins with which it interacts. Recent evidence indicates that many regulatory proteins from complexes: for example, heterodimers via leucine zipper or helix–loop–helix motifs (Landschulz *et al.* 1989; Murre *et al.* 1989*a*). Such interactions can either promote or inhibit the function of a regulator. For instance, the transcription factor MyoD requires E12 in order to bind DNA efficiently (Murre *et al.* 1989*b*), but is prevented from binding DNA when complexed to the protein Id (Benezra *et al.* 1990). The transcription factor NF-κB is inhibited from entering the nucleus and is therefore inactive when it is complexed to I-κB in the cytoplasm (Baeuerle and Baltimore 1988*a,b*). The complexity of these interactions increases as the number of different partners with which a protein can associate increases, as is the case in intact cells (Peterson *et al.*

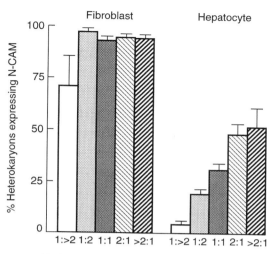

Fig. 1.2. Effect of nuclear ratio on 5.1H11 expression in heterokaryons containing different cell types. Individual heterokaryons containing nuclei from muscle and from either lung fibroblasts or hepatocytes were analysed for nuclear composition and the expression of N-CAM 6 days after fusion. Data were grouped into five ratios of muscle:non-muscle nuclei. Error bars indicate the standard error of the proportion calculated from the standard binomial equation. Reproduced with permission from Blau *et al.* (1985).

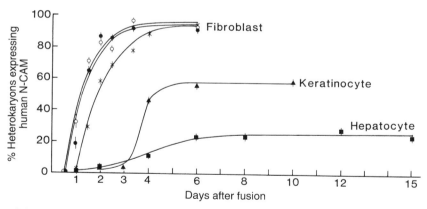

Fig. 1.3. Kinetics of expression of human muscle N-CAM in heterokaryons containing different cell types. Individual heterokaryons containing nuclei from muscle and from either MRC5 lung fibroblasts (∗), fetal skin fibroblasts (●), adult skin fibroblasts (○), keratinocytes (▲), or HepG2 hepatocytes (■) were analysed at the single cell level for nuclear composition using Hoechst 33258 to distinguish mouse and human nuclei, and for expression of human muscle N-CAM using monoclonal antibody 5.1H11 and immunofluorescence. Results for primary fetal hepatocytes were similar to the results obtained for HepG2. Reproduced with permission from Blau (1989).

1990; Schäfer *et al.* 1990). Clearly, in addition to abundance, the relative affinity and co-operative interactions of regulators at DNA binding sites will have a profound impact on gene expression. Recent evidence that synergism between diverse transcriptional regulators occurs even at concentrations at which their DNA binding sites are saturated, suggests that regulators have co-operative effects not just as heterodimers, but also as multimeric complexes (Carey *et al.* 1990; Lin *et al.* 1990). Because proteins act in combinations, small changes in the relative concentrations of a single regulator can have large effects on the expression of the cell's differentiated state, by shifting a critical balance, reaching a threshold, and setting off a cascade of subsequent events. These predictions are borne out *in vivo*. The dosage of genes encoding the helix–loop–helix proteins *daughterless*, *hairy*, and *achaete-scute* determines sex in *Drosophila* (Parkhurst *et al.* 1990). Gene dosage also determines sex in *Caenorhabditis elegans* (Hodgkin 1990) and neurosensory cells in *Drosophila* (Botas *et al.* 1982), and is responsible for several hereditary developmental disorders in man (Epstein 1986).

How are thresholds of regulators stably maintained ensuring that commitment to a differentiated state is propagated to cell progeny? Is the entire hierarchy of regulators that led to the establishment of each distinct differentiated state continuously required to maintain it? Feedback loops provide a mechanism for circumventing the regulatory hierarchy. Yet, to date few have been defined. Autoregulation of transcription factors that act on the promoters of genes that encode them can maintain protein levels in the absence of early steps. This mechanism appears to be employed by phage lambda repressor, some of the *Drosophila* homeotic selector gene products, the signal transducer c-Jun and the helix–loop–helix family of myogenic regulators (Ptashne 1986; Angel *et al.* 1988; Kuziora and McGinnis 1988; Thayer *et al.* 1989). Once activated, autoregulation of genes encoding nodal regulators could serve to maintain these proteins at a critical threshold concentration, providing both stability and memory. Another cellular feedback mechanism involves autocatalytic calcium/calmodulin dependent protein kinases which have been proposed as effective mediators of long-term storage by virtue of their multisubunit holoenzyme structure (Lisman and Goldring 1988). The extracellular matrix components secreted by differentiated cell types such as bone and cartilage act, in turn, to reinforce and stabilize those differentiated states (Greenburg and Hay 1988; Streuli *et al.* 1991). By circumventing the regulatory hierarchy, such feedback mechanisms limit the number of regulators required at any given time and play a central role in maintaining the differentiated state. In years to come novel feedback loops are likely to be discovered that operate in stabilizing the differentiated state by preventing proliferation.

Approaches to identifying novel regulators and feedback mechanisms

Currently, the most widely used approaches to cloning genes encoding mammalian tissue-specific regulators are based on biochemical purification of DNA binding proteins or on homology with known genes in other species (He *et al.* 1989; Benezra *et al.* 1990; Johnson *et al.* 1990). The biochemical approach entails the purification of proteins that bind to *cis*-acting DNA sequence elements that confer tissue-specific expression of a linked gene in transfection experiments. A determination of the amino acid sequence allows the design of oligonucleotide probes used to clone the corresponding genes (Kadonaga and Tjian 1986). It has also been possible to clone genes encoding sequence-specific DNA binding proteins by directly screening bacterial expression libraries with the DNA sequence of interest (Singh *et al.* 1988; Vinson *et al.* 1988). Such biochemical binding assays have led to the efficient isolation of a number of transcription factors, some of which have proven to be tissue-specific (Sen and Baltimore 1986; Scheidereit *et al.* 1987; Ingraham *et al.* 1988). However, in certain cases, the tissue specificity of factors is not readily apparent. Factors that appear ubiquitous by binding assays such as footprinting and mobility shift, only appear tissue-specific in a functional assay such as *in vitro* transcription (Mizusima-Sugano and Roeder 1986). Tissue specificity may be conferred on ubiquitous factors by cooperative interactions with other proteins, as documented for adenovirus E1A, for yeast Gal 80, and for chicken Coup (Yoshinaga *et al.* 1986; Ma and Ptashne 1987; Tsai *et al.* 1987; Diamond *et al.* 1990). In addition, modifications of regulatory proteins, such as phosphorylation (Ninfa *et al.* 1987; Yamamoto *et al.* 1988) are likely to play a critical role in their activity. Numerous protein kinases and phosphatases have been identified; however, they are generally promiscuous *in vitro* and identification of their true substrates may depend on knowledge of their intracellular compartmentation. Taken together, these findings underscore the importance of complementing DNA binding assays with assays based on function.

A genetic approach provides a means for cloning novel regulators based on function. This approach has been extensively exploited in the identification of hierarchies of regulatory genes in non-mammalian systems, for instance, lysogenic development in bacteriophages (Ptashne *et al.* 1980; Herskowitz 1985), certain homeotic genes in *Drosophila* (Hogness *et al.* 1985; Scott and O'Farrell 1986; Gehring 1987), vulval development in nematodes and sex determination in yeast, nematodes, and *Drosophila* (Belote *et al.* 1985; Herskowitz 1985; Ferguson *et al.* 1987; Hodgkin 1987). Since it has been possible to isolate recessive mutants in most of these cases, regulatory genes have been identified by a loss of function which could be corrected by introduction of the missing gene. However, in many cases in yeast and in *Drosophila*, regulatory genes with suppressor functions have also been isolated by genetic complementation downstream of dominant negative mutations (Surana *et al.* 1991).

In theory, a genetic approach should prove useful for identifying regulatory genes based on function in mammalian systems. This approach entails introduction of DNA from a donor cell into recipient cells which are then assayed for the heritable expression of novel gene products in tissue culture. Numerous genes encoding cytokines, cell surface receptors, or ion channels have been cloned in this manner (Kavathas *et al.* 1984; Yokata *et al.* 1984; Chao *et al.* 1986; Lubbert *et al.*

1987; Seed and Aruffo 1987). For cloning regulators of gene expression, a particularly attractive feature of this approach is that it permits identification of genes encoding proteins that act indirectly to induce tissue-specific gene expression. Unlike screens for DNA binding factors or searches for genes based on homology, this approach does not entail preconceptions regarding the nature of the regulators that will be cloned. The power of the genetic approach is the ability to clone diverse regulators including previously unidentified genes. The challenge of this approach is elucidating the role of unknown cloned genes.

In a few cases, a genetic complementation approach using DNA transfection has led to the isolation of regulatory genes in mammalian cell culture systems. In these cases, complementation was usually achieved by a gain of function, since mammalian cell mutants are unlikely to be double mutants of autosomal genes. This should not pose a problem. Given the profound effect of gene dosage on the expression of mammalian genes (Diamond et al. 1990; Blau and Baltimore 1991), it should be possible, as in other species, to clone regulatory genes with suppressor functions by rescuing dominant negative mutations. Among the genes identified by genetic complementation downstream of the primary defect are tumour suppressor genes (Kitayama et al. 1989; Schatz et al. 1989; Eiden et al. 1991), a cell cycle component (Greco et al. 1987), and a VD–J recombinase that controls immunoglobulin production (Schatz and Baltimore 1988). However, the number of mammalian genes cloned by this approach is relatively small. Critical to the success of this approach appears to be the choice of the recipient cell to be complemented. Some cells may repress or interfere with the expression of the transfected gene, whereas others may lack a combination of multiple components required to see its effect (Land et al. 1983). In these cases, complementation by a single gene product will not suffice. Another feature critical to the success of this approach is recovery of the complementing DNA, which has often been problematic. Even when successful, recovery of the genetic material often required a combination of luck and brute force. Below I present a less arbitrary means of identifying recipient cells and a method for recovering the complementing DNA.

Mutant myogenic cell lines as test cells for genetic complementation

The selection of a cell type to be complemented may be of key importance to the success of the genetic approach. Even in a well-defined test cell with a promoter–reporter gene construct that can be activated in heterokaryons, the genetic approach may fail. For instance, if more than one muscle regulatory factor is required that is absent in a recipient cell, such as a fibroblast, this approach is unlikely to work. Alternatively, the recipient cell may contain negative regulators of muscle genes that are not overridden by a single introduced muscle gene. To overcome these problems I have generated mutant myogenic cell lines. The rationale behind this approach is as follows. These cells were at one time capable of expressing the gene in question. Mutagenesis at low doses reduces the likelihood of disrupting more than one gene in a limited regulatory pathway, for instance the network of regulators mediating the activation of the α-cardiac actin promoter. Thus, the chances of complementing a defect, even a dominant one, with a single cDNA are quite high. Moreover, the complementing DNA need not correct the primary defect in the mutant, but can complement regulatory steps downstream of that defect. This approach has been amply exploited in studies of yeast regulatory genes (Surana *et al.* 1991). I propose that it can now be fruitfully exploited in the cloning of mammalian genes involved in differentiation.

Critical to this approach is the identification of mutant cells of interest. A simple screen must be designed. For our myogenic mutants this proved to be indefinite proliferation and lack of response to differentiation inducing cues in the medium. Once selected, the mRNA and protein profile of the mutant relative to the parental cell type must be assessed. This allows identification of those genes that are no longer expressed. The most useful mutants are likely to be those in which more than one myogenic gene fails to be expressed. Of greatest importance in determining that a cell line with a regulatory mutation has been obtained is an analysis of its pattern of gene expression following fusion with a wild-type muscle cell of another species. A species difference is essential in order to identify the gene products encoded by each cell type after fusion. Some mutant lines will have defects in structural genes such as actins and myosins and these must be eliminated. A second type of mutant will repress myogenic genes in wild-type cells in heterokaryons and is a source of *trans*-acting negative regulators of myogenesis (Peterson *et al.* 1990). A third type of mutant can activate its muscle genes in response to myogenic factors contributed by wild-type cells. Such mutants can be used to assay for positive regulators of myogenesis.

Recovery of the complementing DNA

A major problem with genetic complementation in mammalian cells has been recovering the complementing DNA. An approach that has been successfully used recently is to complement cells with cDNAs that remain episomal (Pan *et al.* 1992). Conventional cDNA expression vectors are problematic, since they require integration into the host cell genome in order to be retained and achieve long-term stable expression. This results both in low efficiencies of gene transfer and in difficulty in recovery of the transfected gene. To facilitate the transfer and recovery of the muscle cDNA library, a DNA fragment (EBO) containing an Epstein–Barr virus (EBV) replicon plus hygromycin resistance gene can be incorporated into the pcD vector (Margolskee *et al.* 1988). This modification produces an episomal shuttle vector (EBO-pcD) which is capable of autonomous replication. Since chromosomal integration of EBO-pcD plasmids is not required, stable transfection appears to be limited only by the ability of host cells to take up DNA. Thus, EBV-based plasmids give stable transfection frequencies that approximate transient transfection frequencies, several orders of magnitude higher than plasmids that must integrate (Yates *et al.* 1985; Kioussis *et al.* 1987; Lutfalla *et al.* 1989). Extrachromosomal replication of circular EBO-pcD plasmids also facilitates their recovery in low molecular weight DNA (Hirt 1967) and subsequent introduction into bacterial strains deficient in methylcytosine-dependent restriction of DNA allows for large-scale amplification. Using this approach, full-length cDNAs encoding neural cell adhesion molecules were readily isolated (Pan et al. 1992).

Conclusion

Regulatory feedback loops are likely to play a critical role in maintaining the necessary balance of regulators to perpetuate a stable differentiated state. The goal of the genetic approach is to clone and identify novel regulators and determine their role in initiating and maintaining myogenic and other differentiation pathways. Knowledge of such feedback mechanisms may allow intervention in disease states such as cancer where inhibition of proliferation and promotion of differentiation are desirable.

References

Angel, P., Hattori, K., Smeal, T., and Karin, M. (1988). The jun proto-oncogene is positively autoregulated by its product, jun/AP-1. *Cell*, **55**, 875–85.

Baeuerle, P. A. and Baltimore, D. (1988*a*). IκB: a specific inhibitor of the NF-κB transcription factor. *Science*, **242**, 540–6.

Baeuerle, P. A. and Baltimore, D. (1988*b*). Activation of DNA-binding activity in an apparently cytoplasmic precursor of the NF-κB transcription factor. *Cell*, **53**, 211–17.

Baron, M. H. and Maniatis, T. (1986). Rapid reprogramming of globin gene expression in transient heterokaryons. *Cell*, **46**, 591–602.

Belote, J. M., McKeown, M. B., Andrew, D. J., Scott, T. N., Wolfner, M. F., and Baker, B. S. (1985). Control of sexual differentiation in *Drosophila melanogaster*. *Cold Spring Harbor Symp. Quant. Biol., 50*, 605–14.

Benezra, R., Davis, R. L., Lockshon, D., Turner, D. L., and Weintraub, H. (1990). The protein Id: a negative regulator of the helix–loop–helix DNA binding proteins. *Cell*, **61**, 49–59.

Blau, H. M. (1992). Differentiation requires continuous active control. *Annu. Rev. Biochem.*, **61**, 1213–30.

Blau, H. M. and Baltimore, D. (1991). Differentiation requires continual regulation. *J. Cell Biol.* **112**, 781–3

Blau, H. M., Chiu, C.-P., and Webster, C. (1983). Cytoplasmic activation of human nuclear genes in stable heterokaryons. *Cell*, **32**, 1171–80

Blau, H. M., Pavlath, G. K., Hardeman, E. C., Chiu, C.-P., Silberstein, L., *et al.* (1985). Plasticity of the differentiated state. *Science*, **230**, 758–66.

Botas, J., Moscoso del Prado, J., and Garcia-Bellido, A. (1982). Gene-dose titration analysis in the search of *trans*-regulatory genes in *Drosophila*. *EMBO J.*, **1**, 307–11.

Brown, D. D. (1984). The role of stable complexes that repress and activate eucaryotic genes. *Cell*, **37**, 359–65.

Carey, M., Lin, Y.-S., Green, M. R., and Ptashne, M. (1990). A mechanism for synergistic activation of a mammalian gene by GAL4 derivatives. *Nature*, **345**, 361–4.

Carlsson, S.-A., Luger, O., Ringertz, N. R., and Savage, R. E. (1974). Phenotypic expression in chick erythrocyte × rat myoblast hybrids and in chick myoblast × rat myoblast hybrids *Exp. Cell Res.*, **84**, 47–55.

Chao, M. V., Bothwell, M. A., Ross, A. H., Koprowski, H., Lanahan, A. A., Buck, C. R., and Sehgal, A. (1986). Gene transfer and molecular cloning of the human NGF receptor. *Science*, **232**, 518–21.

Chiu, C.-P. and Blau, H. M. (1984). Reprogramming cell differentiation in the absence of DNA synthesis. *Cell*, **37**, 879–87.

Diamond, M. I., Miner, J. N., Yoshinaga, S. K., and Yamamoto, K. R. (1990). Transcription factor interactions: selectors of positive or negative regulation from a single DNA element. *Science*, **249**, 1266–72.

DiBerardino, M. A. (1988). Genomic multipotentiality of differentiated somatic cells. *Cell Diff. Dev.*, **25**, 129–36.

DiBerardino, M. A. and Hoffner, N. J. (1983). Gene reactivation in erythrocytes: nuclear transplantation in oocytes and eggs of *Rana*. *Science*, **219**, 862–4.

DiBerardino, M. A., Orr, N. H., and McKinnell, R. G. (1986). Feeding tadpoles cloned from *Rana* erythrocyte nuclei. *Proc. Natl Acad. Sci. USA*, **83**, 8231–4.

Eiden, M. V., MacArthur, L., and Okayama, H. (1991). Suppression of the chemically transformed phenotype of BHK cells by a human cDNA. *Mol. Cell. Biol.*, **11**, 5321–9.

Epstein, C. J. (ed.) (1986). The theoretical mechanisms and issues: the primary and secondary effects of aneuploidy. In *The consequences of chromosome imbalance: principles, mechanisms, and models*, pp. 65–79. Cambridge University Press.

Ferguson, E. L., Sternberg, P. W., and Horvitz, H. R. (1987). A genetic pathway for the specification of the vulval cell lineages of *Caenorhabditis elegans*. *Nature*, **326**, 259–67.

Gehring, W. J. (1987). Homeo boxes in the study of development. *Science*, **236**, 1245–52.

Greco, A., Ittman, M., and Basilico, C. (1987). Molecular cloning of a gene that is necessary for G1 progression in mammalian cells. *Proc. Natl Acad. Sci. USA*, **84**, 1565–9.

Greenburg, G. and Hay, E. D. (1988). Cytoskeleton and thyroglobulin expression change during transformation of thyroid epithelium to mesenchyme-like cells. *Development*, **102**, 605–22.

Gurdon, J. B. (1962). The developmental capacity of nuclei taken from intestinal epithelium cells of feeding tadpoles. *J. Embryol. Exp. Morph.*, **10**, 622–40.

Gurdon, J. B., Laskey, R. A., and Reeves, O. R. (1975). The developmental capacity of nuclei transplanted from keratinized skin cells of adult frogs. *J. Embryol. Exp. Morphol.*, **34**, 93–112.

Harris, H. (1988). The analysis of malignancy by cell fusion: the position of 1988. *Cancer Res.*, **48**, 3302–6.

Harris, H. and Klein, G. (1969). Malignancy of somatic cell hybrids. *Nature*, **224**, 1314–16.

Harris, H. and Watkins, J. F. (1965). Hybrid cells derived from mouse and man: artificial heterokaryons of mammalian cells from different species. *Nature*, **205**, 640–6.

Harris, H., Sidebottom, E., Grace, D. M., and Bramwell, M. E. (1969). The expression of genetic information: a study of hybrid animal cells. *J. Cell Sci.*, **4**, 499–525.

He, X., Treacy, M. N., Simmons, D. M., Ingraham, H. A., Swanson, L. W., and Rosenfeld, M. G. (1989). Expression of a large family of POU-domain regulatory genes in mammalian brain development. *Nature*, **340**, 35–42.

Herskowitz, I. (1985). Master regulatory loci in yeast and lambda. *Cold Spring Harbor Symp. Quant. Biol.*, **50**, 565–74.

Hirt, B. (1967). Selective extraction of polyoma DNA from infected mouse cultures. *J. Mol. Biol.*, **26**, 365–9.

Hodgkin, J. (1987). Sex determination and dosage compensation in *Caenorhabditis elegans*. *Annu. Rev. Genet.*, **21**, 133–54.

Hodgkin, J. (1990). Sex determination compared in *Drosophila* and *Caenorhabditis*. *Nature*, **344**, 721–8.

Hogness, D. S., Lipshitz, H. D., Beachy, P. A., Peattie, D. A., Saint, R. B., Goldschmidt-Clermont, M., *et al.* (1985). Regulation and products of the Ubx domain of the bithorax complex. *Cold Spring Harbor Symp. Quant. Biol.*, **50**, 181–94.

Holtzer, H., Biehl, J., Payette, R., Sasse, J., Pacifici, M., and Holtzer, S. (1983). Cell diversification: differing roles of cell lineages and cell–cell interactions. In *Limb development and regeneration*, (ed. R. O. Kelley, P. F. Goetinck, and J. A. MacCabe), pp. 271–80. A. R. Liss Inc., New York.

Ingraham, H. A., Chen, R., Mangalam, H. J. Elsholtz, H. P., Flynn, S. E., Lin, C. R., *et al.* (1988). A tissue-specific transcription factor containing a homeodomain specifies a pituitary phenotype. *Cell*, **55**, 519–29.

Jacob, F. and Monod, J. (1961). Genetic regulatory mechanisms in the synthesis of proteins. *J. Mol. Biol.*, **3**, 318–56.

Johnson, J. E., Birren, S. J., and Anderson, D. J. (1990). Two rat homologues of *Drosophila* achaete-scute specifically expressed in neuronal precursors. *Nature*, **346**, 858–61.

Kadonaga, J. T. and Tjian, R. (1986). Affinity purification of sequence-specific DNA binding proteins. *Proc. Natl Acad. Sci. USA*, **83**, 5889–93.

Kavathas, P., Sukhatma, V. P., Herzenberg, L. A., and Pames, J. R. (1984). Isolation of the gene encoding the human T-lymphocyte differentiation antigen Leu2 (T8) by gene transfer and cDNA subtraction. *Proc. Natl Acad. Sci. USA*, **81**, 7688–92.

Kioussis, D., Wilson, F., Daniels, C., Boyeton, C., Taverne, J., and Playfair, J. H. L. (1987). Expression and rescuing of a cloned human tumor necrosis factor gene using an EBV-based shuttle cosmid vector. *EMBO J.*, **6**, 355–61.

Kitayama, H., Sugimoto, Y., Matsuzaki, T., Ikawa, Y., and Noda, M. (1989). A *ras*-related gene with transformation suppressor activity. *Cell*, **56**, 77–84.

Konieczny, S. F., Lawrence, J. B., and Coleman, J. R. (1983). Analysis of muscle protein expression in polyethylene glycol-induced chicken: rat myoblast heterokaryons. *J. Cell Biol.*, **97**, 1348–55.

Kuziora, M. A. and McGinnis, W. (1988). Autoregulation of a *Drosophila* homeotic selector gene. *Cell*, **55**, 477–85.

Land, H., Parada, L. F., and Weinberg, R. A. (1983). Tumorigenic conversion of primary embryo fibroblasts requires at least two cooperating oncogenes. *Nature*, **304**, 596–602.

Landschulz, W. H., Johnson, P. F., and McKnight, S. L. (1989). The DNA binding domain of the rat liver nuclear protein C/EBP is bipartite. *Science*, **243**, 1681–8.

Lawrence, J. B. and Coleman, J. R. (1984). Extinction of muscle-specific proterites in somatic cell heterokaryons. *Dev. Biol.*, **101**, 463–76.

Lin, Y.-S., Carey, M., Ptashne, M., and Green, M. R. (1990). How different eukaryotic transcriptional activators can cooperate promiscuously. *Nature*, **345**, 359–61.

Lisman, J. E. and Goldring, M. A. (1988). Feasibility of long-term storage of graded information by the Ca^{2+}/calmodulin-dependent protein kinase molecules of the postsynaptic density. *Proc. Natl Acad. Sci. USA*, **85**, 5320–24.

Lubbert, H., Hoffman, B. J., Snutch, T. P., Van Dyke, T., Levine, A. J., Hartig, P. R., *et al.* (1987). cDNA cloning of a serotonin 5-HT_{1c} receptor by electrophysiological assays of mRNA-injected Xenopus oocytes. *Proc. Natl Acad. Sci. USA*, **84**, 4332–6.

Lutfalla, G., Armbruster, L., Dequin, S., and Bertolotti, R. (1989). Construction of an EBNA-producinig line of well-differentiated human hepatoma cells and of appropriate Epstein–Barr virus-based shuttle vectors. *Gene*, **76**, 27–39.

Ma, J. and Ptashne, M. (1987). The carboxy-terminal 30 amino acids of GAL4 are recognized by GAL80. *Cell*, **50**, 137–42.

Margolskee, R. F., Kavathas, P., and Berg, P. (1988). Epstein–Barr virus shuttle vector for stable episomal replication of cDNA expression libraries in human cells. *Mol. Cell. Biol.*, **8**, 2837–47.

Miller, S. C., Pavlath, G. K., Blakely, B. T., and Blau, H. M. (1988). Muscle cell components dictate hepatocyte gene expression and the distribution of the Golgi apparatus in heterokaryons. *Genes Dev.*, **2**, 330–40.

Mizushima-Sugano, J. and Roeder, R. G. (1986). Cell-type-specific transcription of an immunoglobulin κ light chain gene *in vitro*. *Proc. Natl Acad. Sci. USA*, **83**, 8511–5.

Murre, C., McCaw, P.S., and Baltimore, D. (1989a). A new DNA binding and dimerization motif in immunoglobulin enhancer binding, daughterless, MyoD, and myc proteins. *Cell*, **56**, 777–83.

Murre, C., McCaw, P. S., Vaessin, H., Caudy, M., Jan, L. Y., Jan, Y. N., *et al.* (1989b). Interactions between heterologous helix–loop–helix proteins generate complexes that bind specifically to a common DNA sequence. *Cell*, **58**, 537–44.

Ninfa, A. J., Reitzer, L. J., and Magasanik, B. (1987). Initiation of transcription of the bacterial glnAp2 promoter by purified *E.coli* components is facilitated by enhancers. *Cell*, **50**, 1039–46.

Pan, L. C., Margolskee, R. F., and Blau, H. M. (1992). Cloning muscle isoforms of neural cell adhesion molecule using an episomal shuttle vector. *Somat. Cell Mol. Genet.*, **18**, 163–77.

Parkhurst, S. M., Bopp, D., and Ish-Horowicz, D. (1990). X:A ratio, the primary sex-determining signal in *Drosophila*, is transduced by helix–loop–helix proteins. *Cell*, **63**, 1179–91.

Pavlath, G. K. and Blau, H. M. (1986). Expression of muscle genes in heterokaryons depends on gene dosage. *J. Cell Biol.*, **102**, 124–30.

Peehl, D. M. and Stanbridge, E. J. (1982). The role of differentiation in the suppression of tumorigenicity in human cell hybrids. *Int. J. Cancer*, **30**, 113–20.

Peterson, C. A., Gordon, H., Hall, Z. W., Paterson, B. M., and Blau, H. M. (1990). Negative control of helix–loop–helix family of myogenic regulators in NFB mutant. *Cell*, **62**, 493–502.

Ptashne, M. (1986). *A genetic switch: gene control and phage* λ. Cell Press and Blackwell Scientific, Cambridge, MA.

Ptashne, M., Jeffrey, A., Johnson, A. D., Maurer, R., Meyer, B. J., Pabo, C. O., *et al.* (1980). How the λ repressor and cro work. *Cell*, **19**, 1–11.

Ringertz, N. R. and Savage, R. E. (1976). *Cell hybrids*. Academic Press, New York.

Ringertz, N. R., Carlsson, S.-A, Ege, T., and Bolund, L. (1971). Detection of human and chick nuclear antigens in nuclei of chick erythrocytes during reactivation in heterokaryons with HeLa cells. *Proc. Natl Acad. Sci. USA*, **68**, 3228–2.

Schäfer, B. W., Blakely, B. T., Darlington, G. D., and Blau, H. M. (1990). Effect of cell history on response to helix–loop–helix family of myogenic regulators. *Nature*, **344**, 454–8.

Schatz, D. G. and Baltimore, D. (1988). Stable expression of immunoglobulin gene V(D)J recombinase activity by gene transfer into 3T3 fibroblasts. *Cell*, **53**, 107–15.

Schatz, D., Oettinger, M. A., and Baltimore, D. (1989). The V(D)J recombination activating gene RAG-1. *Cell*, **59**, 1035–48.

Scheidereit, C., Heguy, A., and Roeder, R. G. (1987). Identification and purification of a human lymphoid-specific octamer-binding protein (OTF-2) that activates transcription of an immunoglobulin promoter *in vitro*. *Cell*, **51**, 783–93.

Scott, M. P. and O'Farrell, P. H. (1986). Spatial programming of gene expression in early drosophila embryogenesis. *Annu. Rev. Cell Biol.*, **2**, 49–80.

Seed, B. and Aruffo, A. (1987). Molecular cloning of the CD2 antigen, the T-cell erythrocyte receptor, by a rapid immunoselection procedure. *Proc. Natl Acad. Sci. USA*, **84**, 3365–9.

Sen, R. and Baltimore, D. (1986). Multiple nuclear factors interact with the immunoglobulin enhancer sequences. *Cell*, **46**, 705–16.

Singh, H., LeBowitz, J. H., Baldwin, A. S. Jr, and Sharp, P. A. (1988). Molecular cloning of an enhancer binding protein: isolation by screening of an expression library with a recognition site DNA, *Cell*, **52**, 415–23.

Spear, B. T. and Tilghman, S. M. (1990). Role of alpha-fetoprotein regulatory elements in transcriptional activation in transient heterokaryons. *Mol. Cell. Biol.*, **10**, 5047–54.

Streuli, C. H., Bailey, N., and Bissell, M. (1991). Control of mammary epithelial differentiation: basement membrane induces tissue-specific gene expression in the absence of cell–cell interaction and morphological polarity. *J. Cell Biol.*, **115**, 1383–95.

Surana, U., Robitsch, H., Price, C., Schuster, T., Fitch, I., Futcher, A. B., and Nasmyth, K. (1991). The role of CDC28 and cyclins during mitosis in the budding yeast *S. cerevisiae*. *Cell*, **65**, 145–61.

Thayer, M. J., Tapscott, S. J., Davis, R. L., Wright, W. E., Lassar, A. B., and Weintraub, H. (1989). Positive autoregulation of the myogenic determination gene *myoD1*. *Cell*, **58**, 241–8.

Tsai, S. Y., Sagami, I., Wang, H., Tsai, M.-J., and O'Malley, B. W. (1987). Interactions between a DNA-binding transcription factor (COUP) and a non-DNA binding factor (S300-II). *Cell*, **50**, 701–9.

Vinson, C. R., LaMarco, K. L., Johnson, P. F., Landschulz, W. H., and McKnight, S. L. (1988). *In situ* detection of sequence-specific DNA binding activity specified by a recombinant bacteriophage. *Genes Dev.*, **2**, 801–6.

Waddington, C. H. (1940). *Organisers and Genes*, pp. 91–3. Cambridge University Press, London.

Weintraub, H. (1985). Assembly and propagation of repressed and derepressed chromosomal states. *Cell*, **42**, 705–11.

Wright, W. E. (1984a). Expression of differentiated functions in heterokaryons between skeletal myocytes, adrenal cells, fibroblasts, and glial cells. *Exp. Cell Res.*, **51**, 55–69.

Wright, W. E. (1984b). Induction of muscle genes in neural cells. *J. Cell Biol.*, **98**, 427–35.

Wright, W. E. and Aronoff, J. (1983). The suppression of myogenic functions in heterokaryons formed by fusing chick myocytes to diploid rat fibroblasts. *Cell Differ.*, **12**, 299–306.

Wu, K. J., Samuelson, L. C., Howard, G., Meisler, M. H., and Darlington, G. J. (1991). Transactivation of pancreas-specific gene sequences in somatic cell hybrids. *Mol. Cell. Biol.*, **11**, 4423–30.

Yamamoto, K. K., Gonzalez, G. A., Biggs, W. H., III, and Montminy, M. R. (1988). Phosphorylation-induced binding and transcriptional efficacy of nuclear factor CREB. *Nature*, **334**, 494–8.

Yates, J. L., Warren, N., and Sugden, B. (1985). Stable replication of plasmids derived from Epstein–Barr virus in various mammalian cells. *Nature*, **313**, 812–15.

Yokata, T., Lee, F., Rennick, D., Hall, C., Arai, N., Mosmann, T., *et al.* (1984). Isolation and characterization of a mouse cDNA clone that expresses mast-cell growth factor activity in monkey cells. *Proc. Natl Acad. Sci. USA*, **81**, 1070–74.

Yoshinaga, S., Dean, N., Han, M., and Berk, A. J. (1986). Adenovirus stimulation of transcription by RNA polymerase III: evidence for an E1A-dependent increase in transcription factor IIIC concentration. *EMBO J.*, **5**, 343–54.

2. The instability of differentiation in hepatomas

S. J. Goss

Hepatomas and the somatic cell genetics of differentiation

In 1965, Henry Harris showed that viable heterokaryons could be made between HeLa cells and normal differentiated cells as distinctive as rat lymphocytes, hen erythrocytes, and rabbit macrophages; and in these heterokaryons he detected reactivation of RNA and DNA synthesis in the previously quiescent nuclei. Cell fusion was rapidly established as a powerful tool in the investigation of differentiation, as is clear from the intense activity in this field in the immediately following years (Ringertz and Savage 1976). In work with growing hybrid cells, it is liver differentiation that has been most extensively studied. Well-differentiated hepatomas are available that maintain a wide range of liver-specific traits even when they are grown rapidly in standard tissue culture media. Their phenotype is generally stable through repeated cloning (Deschatrette and Weiss 1974), and the usefulness of the system has been increased by the development of media and conditions that select either for or against liver-specific traits (Leffert and Paul 1972; Haggerty *et al.* 1975; Choo and Cotton 1977). When a hepatoma is fused to a non-liver cell, the simplest outcome is the cross-activation of liver-specific genes derived from the non-liver parent (Peterson and Weiss 1972; Darlington *et al.* 1974; Malawista and Weiss 1974; Brown and Weiss 1975). Such results provided the first indication of the diffusible hepatic transcription factors that have since been so well documented (Gorski *et al.* 1986; Cereghini *et al.* 1987; Frain *et al.* 1989; Lichsteiner and Schibler 1989; Tronche and Yaniv 1992). Most often, however, the immediate result of cell fusion is a loss of differentiation, i.e. 'extinction' (Schneider and Weiss 1971; Szpirer and Szpirer 1975), and it has also been found that hepatomas can sometimes spontaneously become 'dedifferentiated' (Deschatrette and Weiss 1974). It is these examples of the loss of differentiation that will be discussed in this article. To seek the physiological significance of the failure of any biological process is, of course, a risky undertaking: a timely and apposite caveat has recently been provided by the demonstration that the partial failure of liver differentiation in *lethal-albino* mutant mice is due not to the deletion of a regulatory gene, as many of us had hoped, but rather to a simple enzyme deficiency causing the accumulation of a toxic metabolite (Gluecksohn-Waelsch 1979; Ruppert *et al.* 1992).

Liver-specific transcription factors in extinction and dedifferentiation

Although the loss of differentiation in extinction results from a deliberate and specific experimental manipulation, whereas dedifferentiation is spontaneous, extinction and dedifferentiation nevertheless resemble each other in several fundamental respects. For example, in both instances the loss of differentiation is accompanied by a loss of hepatic transcription factors (e.g. HNF1α and HNF4: Cereghini *et al.* 1987; Kuo *et al.* 1992) and in both instances differentiation and the transcription factors can reappear without the need for tissue-specific induction (Weiss and Chaplain 1971; Deschatrette *et al.* 1980). It is seen that the cells have an underlying memory of their hepatic origin, and this 'determination' is quite stable even in the absence of the transcription factors. In the case of extinction, it has additionally been shown that extinction cannot be attributed simply to the loss of the key factor HNF1α. Bulla *et al.* (1992) transfected hepatomas and their hybrids with an expression vector that provided them with a permanent supply of this factor, and yet it was found that extinction could neither be reversed nor prevented. This would seem to indicate that extinction must involve a component acting in parallel with HNF1α. Until recently, a promising candidate was HNF1β (otherwise known as vAPF or vHNF1; Cereghini *et al.* 1988). HNF1β resembles HNF1α and recognizes the same DNA motif. In both extinction and dedifferentiation, when HNF1α falls, HNF1β rises, and it seemed a reasonable suggestion that HNF1β might act to inhibit liver gene expression. However, transfection experiments have failed to show that HNF1β has any significant repressor activity (Mendel *et al.* 1991; Rey-Compos *et al.* 1991): indeed one group finds that HNF1β actually activates transcription (Rey-Compos *et al.* 1991). In considering what other factors might act in parallel with HNF1α, we should remember also the systems that regulate the *extent* of differentiation rather than specifying the pattern of gene expression. In this context we shall consider both hormonal regulation and the control of cell growth.

The dissection of extinction by use of microcell hybrids

Whole cell hybrids between two tissues potentially combine in one cell not only the factors specifying the two distinct tissue types, but also the receptor and second messenger systems appropriate to each. A dissection of the resulting interactions has been achieved by using mouse fibroblast microcells to transfer just one (or a very few) fibroblast chromosomes into a rat hepatoma. In this way, Killary and Fournier (1984) have established the presence of an extinguishing locus, *tse-1* (tissue-specific extinguisher 1), on mouse chromosome 11. The locus produces profound, though not total extinction of a number of liver-specific enzymes. It has now been shown that this extinction can be reversed by cyclic AMP (Thayer and Fournier 1989) and that *tse-1* codes for the R1α subunit of cyclic AMP-dependent protein kinase (Boshart *et al.* 1991; Jones *et al.* 1991). This result establishes a hormonal response mechanism as contributing significantly to extinction, though it also poses the intriguing problem of why the rat homologue of *tse-1* is apparently silent in the hepatoma and yet the activity of the mouse homologue persists after microcell-mediated transfer (Weiss 1992). Additional factors must operate in whole cell hybrids, where extinction involves a wider range of traits and is less easily reversed. Chin and Fournier (1989), again using microcell fusion, have identified a further locus, *tse-2*, that suppresses albumin production. It will be interesting to see whether *tse-2* acts via tissue-specific transcription factors, as suggested by Mendel et al. (1991), or whether, like *tse-1*, it acts in parallel with them.

The nature of dedifferentiated variants

The interest in dedifferentiated hepatomas lies in their potential use in genetic analysis. Assuming these variants arise by mutations affecting single factors required for differentiation, it should be possible to analyse the mechanism of differentiation by complementation testing. Back-crosses would identify the recessive mutants, inter-crosses would place these in complementation groups, and transfection might be used as a route to clone the genes involved (Fig. 2.1). All these steps have been attempted, but unfortunately they have so far shed little light on the normal processes of differentiation. There are at least three reasons to doubt that suitable recessive mutants are yet available:-

1. *Dedifferentiated variants arise more frequently than recessive mutants of known genes.* Dedifferentiation may even represent one extreme of the discontinuous variation in liver-specific gene expression described by Peterson (1974, 1976). Using immunoperoxidase staining, he found a stepwise fluctuation in the albumin content of single cells that occurred with a frequency of approximately 10^{-2} (Peterson 1979). Unfortunately, a reliable estimate of the absolute frequency of *de novo* dedifferentiation has never been obtained. The first variants were found by a simple visual inspection of hepatoma clones for morphological peculiarities (Deschatrette and Weiss 1974). These variants were clearly both rare

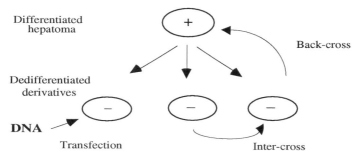

Fig. 2.1. The genetic analysis of differentiation. A scheme for complementation analysis using dedifferentiated hepatomas.

and independent in origin, but, given the circumstances of their isolation, no estimate of their frequency was possible. Subsequent isolations have involved the use of BUdR/light killing or radiolabel-suicide to select against well-differentiated cells, and neither technique is compatible with accurate measurements (Choo and Cotton 1977; Moore and Weiss 1982; Goss 1984*a,b*). Nevertheless, we can note that the apparent frequency of phenylalanine hydroxylase-deficient variants found by Choo and Cotton without prior mutagenesis is about 10^{-4}, which would seem to be on the high side for mutation. These authors were mindful of the anomalously high mutation rates that have been seen in some cell lines, notably in certain Chinese hamster lines that seem to have a widespread functional hemizygosity. However, there is evidence that such circumstances do not apply in the commonly used hepatomas: Moore and Weiss (1982) and Goss (1984*a*) have pointed out that their selective techniques were aimed in each case against the expression of any of several enzymes comprising a metabolic pathway, and yet none of the resulting variants was found to be a mutant lacking just a single enzyme. This was so even after mutagenesis. The outcome of these selections was uniformly a pleiotropic dedifferentiation, which must therefore be considerably more common than genetic mutation. This conclusion is especially obvious in selections against the conversion of ornithine to arginine (Goss 1984*a,b*). Any of four enzymes could have been lost by mutation, two are liver-specific and two are ubiquitous (Fig. 2.2). Loss of the ubiquitous enzymes was never detected, and nor was ornithine transcarbamylase ever lost on its own, even though it is on the X-chromosome which is present in a single active copy.

2. *Most dedifferentiated variants can revert more readily than would be expected by back-mutation.* Many variants are extremely unstable and revert in bulk on continued growth (Moore and Weiss 1982; Goss 1984*a*): naturally, these variants have been very little studied. The remainder are much more stable, but reversion can generally be detected by plating the cells in selective media.

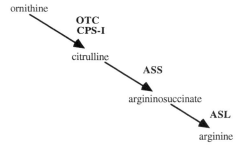

Fig. 2.2. The synthesis of arginine from ornithine. Four enzymes are essential for cells to grow using ornithine in place of arginine (Goss 1984*a*). Two are tissue-specific and are found predominantly in the liver: OTC, ornithine transcarbamylase; and CPS-I, mitochondrial carbamylphosphate synthetase. The other two are ubiquitous: ASS, argininosuccinate synthetase; and ASL, argininosuccinate lyase.

Deschatrette et al. (1980), selecting for the recovery of gluconeogenesis, reported frequencies as low as 10^{-8}, which could well represent back-mutation. However, in that case, one might expect mutagenesis to increase the rate of reversion, and yet no such increase could be demonstrated. In view of the difficulty of ensuring a high cloning efficiency in liver-selective media, these low reversion frequencies should perhaps be interpreted with some caution. It is quite possible, for instance, that hormonal stimulation would have increased the apparent rate of reversion. In the presence of dexamethasone, which stimulates gluconeogenesis, the most stable variant produced in this laboratory gives a spontaneous reversion frequency of 10^{-6}, which is too high to be attributed to back-mutation (Goss 1984*a*); and in other experiments, we found that some variants that had otherwise seemed totally stable were induced to redifferentiate *en masse* by medium containing cyclic AMP and dexamethasone (Goss 1984*b*). There would appear to be a hormonal involvement in dedifferentiation just as there is in extinction. A final complication is that different traits can reappear independently, and then the reversion frequencies that are detected are exquisitely sensitive to the sequence of selection (Goss 1984*a*). The same considerations apply to hepatomas that have become dedifferentiated *in vivo* (Farmer and Goss 1991). Stringent testing of the phenylalanine hydroxylase deficiency in the mouse hepatoma, BWTG3, revealed a reversion frequency of $3-5 \times 10^{-5}$, though this hepatoma had previously been considered totally stable and had been used in transfection assays in an attempt to detect complementation (Faust and Weiss 1990). A few variants are known in which spontaneous reversion seems not to occur. These include one arising *in vitro*, H5 (Deschatrette and Weiss, 1974), and the HTC tumour, which apparently lost most of its liver-specific traits *in vivo* (see below).

At this point, it is appropriate to call to question the suggestion that a variant commonly used in molecular biological experiments, Faof1-C2, is a 'differentiation mutant' (Mendel et al. 1991; Kuo et al. 1992). These authors cite 10^{-6}–10^{-7} as the frequency of generation of such a cell, yet it is derived from one of the variants that were discovered by visual inspection of growing clones of cells (Deschatrette and Weiss, 1974). In that case, it is difficult to see how so low a frequency could reasonably be defended. The frequency cited for the reversion of this cell, around 10^{-9}, relates to the recovery of gluconeogenesis in the absence of hormones. Faof1-C2 can in fact be *induced* to regain albumin synthesis by growth for 2 days as cell aggregates in suspension. The frequency of this change is greater than 1 in 10, and the subsequent recovery of gluconeogenesis (still measured in the absence of hormones) is then apparently raised in some instances to 10^{-6} (Deschatrette 1980). Given the general belief that differentiation is epigenetic, there might be some advantage in considering epigenetic change rather than mutation to explain the behaviour of this cell.

3. *Dedifferentiation appears dominant in back-crosses with differentiated cells.* This finding (Deschatrette and Weiss 1975; Deschatrette *et al.* 1979) may well explain the lack of complementation in inter-crosses between dedifferentiated

hepatomas (Levilliers and Weiss, 1983). Farmer and Goss (1991) have argued that the ornithine transcarbamylase deficiency in BWTG3 hepatoma cells is recessive in a back-cross, but there is no proof that the reactivation of the BWTG3 ornithine transcarbamylase gene in such hybrids represents true genetic complementation. BWTG3 is somewhat unusual in that, apart from a few deficiencies that are mostly extremely labile, this hepatoma is in other respects a relatively well-differentiated cell: it is sometimes considered to represent an immature liver cell. If we then set this example aside, it is seen that typical pleiotropic dedifferentiation involves the acquisition of the ability to suppress differentiation, rather than the simple loss of some regulatory factor.

These considerations lead us to conclude that in most, if not all instances, the basis of dedifferentiation is not genetic mutation. This view will, of course, have implications for the way in which we plan both cell fusion experiments and DNA transfections involving the use of dedifferentiated variants. This is not to say that dedifferentiation is not a genetic phenomenon; for example, gene dosage effects could explain all that is set out above. But equally there is the possibility that epigenetic mechanisms might play a role. These would include not only the *cis*-regulation of gene activity by DNA methylation, heterochromatinization, or some other form of imprinting, but also, for instance, self-maintained states of gene expression dependent on positive feedback by diffusible molecules. It is hardly helpful to be left with so many diverse possibilities, but we could perhaps note that these are mechanisms that might adjust the level of some cellular component rather than changing its nature or abolishing its expression entirely. The kind of cellular component that is then implicated is one whose effect on the cell depends critically on *level* rather than on absolute presence. The involvement of hormonal signalling in some cases of dedifferentiation has been mentioned above. Another sphere of cellular control where the levels of mediators are likely to be crucial is the regulation of cell growth. The interaction between cell growth control and differentiation will be discussed next.

Cell growth control and differentiation

The reciprocal regulation of cell growth and differentiation is obvious and unavoidable in, for example, myoblasts, erythroleukaemia cells, and epidermal cells, and it is equally prominent in normal hepatocytes, both *in vivo* and *in vitro* (Leffert *et al.* 1982). However, it is traditionally ignored in the somatic cell genetics of hepatomas. It has long been known that oncogenes can suppress differentiation, and that, just as in extinction and spontaneous dedifferentiation, the underlying determination of the cells is unaffected. This is clear from the reversibility of the effects of temperature-sensitive oncogenes (in melanoblasts (Boettiger *et al.* 1977), chondroblasts (Pacifici *et al.* 1977), and fetal hepatocytes (Schlegel-Haueter *et al.* 1980)). Another reminder of extinction and dedifferentiation is seen when the oncogene *ras* or *fos* suppresses myogenesis and produces a concomitant loss of the muscle-specific transcription factor, MyoD1 (Lassar et al. 1989). On the other hand, it is well known that the activity of tumour suppressor genes is associated with the promotion of differentiation (Peehl and Stanbridge 1982; Harris 1985). The molecular mechanisms of these opposing actions are beginning to be unravelled (Yang-Yen et al. 1990; Baichwal et al. 1992; Zambetti et al. 1992). Given this background, we should perhaps consider the extent of oncogene involvement in hepatoma dedifferentiation. This approach could help explain the peculiarities of dedifferentiated variants described in the preceding section. For example, if dedifferentiated variants were in some respect 'more transformed', they might possess a selective advantage that would increase the frequency with which they were detected. Furthermore, a high rate of *de novo* dedifferentiation (and of reversion) could be generated by changes in the balance of oncogenes resulting, for instance, from karyotypic evolution and alterations in DNA methylation. Finally, in cell hybrids, depending on the conditions of their isolation, dedifferentiation through oncogenic imbalance could well appear to be dominant. Some support for this analysis is provided by the experiments described below.

An association between mitogen independence and the loss of a urea-cycle enzyme in a hepatoma and its hybrids

Experiments currently in progress in this laboratory are providing clear evidence of an association between increased transformation and dedifferentiation. The rat hepatoma, HTC, is poorly differentiated. It is deficient in most liver-specific traits and seems to be quite stable: it has never yielded revertants in any selective medium so far tried in this laboratory. The hepatocellular origin of HTC is nevertheless clear from its hormone-inducible expression of tyrosine aminotransferase (Thompson et al. 1966). HTC has been fused in this laboratory to several well-differentiated hepatomas, and in agreement with the literature, in standard conditions, dedifferentiation has appeared dominant. The circumstances under which some liver-specific traits can be maintained or re-induced in these hybrids is currently under investigation, and the simplest example, that of the urea-cycle enzyme carbamylphosphate synthetase (CPS-I), will be described here. In these experiments, HTC TG3^{neo}, a subclone of HTC resistant to 6-thioguanine and the antibiotic G418, was fused using Sendai virus to another rat hepatoma, Fu5-t. This is a well-differentiated stock of Fu5 (Deschatrette and Weiss 1974) which had previously been grown in medium without tyrosine (to select for the liver-specific enzyme phenylalanine hydroxylase). Hybrids were selected in the usual way with HAT and G418 (Table 2.2), and were grown as pools rather than clones, so biasing the results in favour of the most rapidly growing cells. Tables 2.1 and 2.2 show the results of CPS-I assays on the parental cells and on freshly isolated hybrids, and Fig. 2.3 shows the morphology of the cells.

HTC TG3^{neo} has only trace amounts of CPS-I, at the limit of detection of the assay. No medium we have tried has caused any significant induction of CPS-I in this cell. In contrast, Fu5-t has a high basal activity of CPS-I (9 units) and can be

Table 2.1. Expression of CPS-I in the parental hepatomas

Cell	Medium supplements		CPS-I (nmol/min/mg)
HTC TG3^{neo}	FCS ± dex^{*a}		0.2 ± 0.02 (n = 9)
Fu5-t	FCS		9.0 ± 0.9 (n = 7)
	FCS + dex	(6 days)	25.8
	FCS + iR.dex	(6 days)	15.0
	HS + dex	(5 days)	17.8
	serum-free + dex	(16 days)	35.8

Specific activities of CPS-I were determined as described by Goss (1984a). An enriched form of Eagle's minimal essential medium was used throughout (MEM', i.e. 'arginine medium' of Goss (1984a) without N-acetylglutamate, plus 0.5 mM glycine, 1 mM alanine, and 33 mM glucose. Fetal calf serum (FCS) and horse serum (HS) were used at 5% v/v; dexamethasone (dex) at 17 nM; insulin and retinoic acid (iR) at 10 nM and 2.5 μM respectively. Serum-free medium additionally contained 1.4 mM choline, 30 nM sodium selenite 1 μM putrescine, and 10 μg/ml transferrin. Serum-free cultures were plated in plastic tissue-culture flasks that had first been protein-coated by overnight incubation with 3 ml phosphate-buffered saline + FCS (5% v/v): these flasks were rinsed exhaustively before use. Assays of basal activity were repeated (s.e.m. and n given). (a) The result for HTC TG3^{neo} includes five assays with and four without dexamethasone: CPS-I is not inducible in this cell. Assays after induction are single results: the duration of induction is given in brackets.

induced twofold or more, to near *in vivo* levels (Goss 1984a), by physiological stimulation with glucocorticoid (17 nM dexamethasone). Table 2.1 shows this induction in media with different mitogenic potentials. As expected, the growth of Fu5-t is slowed by dexamethasone (Barnett *et al.* 1979). In particular, in medium with horse serum and dexamethasone the passage time is typically doubled, and in serum-free medium with dexamethasone cell growth is virtually stopped (hence the delay in obtaining this assay result: Table 2.1). This behaviour contrasts markedly with that of HTC TG3^{neo} which grows well in all these media, and which even grows vigorously in serum-free medium supplemented with 1 μM dexamethasone.

Table 2.2. Expression of CPS-I in HTC TG3^{neo} x Fu5-t hybrids

Medium supplements		CPS-I (nmol/min/mg)
FCS + iR		9.6 ± 1.1 (n = 6)
FCS + iR.dex	(7 days)	19.2
HS + dex	no growth	not assayed
HS + iR.dex	(9 days)	26.4
serum-free + dex	no growth	not assayed
serum-free + iR.dex	(6 days)	30.6

Hybrids were isolated in MEM' + 5% FCS + HAT (i.e. 300 μM hypoxanthine, 0.9 μM aminopterin and 50 μM thymidine + the antibiotic G418 at 1 mg/ml (G418 was omitted 2 weeks after isolation). This table shows results obtained on newly isolated cells (grown for less than 3 weeks after removal of G418). Assays on induced cells are single results, but see Fig. 2.4 for reproducibility. Abbreviations, as in Table 2.1.

The two hepatomas thus differ in their growth control as well as in the extent of their differentiation. Furthermore, they have radically different morphologies: HTC TG3^{neo} looks distinctly transformed in comparison with Fu5-t, which forms a well-ordered pavement of epithelial cells with prominent nucleoli and phase-dark cytoplasm typical of normal hepatocytes (Fig. 2.3, cf. panels (a) and (b)).

Hybrids isolated with 5 per cent fetal calf serum as the only mitogen have proved difficult to grow, and they have been little studied. Their basal CPS-I was less than 1 unit, and dexamethasone raised this only to 3 units: dedifferentiation appeared to be dominant. In contrast, as shown in Table 2.2, in hybrids isolated in medium supplemented with 5 per cent fetal calf serum, insulin, and retinoic acid (an effective mitogenic combination for these cells), the basal and induced levels of CPS-I resemble those in the Fu5-t parent: dedifferentiation now appears to be recessive. Table 2.2 also indicates that these differentiated hybrids have a greater dependence than either parental cell on exogenous mitogens. The hybrids have an absolute requirement for the insulin/retinoate supplement for growth in media containing dexamethasone. This applies not only in serum-free medium, but even in the presence of horse serum. A similar antagonism between physiological levels of dexamethasone and insulin has been described in normal hepatocytes cultured with epidermal growth factor as the primary mitogen (Baribault and Marceau 1986). In this respect, the hybrids more closely resemble normal hepatocytes than does either parental cell: there has apparently been some kind of complementation between the parental cells that has resulted in more normal growth control in the hybrids.

The subsequent evolution of these hybrids emphasizes the correlation between dedifferentiation and transformation. Figure 2.4 shows that the induction of CPS-I

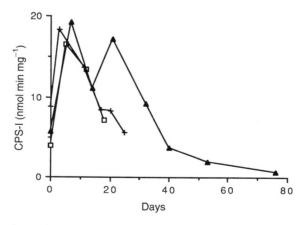

Fig. 2.4. Induction and subsequent loss of CPS-I in HTC TG3^{neo} × Fu5-t hybrids. Newly produced stocks of hybrid cells growing in HAT medium with 5% FCS, insulin, and retinoate (see Table 2.2) were induced by the inclusion of 17 nM dexamethasone from day 0. The cells were assayed for CPS-I at intervals, as they were sub-cultured. Each point is a single assay result: three independent experiments are shown.

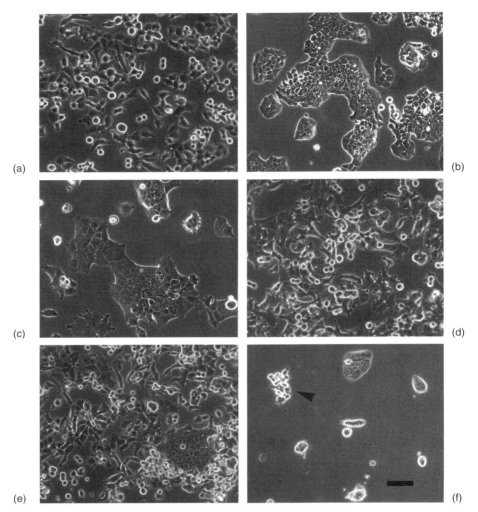

Fig. 2.3. Growth and morphology of hepatomas and their hybrids. The cultures shown were all seeded at one-tenth confluency and photographed under phase-contrast after 2 days (a)–(d) or 4 days (e) and (f). The media are explained in Tables 2.1 and 2.2. (a) HTC TG3^{neo} in MEM' + 5% FCS. (b) Fu5-t in MEM' + 5% FCS: this morphology is typical of well-differentiated cells. (c) Newly isolated hybrid cells in HAT + 5% FCS, insulin, and retinoate. (d) Dedifferentiated hybrids in HAT + 5% FCS, insulin, retinoate, and dexamethasone (after a total of 62 days' growth in this medium). (e) Dedifferentiated hybrids, as shown in (d), but plated in serum-free medium + dexamethasone: most cells continue to multiply, but note the aggregate of stationary differentiated cells. (f) Mitogen-dependent cells partly purified from culture (e) by BUdR/light treatment (see text), grown up in HAT + 5% FCS, insulin, and retinoate, and then replated in serum-free medium + dexamethasone: note just one clone of dividing cells (arrow) and several aggregates of arrested cells. Scale bar = 80 μm.

by dexamethasone is not sustained. Within one month the cells become refractory to induction, their appearance is radically changed (Fig. 2.3, cf. panel (d) and (c)), and most of the cells acquire the ability to grow in serum-free medium with dexamethasone (Fig.2.3(e)). CPS-I is barely detectable in these cultures. There remains, however, a minority of cells whose growth is arrested in this medium. These cells, which form small, morphologically differentiated aggregates (Fig. 2.3(e)), have been partly purified by using the BUdR/light technique of Puck and Kao (1967) to kill the growing cells. The surviving cells, rescued in dexamethasone-free medium supplemented with 5% fetal calf serum, insulin, and retinoate, were found to look similar to the original hybrids (not shown). When these cells were returned to serum-free medium with dexamethasone, as expected, they stopped growing (Fig. 2.3(f)), and once again they accumulated high levels of CPS-I (>20 units after 4 days). In summary, two types of hybrid cell have been identified. Some are mitogen-dependent, and these show both basal and inducible CPS-1, whilst the others are mitogen-independent and dedifferentiated.

These results illustrate a link between differentiation and cell growth control. The progressive loss of CPS-I and the associated increase in transformation seen in these hybrids may provide a convenient *in vitro* model for some aspects of tumour progression. It was expected that this phenotypic shift would simply be due to chromosome loss tending to restore the hybrids towards the HTC karyotype. However, perhaps because the hybrids are intraspecific, their karyotype seems to be remarkably stable. A preliminary analysis has shown that the dedifferentiated hybrids maintain a modal chromosome number close to that expected for complete 1:1 hybrids. The basis of the phenotypic conversion in these hybrids needs further investigation in cloned populations of cells.

Summary

There remain considerable uncertainties about the changes responsible for the loss of hepatoma differentiation in extinction and in spontaneous dedifferentiation. In both instances, cellular determination appears unaffected despite radical changes in the pattern of expression of tissue-specific transcription factors. Both extinction and dedifferentiation are generally reversible, and both can involve a significant contribution from hormonal response mechanisms. Spontaneous dedifferentiation is generally too frequent and too readily reversible to be attributed to genetic mutation. Dedifferentiation is usually dominant in cell hybrids, though that may be dependent on the conditions of hybrid isolation. The lack of recessive mutants affecting differentiation has so far frustrated attempts at complementation analysis. Finally, it is suggested that changes in oncogenic balance might make significant contributions to the instability of hepatoma differentiation. This view is supported by the correlation between dedifferentiation and mitogen-independent growth seen in certain hepatoma hybrid cells. A better understanding of dedifferentiation *in vitro* may further our understanding of tumour progression *in vivo*.

References

Baichwall, V. R., Park, A., and Tjian, R. (1992). The cell-type-specific activator region of c-Jun juxtaposes constitutive and negatively regulated domains. *Genes Dev.*, **6**, 1493–502.

Baribault, H. and Marceau, N. (1986). Dexamethasone and dimethylsulphoxide as distinct regulators of growth and differentiation of cultured suckling rat hepatocytes. *J. Cell. Physiol.*, **129**, 77–84.

Barnett, C. A., Barnhorst, M., Fooshee, C. M., and Saneto, R. P. (1979). Selection of a dexamethasone-resistant H4-IIE-C3 rat hepatoma tissue-culture line. *In Vitro*, **15**, 128–37.

Boettiger, D., Roby, K., Brumbaugh, J., Biehl, J., and Holtzer, H. (1977). Transformation of chicken embryo retinal melanoblasts by a temperature-sensitive mutant of Rous sarcoma virus, *Cell*, **11**, 881–90.

Boshart, M., Weih, F., Nichols, M. and Schütz, G. (1991). The tissue-specific extinguisher locus *Tse-1* encodes a regulatory subunit of cAMP-dependent protein kinase. *Cell*, **66**, 849–59.

Brown, J. E., and Weiss, M. C. (1975). Activation of production of mouse liver enzymes in rat hepatoma–mouse lymphoid cell hybrids. *Cell*, **6**, 481–94.

Bulla, G. A., De Simone, V., Cortese, R., and Fournier, R. E. (1992). Extinction of alpha-1-antitrypsin gene expression in somatic cell hybrids: evidence for multiple controls. *Genes Dev.*, **6**, 316–27.

Cereghini, S., Raymondjean, M., Carranca, A. G., Herbomel, P., and Yaniv, M. (1987). Factors involved in tissue-specific expression of albumin gene. *Cell*, **50**, 627–38.

Cereghini, S., Blumenfeld, M., and Yaniv, M. (1988). A liver-specific factor essential for albumin transcription differs between differentiated and dedifferentiated rat hepatoma cells. *Genes Dev.*, **2**, 957–74.

Chin, S. C. and Fournier, R. E. K. (1989). *Tse-2*: a *trans*-dominant extinguisher of albumin gene expression in hepatoma hybrid cells. *Mol. Cell. Biol.*, **9**, 3736–43.

Choo, K. H. and Cotton, R. G. H. (1977). Genetics of the mammalian phenylalanine hydroxylase system: 1. Isolation of phenylalanine hydroxylase-deficient tyrosine auxotrophs from rat hepatoma cells. *Somat. Cell Genet.*, **3**, 457–70.

Darlington, G. J., Bernhard, H. P., and Ruddle, F. H. (1974). Human serum albumin phenotype activation in mouse hepatoma–human leukocyte hybrids. *Science*, **185**, 859–62.

Deschatrette, J. (1980). Dedifferentiated variants of a hepatoma cell: partial reversion induced by cell aggregation. *Cell*, **22**, 501–11.

Deschatrette, M. and Weiss, M. C. (1974). Characterization of differentiated and dedifferentiated clones from a rat hepatoma. *Biochimie*, **56**, 1603–11.

Deschatrette, J. and Weiss, M. C. (1975). Extinction of liver-specific functions in hybrids between differentiated and dedifferentiated hepatoma cells. *Somat. Cell Genet.*, **1**, 279–92.

Deschatrette, J., Moore, E. E., Dubois, M., Cassio, D., and Weiss, M. C. (1979). Dedifferentiated variants of a rat hepatoma: analysis by cell hybridisation. *Somat. Cell Genet.*, **5**, 697–718.

Deschatrette, J., Moore, E. E., Dubois, M., and Weiss, M. C. (1980) Differentiated variants of a rat hepatoma: reversion analysis. *Cell*, **19**, 1043–51.

Farmer, A. A., and Goss, S. J. (1991). BWTG3 hepatoma cells can acquire phenylalanine hydroxylase, cystathionine synthase and CPS-I without genetic manipulation, but activation of the silent OTC gene requires cell fusion with hepatocytes. *J. Cell Sci.*, **98**, 533–8.

Faust, D. M., Imaizumi-Scherrer, T., Fulchignoni-Lataud, M. C., Catherin, A. M., Iost, I., and Weiss, M. C. (1990). Activation of phenylalanine hydroxylase expression following genomic DNA transfection into hepatoma cells. *Differentiation*, **44**, 74–9.

Frain, M., Swart, G., Monaci, P., Nicosia, A., Stampfli, S., Frank, R., *et al.* (1989). The liver-specific transcription factor LF-B1 contains a highly diverged homeobox DNA binding domain. *Cell*, **59**, 145–57.

Gluecksohn-Waelsch, S. (1979). Genetic control of morphogenetic and biochemical differentiation: lethal albino deletions in the mouse. *Cell*, **16**, 225–37.

Gorski, K., Carneiro, M., and Schibler, U. (1986). Tissue-specific *in vitro* transcription from the mouse albumin promoter. *Cell*, **47**, 767–76.

Goss, S. J. (1984*a*). Arginine synthesis by hepatomas *in vitro* I. The requirements for cell growth in medium containing ornithine in place of arginine, and the isolation and characterization of variant hepatomas auxotrophic for arginine. *J. Cell Sci.*, **68**, 285–303.

Goss, S. J. (1984*b*). Arginine synthesis by hepatomas *in vitro* II. Isolation and characterization of Morris hepatoma variants unable to convert ornithine to arginine, and modulation of urea-cycle enzymes by dexamethasone and cyclic-AMP. *J. Cell Sci.*, **68**, 305–19.

Haggerty, D. F., Young, P. L., and Buese, J. V. (1975). A tyrosine-free medium for the selective growth of cells expressing phenylalanine hydroxylase activity. *Dev. Biol.*, **44**, 158–68.

Harris, H. (1965). Behaviour of differentiated nuclei in heterokaryons of animal cells from different species. *Nature*, **206**, 583–8.

Harris, H. (1985). Suppression of malignancy in hybrid cells: the mechanism. *J. Cell Sci.*, **79**, 83–94.

Jones, K. W., Shapero, M. H., Chevrette, M., and Fournier, R. E. K. (1991). Subtractive hybridization cloning of a tissue-specific extinguisher: *Tse-1* encodes a regulatory subunit of protein kinase A. *Cell*, **66**, 861–72.

Killary, A. M. and Fournier, R. E. K. (1984). A genetic analysis of extinction: transdominant loci regulate expression of liver-specific traits in hepatoma hybrid cells. *Cell*, **38**, 523–34.

Kuo, C. J., Conley, P. B., Chen, L., Sladek, F. M., Darnell, J. E., and Crabtree, G. R. (1992). A transcriptional hierarchy involved in mammalian cell-type specification. *Nature*, **355**, 457–61.

Lassar, A. B., Thayer, M. J., Overell, R. W., and Weintraub, H. (1989). Transformation by activated *ras* or *fos* prevents myogenesis by inhibiting expression of MyoD1. *Cell*, **58**, 659–67.

Leffert, H. L. and Paul, D. (1972). Studies on primary cultures of fetal liver cells. *J. Cell Biol.*, **52**, 559–68.

Leffert, H. L., Koch, K. S., Lad, P. J., Shapiro, I. P., Skelly, H., and de Hemptine, B. (1982). Hepatocyte regeneration, replication and differentiation. In *The liver: biology and pathobiology*, (ed. I. Arias, H. Popper, D. Schacter, and D. Schafritz), pp. 833–50. Raven Press, New York.

Levilliers J. and Weiss, M. C. (1983). Differentiation is not restored in hybrids between independent variants of a rat hepatoma. *Somat. Cell Genet.*, **9**, 407–13.

Lichtsteiner, S. and Schibler, U. (1989). A glycosylated liver-specific transcription factor stimulates transcription of the albumin gene. *Cell*, **57**, 1179–87.

Malawista, S. E. and Weiss, M. C. (1974). Expression of differentiated functions in hepatoma cell hybrids. High frequency of induction of mouse albumin production in rat hepatoma × mouse lymphoblast hybrids. *Proc. Natl. Acad. Sci. USA*, **71**, 927–31.

Mendel, D. B., Hansen, L. P., Graves, M. K., Conley, P. B., and Crabtree, G. R. (1991). HNF-1α and HNF-1β (vHNF-1) share dimerization and homeo domains, but not activation domains, and form heterodimers *in vitro*. *Genes Dev.*, **5**, 1042–56.

Moore, E. E. and Weiss M. C. (1982). Selective isolation of stable and unstable dedifferentiated variants from a rat hepatoma cell line. *J. Cell Physiol.*, **111**, 1–8.

Pacifici, M., Boettiger, D., Roby, K., and Holtzer, H. (1977). Transformation of chondroblasts by Rous sarcoma virus and synthesis of sulfated proteoglycan matrix. *Cell*, **11**, 891–9.

Peehl, D. M. and Stanbridge, E. J. (1982). The role of differentiation in the suppression of tumorigenicity in human hybrid cells. *Int. J. Cancer*, **30**, 113–20.

Peterson, J. A. (1974). Discontinuous variability, in the form of a geometric progression, of albumin production in hepatoma and hybrid cells. *Proc. Natl. Acad. Sci. USA*, **71**, 2062–6.

Peterson, J. A. (1976). Clonal variation in albumin messenger RNA activity in hepatoma cells. *Proc. Natl. Acad. Sci. USA*, **73**, 2056–60.

Peterson, J. A. (1979). Analysis of discontinuous variation in albumin production by hepatoma cells at the cellular level. *Somat. Cell Genet.*, **5**, 641–51.

Peterson, J. A. and Weiss, M. C. (1972). Expression of differentiated functions in hepatoma cell hybrids: induction of mouse albumin production in rat hepatoma–mouse fibroblast hybrids. *Proc. Natl. Acad. Sci. USA*, **69**, 571–5.

Puck, T. T. and Kao, F.-T. (1967). Genetics of somatic mammalian cells. V. Treatment with 5-bromodeoxyuridine and visible light for isolation of nutritionally deficient mutants. *Proc. Natl. Acad. Sci. USA*, **58**, 1227–34.

Rey-Compos, J., Chouard, T., Yaniv, M., and Cereghini, S. (1991). vHNF-1 is a homeoprotein that activates transcription and forms heterodimers with HNF-1. *EMBO J.*, **10**, 1445–57.

Ringertz, N. R. and Savage, R. E. (1976). *Cell hybrids*, pp. 180–212. Academic Press, New York.

Ruppert, S., Kelsey, G., Schedl, A., Schmidt E., Thies, E.,and Schütz, G. (1992). Deficiency of an enzyme of tyrosine metabolism underlies altered gene expression in newborn liver of lethal albino mice. *Genes Dev.*, **6**, 1430–43.

Schlegel-Haueter, S., Schlegel, W., and Chou, J. Y. (1980). Establishment of a fetal rat liver cell line that retains differentiated functions. *Proc. Natl. Acad. Sci. USA*, **77**, 2731–4.

Schneider, J. and Weiss, M. C. (1971). Expression of differentiated functions in hepatoma cell hybrids. 1. Tyrosine aminotransferase in hepatoma–fibroblast hybrids. *Proc. Nat. Acad. Sci. USA*, **68**, 127–31.

Szpirer, J. and Szpirer, C. (1975). The control of serum protein synthesis in hepatoma–fibroblast hybrids. *Cell*, **6**, 53–60.

Thayer, M. J. and Fournier, R. E. (1989). Hormonal regulation of *Tse-1*-repressed genes: evidence for multiple genetic controls in extinction. *Mol. Cell. Biol.*, **9**, 2837–46.

Thompson, E. B., Tompkins, G. M., and Curran, J. F. (1966). Induction of tyrosine-α-ketoglutarate transaminase by steroid hormones in a newly established tissue-culture cell line. *Proc. Natl. Acad. Sci. USA*, **56**, 296–303.

Tronche, F. and Yaniv, M. (1992). HNF1, a homeoprotein member of the hepatic transcription regulatory network. *BioEssays*, **14**, 579–87.

Weiss, M. C. (1992). Extinction by indirect means. *Nature*, **355**, 22–3.

Weiss, M. C. and Chaplain, M. (1971). Expression of differentiation functions in hepatoma hybrids: reappearance of tyrosine aminotransferase inducibility after loss of chromosomes. *Proc. Natl. Acad. Sci. USA*, **68**, 3026–30.

Yang-Yen, H.-F., Chambard, J.-C., Sun, Y.-L., Smeal, T., Schmidt, T. J., Dronin, J., *et al.* (1990). Transcriptional interference between c-Jun and the glucocorticoid receptor: mutual inhibition of DNA binding due to direct protein–protein interaction. *Cell*, **62**, 1205–15.

Zambetti, G. P., Bargonetti, J., Walker, K., Prives, C., and Levine, A. J. (1992). Wild-type p53 mediates positive regulation of gene expression through a specific DNA sequence element. *Genes and Dev.*, **6**, 1143–52.

3. Nuclear protein sorting in heterokaryons and homokaryons

Nils R. Ringertz

Introduction

Our interest in nuclear protein sorting began 25 years ago and was triggered by Professor Harris' report that it was possible to reactivate RNA and DNA synthesis in inactive chick erythrocyte (CE) nuclei by Sendai virus-induced fusion of chicken red cells with human HeLa cells (Harris 1965, 1967). This finding suggested that human tumour cells contain factors capable of activating chick genes and raised a number of interesting questions about the nature of cytoplasmic signals regulating transcription and replication. In collaboration with Professor Harris we carried out a quantitative cytochemical study of chromatin changes during nuclear reactivation in CE × HeLa heterokaryons. Individual nuclei in fixed heterokaryons were analysed using microspectrophotometry, microfluorimetry, and microinterferometry (Bolund et al. 1969). During early stages of reactivation the chick chromatin underwent drastic biophysical changes resulting in an increased ability to bind intercalating fluorochromes to DNA, altered thermal stability of the DNA, etc. These changes were followed by a marked dispersion of the condensed erythrocyte chromatin. Model experiments with isolated CE nuclei suggested that the biophysical changes, which we named 'chromatin activation' (Ringertz 1969), were triggered by ionic signals causing conformational changes in the deoxyribonucleoprotein complexes. The drastic increase in nuclear size noted by Harris (1967) and illustrated in Fig. 3.1, was paralleled by a dramatic increase in nuclear dry mass and protein content. An interesting aspect was that the nuclei grew in size and weight *before* there was any significant increase in chick RNA and chick proteins. Using species-specific antibodies to nuclear proteins (Ringertz et al. 1971) we subsequently found that the chick nuclei grow because they selectively take up and concentrate human nuclear proteins contributed by the HeLa fusion partner. This led us to examine which human proteins entered the chick nuclear compartment and which chick proteins, if any, were lost from the inactive CE chromatin. We also tested if different cytoplasmic environments, i.e. different fusion partners, resulted in activation of the same or of different genes.

As a tribute to Professor Harris I have decided to devote the first part of this article to a review of past work on nucleocytoplasmic protein exchange and gene expression in CE heterokaryons. The second part of the article will be devoted to more recent work on nuclear protein sorting in multinucleate myotubes. The

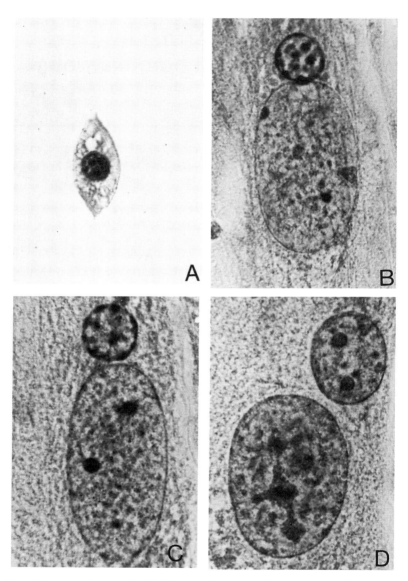

Fig. 3.1. UV microphotographs (λ 260 nm) illustrating the appearance of (A) an unfused chick erythrocyte (CE) of the definitive lineage (13 day embryo), (B) heterokaryon containing one small CE nucleus and a large rat fibroblast nucleus at 8 h, (C) 24 h, and (D) 72 h after fusion. The chromatin undergoes a gradual dispersion as the nucleus expands. At 72 h the CE nucleus contains two newly developed nucleoli but is still smaller than the rat nucleus.

problems there are somewhat different and deal mainly with the role of topological factors in nuclear specialization.

Chick erythrocyte heterokaryons

A summary of the early steps in the reactivation of CE nuclei after fusion with mammalian cells is given in Fig. 3.2. For detailed references to the older literature the reader is referred to reviews by Harris (1968, 1970), Appels and Ringertz (1975), and Ringertz and Savage (1976).

Using immune cytochemistry with species-specific antibodies to *mammalian* nuclear and cytoplasmic antigens we have shown that chick nuclei can take up and selectively concentrate a variety of mammalian nuclear proteins (Table 3.1). Mammalian cytoplasmic proteins are excluded from the CE nuclear compartment. Mammalian nucleolar, nuclear envelope, and nucleoplasmic antigens appeared in the corresponding subcompartments of the chick nuclei. The translocation of mammalian nuclear proteins into the chick nuclei was verified biochemically by isolating chick nuclei from heterokaryons and analysing them for mammalian proteins (Appels *et al.* 1974a,b, 1976; Appels and Ringertz 1975). In addition to uptake of mammalian proteins we found that chick H5 linker histone was lost from erythrocyte chromatin. One of the mammalian proteins concentrated by CE nuclei is RNA polymerase I

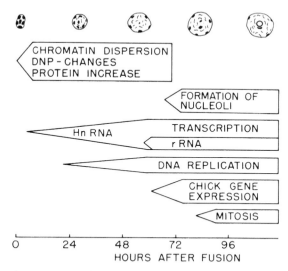

Fig. 3.2. Schematic summary of nuclear events during chick nuclear reactivation in heterokaryons made with HeLa cells and other transcriptionally active mammalian cells (based on data summarized by Harris (1967, 1968, 1970) and Ringertz and Savage (1976)).

Table 3.1. Uptake and loss of proteins by reactivating CE nuclei

A. Protein/antigen taken up	Reference
RNA polymerase I	Scheer et al. (1983)
RNA polymerase II	Zuckerman et al. (1982)
DNA polymerase alpha	Nakamura, Morita and Ringertz, N.R (unpublished observations)
SV40T	Rosenqvist et al. (1975) Kellermayer et al. (1978) Dubbs et al. (1978)
DNA repair factors	Darzynkiewicz and Chelmicka-Szorc (1972) van der Veer and Bootsma (1982)
snRNP antigens	Bergman et al. (1990)
La (RNA pol III termination factor)	Bergman et al. (1990)
Nuclear matrix proteins	Lafond et al. (1983) Lafond and Woodcock (1983)
Non-histone nuclear proteins	Appels et al. (1974a, 1976)
Nucleolar antigens	Ringertz et al. (1971)
Nuclear envelope proteins/antigens	Ringertz et al. (1971) Jost et al. (1979) Nyman et al. (1984)
B. Loss of chick nuclear components	Reference
H5 linker histone	Appels et al. (1974a) Linder et al. (1982b)
Polyamines	Hougaard et al. (1987)
DNA synthesis inhibitor	Ringertz et al. (1985)

(Scheer et al. 1983) (Fig. 3.3). Other studies demonstrated uptake of mammalian RNA polymerase II (Zuckerman et al. 1982), small nuclear ribonucleoprotein (snRNP) complexes required for splicing (Bergman et al. 1990), DNA polymerase α (H. Nakamura, T. Morita, and N. Ringertz, unpublished observations) and DNA repair enzymes (Darzynkiewicz and Chelmicka-Szorc 1972). Using α-amanitin resistant mammalian cells with a mutant RNA polymerase II (Zuckerman et al. 1982) we showed that the chick DNA is transcribed by the mutant mammalian enzyme. Other groups used temperature-sensitive replication mutants as fusion partners for the CE to show that replication too is by the imported mammalian enzyme (Dubbs and Kit 1977; Dubbs et al. 1978; Floros, et al. 1978; Tsutsui et al. 1978; Floros and

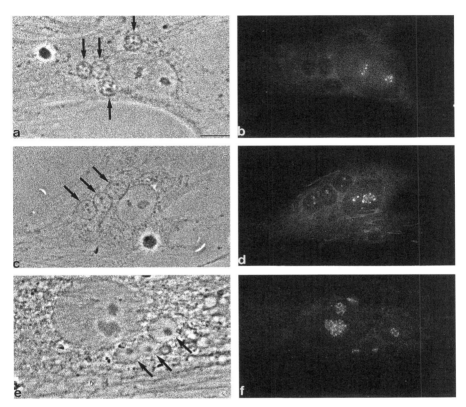

Fig. 3.3. Rat myoblast × CE heterokaryons photographed in the fluorescence microscope after immune staining with rabbit antiserum to RNA polymerase I (b, d, and f) and in the phase contrast microscope (a, c, and e). At 28 h after fusion nucleoli in the rat nuclei of heterokaryons show a punctuate fluorescence pattern (b). Chick erythrocyte nuclei (arrows in (a)) are negative and there are no distinct nucleoli. Seventy-two hours after fusion the newly formed nucleoli of CE nuclei are stained (d). This immunofluorescence becomes stronger with time. At 190 h several small foci can be distinguished inside the reactivated CE nuclei (f). Consequently, rat RNA polymerase I is taken up by the chick nuclei and takes part in their assembly of nucleoli. (Reproduced from Scheer *et al.* (1983) with the permission of Academic Press.)

Baserga 1980). Formation of a nuclear matrix is an early event in CE nuclear reactivation and uses imported, preformed mammalian proteins (Lafond and Woodcock 1983; Lafond *et al.* 1983; Woodcock and Woodcock 1986).

Three days after fusion a number of chick proteins are made. Among these are chick nucleolar proteins, which appear not only in the newly formed nucleoli of the chick nuclei but also in the nucleoli of the mammalian fusion partner (Ringertz *et al.* 1971). Taken together these observations show that in heterokaryons there is

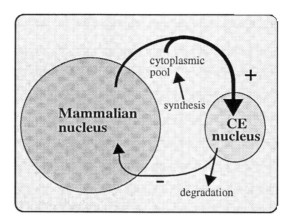

Fig. 3.4. Schematic summary of nucleocytoplasmic protein exchange in CE × mammalian heterokaryons. The CE nucleus takes up a variety of mammalian nuclear proteins while at the same time excluding mammalian cytoplasmic proteins. Among the proteins taken up are RNA and DNA polymerases and factors which activate (+) transcription and replication in the chick nuclei. The CE nuclei also release some proteins bound to the inactive CE chromatin. These proteins tend to suppress (–) functions in the mammalian nucleus and may be responsible for a suppression of some nuclear antigens.

a drastic redistribution of proteins between the mammalian and the chick nuclear compartments and that this redistribution is, most likely, an important part of the mechanisms causing the reactivation (Fig. 3.4)

Nuclear competition phenomena

Partial reactivation of CE nuclei occurs also in fusions with enucleated mammalian myoblasts (Ege et al. 1973, 1975) and fibroblasts (Lipsich et al. 1978; Woodcock and Woodcock 1986), but the degree of nuclear activation never reaches the levels seen in heterokaryons. Analysing heterokaryons containing different proportions of mammalian and chick nuclei it has been found (Ege et al. 1971; Carlsson et al. 1973) that the rate of nuclear growth, the rate at which mammalian nuclear antigens are taken up, and the rate of [^{3}H]uridine incorporation into RNA are affected by the ratio of mammalian to chick nuclei. The fastest reactivation is seen in heterokaryons containing one CE nucleus and one mammalian nucleus. The slowest is in heterokaryons containing a high ratio of CE to mammalian nuclei. If the heterokaryons contain many chick nuclei, transcription in the mammalian nucleus is slowed in direct proportion to the number of CE nuclei present. The total RNA synthesis of the heterokaryon (mammalian plus chick nuclei) remains unaffected. These observations suggest that the nuclei compete with one another for a limited amount of factors necessary for transcription. The CE nuclei may actually 'steal' nuclear proteins from the mammalian nuclei. This could explain why the expression of snRNP antigens decreases in the mammalian nuclei at the same time as the CE nuclei become positive for the same antigens (Fig. 3.5). This type of phenomenon could, however, also be due to conformational changes and does not necessarily reflect a redistribution of molecules. Conformational changes, most likely, explain some results obtained with DNA antibodies (Fig. 3.6). At the same time as the CE nucleus becomes strongly reactive, the mammalian nucleus of the heterokaryon becomes less reactive (Nyman *et al.* 1984).

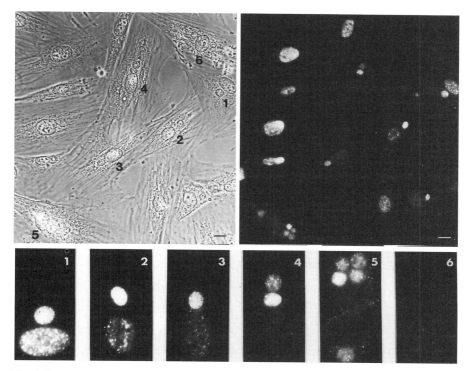

Fig. 3.5. Chick erythrocyte × rat L6 myoblast heterokaryons 24 h after fusion. Immune fluorescence staining with monoclonal Y12 snRNP antibodies. Uptake of mammalian snRNP antigens by chick nuclei is accompanied in most heterokaryons by a decrease or complete loss of snRNP antigens from the mammalian nucleus. Upper left: phase contrast picture in which six heterokaryons (numbered 1–6) are present. Upper right: fluorescence photomicrograph of the same field. Inserts 1–6 at the bottom show magnifications of heterokaryons 1–6. In number 1 both the CE and the mammalian nucleus are antigen positive. Heterokaryons 2–5 show different degrees of suppression of the snRNP antigens in the mammalian nucleus. Heterokaryons no. 6 is completely negative in both nuclei (bar = 10 μm).

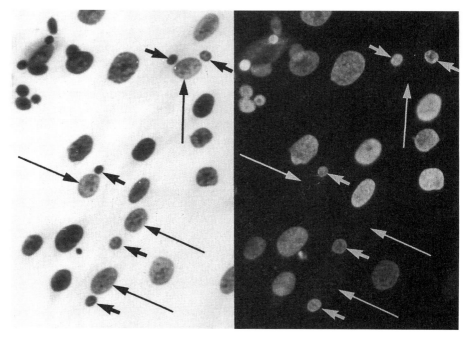

Fig. 3.6. Chick erythrocyte × rat L6 myoblast heterokaryons 24 h after fusion. Right: immune staining with human autoantibody to double-stranded DNA (P22). Left: Feulgen staining of the same field performed after immune staining had been photographed. DNA antigenicity is suppressed in mammalian nuclei (long arrows) present in CE heterokaryons. The positions of CE nuclei are marked by short arrows.

Inhibitory factors released by chick erythrocyte nuclei

Carlsson et al. (1973) noted that transcription was suppressed in the rat nuclei of rat epithelial cell × CE heterokaryons. The degree of suppression was directly related to the ratio of CE to rat nuclei. Similar observations were made by Darzynkiewicz et al. (1974). These authors also found that protease-sensitive suppressors are released from CE nuclei and taken up by the mammalian nucleus. Marked inhibition of DNA synthesis was noted in quail nuclei after fusion of quail fibroblasts with chick red cells. As the chick nuclei initiated DNA synthesis the quail nuclei were blocked in mid-S-phase (Ringertz et al. 1985). Fusion with nucleated CEs inhibited replication in rat L6 myoblast nuclei while no such effect was observed in fusions with mouse erythrocytes, which lack nuclei. Suppression of repair DNA synthesis was observed in human nuclei in human fibroblast × CE heterokaryons by van der Veer and Bootsma (1982). The inhibitory effect was again strongest in heterokaryons containing a high proportion of CE nuclei.

Linder et al.(1982b) confirmed the previous finding of Appels et al. (1974a) that H5 histone is lost from the CE nuclei during reactivation. They also observed that

H5 antigens sometimes appear in mammalian nuclei in CE × mammalian heterokaryons. Microinjection of chick H5 histone into proliferating mammalian cells results in transcriptional inactivation and nuclear condensation (Bergman et al. 1988). One of the factors responsible for the suppression of mammalian nuclei in CE × mammalian heterokaryons could be the H5 histone.

Mechanism of nucleocytoplasmic protein transport and sorting

Several reviews (e.g. Nigg 1990; Silver 1991) have summarized mechanisms involved in the import of protein from the cytoplasm into the nucleus. Since this article is focused on protein sorting in hetero- and homokaryons only a few brief comments will be made. Previous work has shown that large nuclear proteins have nuclear localization signals which target them for nuclear import. Translocation across the nuclear membrane takes place at the nuclear pores by a process which involves binding to receptors and an energy-dependent translocation step. Once inside the nucleus there must be further sorting and assembly processes of which we know very little. In the case of the nucleolar domain, specific nucleolar targeting signals have been established. Some proteins tend to remain for long periods of time in the nuclear compartment while others shuttle back and forth between the nucleus and the cytoplasm. The mechanisms controlling this traffic are unknown. At mitosis there is a major reorganization of nuclear proteins. A broad spectrum of proteins is dispersed into the cytoplasm at prophase and then concentrated and reassembled into nuclei during telophase and early G_1.

The fact that mammalian nuclear antigens are concentrated in chick nuclei and integrated into appropriate nuclear subcompartments suggests that the nuclear localization signals, the receptors at the nuclear pores, and the intranuclear targeting signals must be conserved during evolution. Binding of imported proteins to pre-existing nuclear structures may be an important element in the reactivation of CE nuclei. If for instance actinomycin is bound to DNA at concentrations which inhibit ribosomal gene transcription, mammalian nucleolar antigens enter the CE nuclear compartment but fail to get organized into nucleoli (Ege et al. 1971). Instead the mammalian nucleolar antigens are found throughout the nucleoplasm. This means that the nuclear import mechanism functions but that the intranuclear binding and assembly of the antigenic molecules have been perturbed. For a discussion of intranuclear protein sorting see Ringertz (1992).

The loss of H5 histone from the CE nuclei may be due to this histone being replaced by mammalian H1 linker histone (Appels et al. 1974a), but may also be a consequence of histone modification reactions. Pfeffer et al. (1988) found hyperacetylation of histones in CE nuclei undergoing reactivation in heterokaryons.

The swelling of the CE nucleus during reactivation is independent of chick nuclear transcription and takes place also in UV-irradiated CE nuclei unable to synthesize RNA. Autoradiography of heterokaryons formed by fusion with cells prelabelled with radioactive amino acids shows that many of the proteins concentrated by the CE nuclei must have been made before fusion. Further support for this conclusion is the finding that the nuclei swell also if new synthesis of proteins is blocked by puromycin. An interesting finding by Kellermayer *et al.* (1978) is that CE nuclear swelling, import of SV40T antigen, and DNA replication are blocked by colchicine. This suggests that CE nuclear reactivation depends on the integrity of the microtubular system. Uptake of nuclear proteins by the CE nucleus may therefore be not by random diffusion but by a vectorial transport system involving microtubules.

Does nucleocytoplasmic protein exchange reprogram the CE nucleus?

Since CE nuclei take up functionally important mammalian RNA and DNA polymerases it was of considerable interest to examine if, and to what extent, the CE nuclei retained their programming for erythroid gene expression. Chick erythrocytes of the definitive type were fused to rat L6 myoblasts, hamster BHK cells, and mouse neuroblastoma cells (Linder *et al.* 1982*a*). Immunopurified lysates from [^{35}S]methionine-labelled heterokaryons were analysed by two-dimensional gel electrophoresis. Subsequent to the reactivation of the CE nuclei, at least 40 new chick proteins appeared and increased several-fold with time after fusion. The pattern of chick polypeptide synthesis was very similar in the three types of heterokaryons. Three polypeptides synthesized by non-erythroid chicken cells, but less so in embryonic erythrocytes, were conspicuous in heterokaryons. Chicken adult globins α^A, α^D, and β were synthesized in several types of heterokaryons but the level of expression varied (Linder *et al.* 1981).

In later work we used nucleic acid hybridization with cDNA probes to analyse chick gene expression (Bergman and Ringertz 1990). Previously work by Gariglio *et al.* (1972) had shown that in spite of the inactive state of nuclei in mature erythrocytes, RNA polymerases remain attached to some genes and will produce transcripts in run-off assays. Northern blot analysis of chick transcripts present in CE × rat myoblast heterokaryons examined shortly after fusion (6 h) showed degradation of pre-existing RNA species brought into the heterokaryons by the erythrocyte fusion partner. Beginning at approximately 20 h post-fusion, and increasing with time after fusion, newly made chick RNA species were detected. These RNA molecules failed to appear if the CEs were irradiated with UV light prior to fusion. This treatment prevents transcriptional reactivation after fusion. It was concluded, therefore, that in normal heterokaryons, the new RNA species detected from 20 h after fusion were due to the reactivation of CE nuclei.

The expression of individual chick genes was probed by Northern blot hybridization with cDNA probes (Bergman and Ringertz 1990) and found to fall into five categories. Firstly, some genes were capable of transcription in the run-off assay with isolated CE nuclei which continue to be transcribed in heterokaryons. Examples of this pattern were the H5 histone, adult α-globin, and CAII (carbonic anhydrase, erythroid-specific isoform II) genes. Secondly, there were genes that were somewhat active in the run-off assay, but which were suppressed in the heterokaryons (Eryf1, band 4.1, and band 3 genes). Thirdly, some genes were inactive in the run-off assay, but were reactivated in the heterokaryons. The c-*myb* and the ALA-S (D-aminolevulinate synthase) genes belong in this category. Fourthly, the embryonic β-globin gene represents an example of a gene that was inactive in the run-off assay with nuclei from mature erythrocytes but which was transcriptionally activated in at least one type of heterokaryon. The corresponding protein, however, was not seen in two-dimensional gel analysis of cell lysates (Lanfranchi *et al.* 1984). The fifth category is represented by the chick acetylcholine receptor and ovalbumin genes. Neither of these genes was in an active state in erythrocytes and neither was reactivated after fusion.

The main conclusion from our analysis of chick gene expression in heterokaryons is that some genes that are active during erythropoiesis (e.g. globin, H5 linker histone, ALA-S, and CAII) are expressed in heterokaryons. Globin gene expression is strong in fusions with K562 human fetal erythroleukaemia cells and rat L6 myoblasts but barely detectable in fusions with mouse A9 fibroblasts and A9 $\times$ L6 hybrid cells. It is possible, therefore, that some types of mammalian fusion partners contain transcriptional regulators that suppress chicken globin gene expression. We have been unable to obtain convincing evidence that myoblast fusion partners reprogram CE nuclei to express chick muscle genes (Carlsson *et al.* 1974*b*). In this respect our results differ from those of Chiu and Blau (1984) and Blau *et al.* (1985) who found activation of human myogenic gene expression in fusions of a variety of human cell types with a mouse myoblast line. This system differs from ours in that it uses two mammalian cell lines both of which are transcriptionally active at the time of fusion.

Experiments by Diberardino and Hoffner (1983) have showed that frog erythrocyte nuclei, if serially transplanted through oocyte cytoplasms, do become totipotent and capable of supporting development of tadpoles, a developmental stage at which the organism contains several forms of differentiated cells, among them muscle. Again, it can be said that this system is different from ours since it involves longer exposure to an environment that may be effective in reprogramming nuclei. Somatic cell hybridization studies using adult mouse erythroleukaemias and human non-erythroid cells (for references see Baron and Maniatis 1986) have shown that globin gene expression can be activated in a variety of non-erythroid cell types. These results suggest that the genome of differentiated cells can be reprogrammed to express another tissue-specific pattern of genes. The failure to reprogram CE nuclei may be because the phylogenetic difference between the species involved (chick and rat) is greater than that in some of the other combinations mentioned above. Time could be an important factor. The CE nucleus starts from a transcriptionally inactive state and may require a longer exposure to the reprogramming cytoplasm than would a nucleus already active at the time of fusion.

Myoblast homo- and heterokaryons

Rat myoblasts fuse spontaneously to form large multinucleated myotubes (homokaryons) which begin to synthesize contractile proteins and develop into muscle fibres. The main features of this process are summarized in Fig. 3.7. Hybrid myotubes

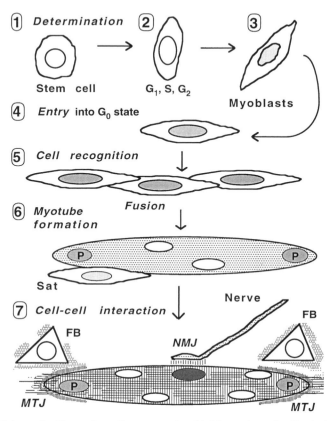

Fig. 3.7. Schematic illustration of myogenesis. (1) Mesenchymal cells derived from the myotome part of the somites are programmed (determined) for the myogenic differentiation pathway, (2) multiply, (3) migrate, and enter a G_0 resting stage (4). Non-replicating G_0 myoblasts undergo cell surface changes and cell recognition phenomena so as to line up into bands (5), and then fuse into multinucleate myotubes (6). In these fibres genes for contractile proteins and other muscle-specific proteins are activated and myofilaments form. Some myoblasts remain in a dormant stage (satellite cells, 'Sat') but may be activated later in regenerative processes. (7) Terminal differentiation of muscle fibres from myotubes involves cell–cell interactions with nerve cells to establish neuromuscular junctions (NMJs), and with fibroblasts to form myotendinous junctions (MTJs) which anchor the cytoskeleton of the muscle fibre to extracellular matrix and collagen. In the text, evidence is discussed that nuclei specialize at the NMJ and the two MTJs with respect to their gene expression patterns. Nuclei at the NMJ express their acetylcholine receptor genes at a higher level than other nuclei (Changeux 1991). Nuclei at the poles (P) of the myotubes show a stronger expression of the RB protein than other nuclei and may be specialized because of the need to form MTJs. At these points a number of specialized proteins are needed to anchor the cytoskeleton to extracellular matrix and collagen. FB denotes a fibroblast.

can be formed by Sendai virus-induced fusion of CEs with rat myoblasts (Carlsson *et al.* 1970). In these myotubes the chick nuclei rapidly take up and concentrate rat nucleoplasmic and nucleolar antigens. Uptake of chick nuclear envelope antigens by adjacent rat nuclei was observed in hybrid myotubes formed by the spontaneous fusion of chick and rat myoblasts (Fig. 3.8). This uptake was slower, however, than the uptake of rat nuclear antigens by the rapidly growing CE nuclei (Carlsson *et al.* 1974*a*). In some myotubes, gradients with respect to the amount of chick antigen taken up by the rat nuclei were observed; the rat nuclei nearest to the chick nuclei were more positive than the more distant rat nuclei. Localization of cytoplasmic muscle gene products in restricted domains has also been described by Pavlath *et al.* (1989) for interspecific human–mouse hybrid myotubes. Ralston and Hall (1989) mixed transfected myoblasts expressing a reporter protein fused to a nuclear localization signal with normal myoblasts. In multinucleate hybrid myotubes the reporter protein was expressed not only by the transfected nuclei but also by neighbouring nuclei within a distance of 50 μM of the source nucleus.

Heterokaryons made by polyethylene glycol induced fusion of human non-muscle cells with mouse myoblasts have been used extensively by Blau and collaborators (Blau *et al.* 1983, 1985; Hardeman *et al.* 1986; Pavlath and Blau 1986) to analyse mechanisms regulating tissue-specific gene expression. The mouse myoblasts induce expression of human muscle genes in human amniocytes. Furthermore, human fibroblasts nuclei after being present for some time in heterokaryons made with mouse myoblasts, will produce *trans*-acting muscle regulators that change the expression of mouse muscle genes. Similar to the CE × mammalian heterokaryons there is a dialogue between nuclei of two different species and tissue types. The difference between the two systems seems to be that fusions of transcriptionally active mammalian cells allow a massive reprogramming of gene expression to occur.

Normal multinucleate myotubes formed by fusion of cloned myoblasts are homokaryons, i.e. all nuclei are believed to be equal in terms of their species of origin, programming for tissue-specific gene expression and cell cycle stage (G_0). Some recent findings obtained in collaboration with George Klein's group (Szekely *et al.* 1993) suggest, however, that nuclei may become specialized *after* fusion because of their geographical positions within myotubes. Fig. 3.9 shows the immunostaining pattern of a rat L6 myotube with antibodies directed against the retinoblastoma (RB) protein. The RB gene is a tumour suppressor gene encoding a 140 kDa nuclear phosphoprotein believed to have a role in controlling cell cycle exit. We found RB antigens to be expressed by mononucleate myoblasts at a stage when they were entering a G_0 state and beginning to fuse. In myotubes the nuclei at the two ends of the myotubes showed a much stronger immunofluorescence than did centrally located nuclei. Similar observations (Szekely *et al.* 1993) were made in experiments in which *human* RB was expressed from a cytoplasmic Semliki Forest virus vector. Using an antibody that detects human but not rat RB protein, we found human RB antigens to be concentrated in nuclei located at the two ends of the rat myotubes. These results suggest that nuclei are specialized within the rat myotubes and that the RB protein, irrespective if its species of origin, is selectively

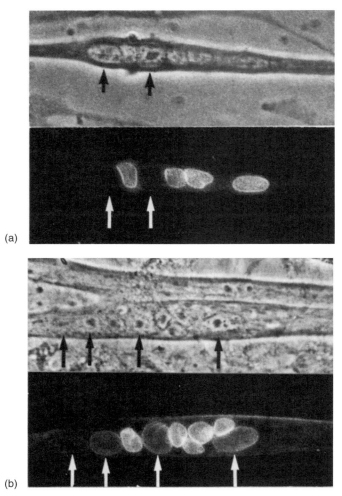

Fig. 3.8. Migration of chick nuclear envelope antigens into rat nuclear envelopes in chick myoblast × rat myoblast hybrid myotubes, (a) 3 days after starting the cultures and (b) at 6 days. Rat nuclei are marked with arrows and are larger than the chick myoblast nuclei. Upper parts of (a) and (b) show phase contrast images; lower parts show immune fluorescence after staining with human antibodies reacting with chick but *not* with rat nuclear envelope antigens. Migration of chick antigens into rat nuclei is seen at 6 days but not at 3 days. In the upper part of (b) there is a myotube that contains only rat nuclei. In some myotubes, gradients with respect to the amount of chick antigen taken up by the rat nuclei were observed; the rat nuclei nearest to the chick nuclei were more strongly positive than the more distant rat nuclei. Note the negative rat nucleus to the far left in (b) at the same time as the rat nuclei close to the (smaller) chick myoblast nuclei are positive.

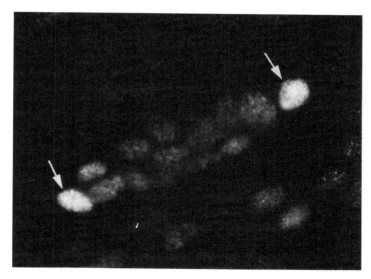

Fig. 3.9. Nuclei occupying positions at the extreme ends of myotubes (arrows) show a much stronger RB antigen expression than do centrally located nuclei. Nuclear specialization within myotubes may be related to particular cytoplasmic and surface specializations such as the formation of the myotendinous (MTJ) and neuromuscular (NMJ) junctions (for explanations see legend to Fig. 3.7).

concentrated in nuclei occupying terminal positions. This expression pattern brings to mind observations made in studies of the neuromuscular junction by Fontaine *et al.* (1988) and Harris *et al.* (1989), and reviewed by Changeux (1991). As nerve endings form stable motor end plates with myotubes, the nuclei and cytoplasm underlying the end plates become specialized. The nuclei in this domain continue to express the acetylcholine receptor gene while expression is reduced in other parts of the myotubes. Although we do not yet understand why the RB protein would be needed at the two ends of the myotubes it could be part of a nuclear specialization that fulfils an important function. At the two ends of the myotube there may be a need for a specialized gene expression in order to produce proteins needed for establishing the myotendinous junctions (Fig. 3.7).

Concluding remarks

When Professor Harris introduced heterokaryons as a system for studying gene regulatory mechanisms in somatic cells he opened a new field. The CE heterokaryons have turned out to be a useful system for the analysis of nucleocytoplasmic protein exchange and its role in gene regulation. One of the advantages of the system is that chick and mammalian cells are sufficiently similar that in fused cells the genome of one species responds to signals from the other species. Yet, chick and mammalian species are different enough to make it easy to obtain antibodies that recognize proteins of one species but not of the other. The same can be said for cDNA probes used to study gene expression. Lastly, the size and DNA content of the nuclei are sufficiently different that one can recognize individual nuclei as being of chick or mammalian origin. The system is easily accessible to immunocytochemistry but it can also be studied by biochemical methods.

The lessons learnt from CE heterokaryons suggest that reactivation of CE nuclei involves several steps, the most important ones being

1. early biophysical changes in deoxyribonucleoprotein complexes triggered by ionic signals;
2. uptake of mammalian nuclear proteins, including RNA and DNA polymerases;
3. loss of suppressors from the inactive CE chromatin;
4. histone modifications;
5. assembly processes resulting in the formation of a nuclear matrix and specialized nuclear domains.

Reactivation of CE nuclei in CE × mammalian heterokaryons suppresses the mammalian nuclei. This could be brought about by competion between nuclei for a limited supply of transcription factors and/or by release of suppressors from the CE nuclei.

Our work on CE × rat myoblast heterokaryons shows that CE nuclei take up rat nuclear antigens during the reactivation process. In spite of this, and unlike the situation in some other hybrid cells, the CE nuclei tend to 'remember' at least part of their erythroid programming when reactivated in mammalian myogenic cells. In chick myoblast × rat myoblast hybrid myotubes there is a slow exchange of nuclear antigens between neighbouring nuclei and coexpression of both chick and rat myogenic differentiation markers.

Myoblast homokaryons formed by fusion of cloned rat myoblasts tend to undergo nuclear specialization *after* fusion. These phenomena appear to be controlled by topological factors and may be important for the formation of the myotendinous junctions.

References

Appels, R. and Ringertz, N. R. (1975). Chemical and structural changes within chick erythrocyte nuclei introduced into mammalian cells by cell fusion. *Curr. Topics Dev. Biol.*, **9**, 137–66.

Appels, R., Bolund, L., and Ringertz, N. R. (1974*a*). Biochemical analysis of reactivated chick erythrocyte nuclei isolated from chick/HeLa heterokaryons. *J. Mol. Biol.*, **87**, 339–55.

Appels, R., Bolund, L., Goto, S., and Ringertz, N. R. (1974*b*). The kinetics of protein uptake by chick erythrocyte nuclei during reactivation in chick–mammalian heterokaryons. *Exp. Cell. Res.*, **85**, 173–81.

Appels, R., Tallroth, E., Appels, D. M., and Ringertz, N. R. (1976). Differential uptake of protein into the chick nuclei of HeLa × chick erythrocyte heterokaryons. *Exp. Cell Res.*, **92**, 79–86.

Baron, M. H. and Maniatis, T. (1986). Rapid reprogramming of globin gene expression in transient heterokaryons. *Cell.*, **46**, 591–602.

Bergman, M. and Ringertz, N. R. (1990). Gene expression pattern of chicken erythroid nuclei in heterokaryons. *J. Cell Sci.*, **97**, 167–75.

Bergman, M., Wawra, E., and Winge, M. (1988). Chicken histone H5 inhibits transcription and replication when introduced into proliferating cells by microinjection. *J. Cell Sci.*, **91**, 201–9.

Bergman, M., Nyman, U., Ringertz, N. R., and Pettersson, I. (1990). Appearance and origin of snRNP antigens in chick erythrocyte nuclei reactivated in heterokaryons. *J. Cell Sci.*, **95**, 361–70.

Blau, H. M., Chiu, C.-P and Webster, C. (1983). Cytoplasmic activation of human nuclear genes in stable heterocaryons. *Cell*, **32**, 1171–80

Blau, H. M., Pavlath, G. K., Hardeman, E. C., Chiu, C.-P., Silberstein, L., Webster, S. G., *et al.* (1985). Plasticity of the differentiated state. *Science*, **230**, 758–66.

Bolund, L., Ringertz, N. R., and Harris, H. (1969). Changes in the cytochemical properties of erythrocyte nuclei reactivated by cell fusion. *J. Cell Sci.*, **4**, 71–87.

Carlsson, S.-A., Savage, R. E., and Ringertz, N. R. (1970). Behaviour of differentiated hen nuclei in the cytoplasm of rat myoblasts and myotubes. *Nature*, **228**, 869–71.

Carlsson, S.-A., Moore, G. P. M., and Ringertz, N. R. (1973). Nucleocytoplasmic protein migration during the activation of chick erythrocyte nuclei in heterokaryons. *Exp. Cell Res.*, **76**, 234–41.

Carlsson, S.-A., Ringertz, N. R., and Savage, R. E. (1974*a*). Intracellular antigen migration in interspecific myoblast heterokaryons. *Exp. Cell Res.*, **84**, 255–66.

Carlsson, S.-A., Luger, O., Ringertz, N. R., and Savage, R. E. (1974*b*). Phenotypic expression in chick erythrocyte × rat myoblast hybrids and in chick × rat myoblast hybrids. *Exp. Cell Res.*, **84**, 47–55.

Changeux, J.-P. (1991). Compartmentalized transcription of acetylcholine receptor genes during motor endplate epigenesis. *New Biol.*, **3**, 413–29.

Chiu, C.-P. and Blau, H. M. (1984). Reprogramming of differentiation in the absence of DNA synthesis. *Cell*, **37**, 879–87.

Darzynkiewicz, Z. and Chelmicka-Szorc, E. (1972). Unscheduled DNA synthesis in hen erythrocytes reactivated in heterokaryons. *Expt. Cell Res.*, **74**, 131–9.

Darzynkiewicz, Z., Chelmicka-Szorc, E., and Arnason, B. G. (1974). Chick erythrocyte nucleus reactivation in heterokaryons: suppression by inhibitors of proteolytic enzymes. *Proc. Natl Acad. Sci. USA*, **71**, 644–7.

Diberardino, M. A. and Hoffner, N.J. (1983). Gene activation in erythrocytes: nuclear transplantation in oocytes and eggs of *Rana*. *Science*, **219**, 862–4.

Dubbs, D. R. and Kit, S. (1977). Initiation of DNA synthesis and uptake of T-antigen by chick erythrocyte nuclei in heterokaryons with SV40 transformed human cells. *Somat. Cell. Genet.*, **3**, 61–9.

Dubbs, D. R., Trkula, D., and Kit, S. (1978). T-antigen and initiation of cell DNA synthesis in a temperature-sensitive mouse line transformed by an SV40tsA mutant and in heterokaryons of the transformed cells and chick erythrocytes. *Somat. Cell Genet.*, **4**, 95–110.

Ege, T., Carlsson, S.-A., and Ringertz, N. R. (1971). Immune microfluorimetric analysis of the distribution of species specific nuclear antigens in HeLa–chick erythrocyte heterokaryons. *Exp. Cell Res.*, **69**, 472–7.

Ege, T., Zeuthen, J., and Ringertz, N. R. (1973). Cell fusion with enucleated cytoplasms. Nobel Symposium 23 In *Chromosome identification*, pp. 189–94. (Ed. T. Caspersson and L. Zech) Academic Press, New York.

Ege, T., Zeuthen, J., and Ringertz, N. R. (1975). Reactivation of chick erythrocyte nuclei after fusion with enucleated cells. *Somat. Cell Genet.*, **1**, 65–80.

Floros, J. and Baserga, R. (1980). Reactivation of G0 nuclei by S-phase cells. *Cell Biol. Int. Rep.*, **4**, 75–82.

Floros, J., Ashihara, T., and Baserga, R. (1978). Characterization of ts13 cells, a temperature-sensitive mutant of the G1 phase of the cell cycle. *Cell Biol. Int. Rep.*, **2**, 259–69.

Fontaine, B., Sassoon, D., Buckingham, M., and Changeux, J.-P. (1988). Detection of the nicotine acetylcholine receptor α-subunit mRNA by *in situ* hybridization at neuromuscular junctions of 15 day old striated muscles. *EMBO J.*, **7**, 603–9.

Gariglio, R., Bellard, M., and Chambon, P. (1972). Clustering of RNA polymerase B molecules in the 5′ moiety of the adult beta-globin gene of hen erythrocytes. *Nucleic Acids Res.*, **9**, 2589–98.

Hardeman, E. C., Chiu, C.-P, Minty, A., and Blau, H. M. (1986). The pattern of actin expression in human fibroblast × mouse muscle heterokaryons suggests that human muscle regulatory factors are produced. *Cell*, **47**, 123–30.

Harris, H. (1965). *Cell fusion*, The Dunham Lectures. Oxford University Press, London.

Harris, H. (1967). The reactivation of the red cell nucleus. *J. Cell Sci.* **2**, 23–32.

Harris, H. (1968). *Nucleus and cytoplasm*. Oxford University Press, London and New York.

Harris, H. (1970). *Cell fusion*, The Dunham Lectures. Oxford University Press, London and New York.

Harris, D. A., Falls, D. L., and Fischbach, G. D. (1989). Differential activation of myotube nuclei following exposure to an acetylcholine receptor-inducing factor. *Nature*, **337**, 173–6.

Hougaard, D. M., Bolund, L., Fujiwara, K., and Larsson, L. I. (1987). Endogenous polyamines are intimately associated with highly condensed chromatin *in vivo*. A fluorescence cytochemical and immunocytochemical study of spermine and spermidine during the cell cycle and in reactivated nuclei. *Eur. J. Cell. Biol.* **44**, 151–5.

Jost, E., d'Arcy, A., and Ely, S. (1979). Transfer of mouse nuclear envelope specific proteins to nuclei of chick erythrocytes during reactivation in heterokaryons with mouse cells. *J. Cell Sci.*, **37**, 97–107.

Kellermayer, M., Jobst, K., and Szües, G. (1978). Inhibition of internuclear transport of SV40 induced T-antigen in heterokaryons. *Cell Biol. Int. Rep.*, **2**, 19–24.

LaFond, R.E. and Woodcock, C. L. F. (1983). Status of the nuclear matrix in mature and embryonic chick erythrocyte nuclei. *Exp. Cell Res.*, **147**, 31–9.

Lafond, R. E., Woodcock, H., Woodcock, C. L. F., Kundahl, E. R., and Lucas, J. J. (1983). Generation of an internal nuclear matrix in avian erythrocyte nuclei during reactivation in cytoplasts. *J. Cell Biol.*, **96**, 1815–19.

Lanfranchi, G., Linder, S., and Ringertz, N. R. (1984). Globin synthesis in heterokaryons formed between chick erythrocytes and human K562 cells or rat L6 myoblasts. *J. Cell Sci.* **66**, 309–19.

Linder, S., Zuckerman, S. H. and Ringertz, N. R. (1981). Reactivation of chicken erythrocyte nuclei in heterokaryons results in expression of adult chicken globin genes. *Proc. Natl Acad. Sci. USA*, **78**, 6286–9.

Linder, S., Zuckerman, S. H., and Ringertz, N. R. (1982*a*). Pattern of chick gene activation in chick erythrocyte heterokaryons. *J. Cell Biol.*, **95**, 885–92.

Linder, S., Zuckerman, S. H., and Ringertz, N. R. (1982*b*). Distribution of histone H5 in chicken erythrocyte-mammalian cell heterokaryons. *Exp. Cell Res.*, **140**, 464–8.

Lipsich, L. A., Lucas, J. J., and Kates, J. R. (1978). Cell cycle dependence of the reactivation of chick erythrocyte nuclei after transplantation into mouse L-929 cytoplasts. *J. Cell. Physiol.*, **97**, 199–208.

Nigg, E. A. (1990). Mechanisms of signal transduction to the nucleus. *Adv. Cancer Res.*, **55**, 271–310.

Nyman, U., Lanfranchi, G., Bergman, M. and Ringertz, N. R. (1984). Changes in nuclear antigens during reactivation of chick erythrocyte nuclei in heterokaryons. *J. Cell. Physiol.*, **120**, 257–64.

Pavlath, G. K. and Blau, H. M. (1986). Expression of muscle genes in heterokaryons depends on gene dosage. *J. Cell Biol.*, **102**, 124–30.

Pavlath, G. K., Rich, K., Webster, S. G. and Blau, H. M. (1989). Localization of muscle gene products in nuclear domains. *Nature*, **337**, 570–3.

Pfeffer, U., Ferrari, N., Tosetti, F., and Vidali, G. (1988). Histone hyperacetylation is induced in chick erythrocyte nuclei during reactivation in heterokaryons. *Exp. Cell Res.*, **178**, 25–30.

Ralston, E. and Hall, Z. W. (1989). Transfer of a protein encoded by a single nucleus to nearby nuclei in multinucleated myotubes. *Science*, **244**, 1066–9.

Ringertz, N. R. (1969). Cytochemical properties of nuclear proteins and deoxyribonucleoprotein complexes in relation to nuclear function. In *Handbook of molecular cytology*, (ed. a. Lima de Faria) pp. 656–84. North Holland, Amsterdam.

Ringertz, N. R. (1992). Intranuclear sorting and assembly of proteins. *Cell Biol. Int. Rep.*, **16**, 799–810.

Ringertz, N. R., Carlsson, S.-A., Ege, T., and Bolund, L. (1971). Detection of human and chick nuclear antigens in chick erythrocyte nuclei during reactivation in heterokaryons with HeLa cells. *Proc. Natl Acad. Sci. USA*, **68**, 3228–32.

Ringertz, N. R. and Savage, R. (1976). *Cell hybrids*. Academic Press, New York.

Ringertz, N. R., Nyman, U., and Bergman, M. (1985). DNA replication and histone H5 exchange during reactivation of chick erythrocyte nuclei in heterokaryons. *Chromosoma (Berlin)*, **91**, 391–6.

Rosenqvist, M., Stenman, S., and Ringertz, N. R. (1975). Uptake of SV40 T-antigen into chick erythrocyte nuclei in heterokaryons. *Exp. Cell Res.*, **92**, 515–18.

Scheer, U., Lanfranchi, G. Rose, K. M., Franke, W. W., and Ringertz, N. R. (1983). Migration of rat RNA polymerase I into chick erythrocyte nuclei undergoing reactivation in chick–rat heterokaryons. *J. Cell Biol.*, **97**, 1641–3.

Silver, P. A. (1991). How proteins enter the nucleus. *Cell*, **64**, 489–97.

Szekely, L., Jiang, W.-Q., Klein, G., Rosén, A., and Ringertz, N. R. (1993). Position dependent nuclear translocation of the retinoblastoma (RB) protein during *in vitro* myogenesis. *J. Cell Physiol.*, **155**, 313–22.

Tsutsui, Y., Chang, S. D., and Baserga, R. (1978). Failure of reactivation of chick erythrocytes after fusion with temperature-sensitive mutants of mammalian cells arrested in G1. *Exp. Cell Res.*, **113**, 359–67.

Van der Veer, E. and Bootsma, D.(1982). Repair DNA synthesis in heterokaryons during reactivation of chick erythrocytes fused with human diploid fibroblasts or HeLa cells. *Exp. Cell. Res.*, **138**, 469–74.

Woodcock, L. F. and Woodcock, H. (1986). Nuclear matrix generation during reactivation of avian erythrocyte nuclei: an analysis of the protein traffic in hybrids. *J. Cell Sci.*, **84**, 105–27.

Zuckerman, S. H., Linder, S., and Ringertz, N. R. (1982). Transcription of chick genes by mammalian RNA polymerase II in chick erythrocyte–mammalian cell heterokaryons. *J. Cell. Physiol.*, **113**, 99–104.

4. Some insights into the replication of damaged DNA in mammalian cells

R. T. Johnson, C. S. Downes, D. B. Godfrey, D. H. Hatton, A. J. Ryan, and S. Squires

Introduction

In a volume devoted to the legacy of cell fusion, this article on cellular responses to DNA damage contains only a single example of the use of the technique, though the information gained thereby has transformed our view of the mutant cell whose phenotype is described at some length below. In the wider sense cell fusion continues to play an invaluable role in the analysis of DNA repair problems, most notably in complementation analysis using heterokaryons (Giannelli *et al*. 1982) or permanent hybrid cells (Johnson *et al*. 1989). In this way the number of genes involved in a repair pathway can be identified, by separating a set of repair-defective mutant cells into different complementation groups. Cell fusion also allows the recessive state of repair-defective mutants to be confirmed in hybrids between mutant and wild-type cells (e.g. Hentosh *et al*. 1990) or, much less commonly, shown as a dominant (A. J. Ryan, C. S. Downes, and R. T. Johnson, unpublished). Complementation of a repair-defective cell by fusion with microcells containing few chromosomes has allowed the identification of the human chromosome carrying the gene for xeroderma pigmentosum complementation group D (chromosome no. 19) (Fleijter *et al*. 1992) or for Bloom's syndrome (no. 15) (McDaniel and Schultz 1992). Finally, metaphase × interphase cell fusion has allowed interphase chromosomes to be visualized as PCC (Johnson and Rao 1970) and in turn this has allowed repair of interphase chromosome damage to be observed and quantified after treatment with ionizing radiation (Hittelman and Rao 1974; Waldren and Johnson 1974; Johnson *et al*. 1982). The sensitivity of the PCC technique enabled Cornforth and Bedford (1985) to show definitively that cells from individuals with the X-ray sensitive condition ataxia telangiectasia were unable to repair as many chromosome breaks as normal cells, a result that predates the demonstration of defective DNA double-strand break repair in these cells (Coquerelle *et al*. 1987; Blocher *et al*. 1991).

One of the dominating themes in Henry Harris' work has been the nature of malignancy in tumour cells. We here deal with a related theme: what is it that causes cells to become malignantly transformed in the first place? It is agreed that DNA damage causes mutations of all sorts, including oncogenic mutations to the transformed state and that DNA repair processes remove the damage and oppose this action. But DNA damage is not inherited by the descendants of a damaged cell; at least, in as far as unrepaired damage is inherited, it must be diluted out in subsequent generations. Mutations, on the other hand, are stably inherited. What we seek, therefore, is a mechanism by which DNA damage can be converted into stable mutations; this is to be found in the processes by which cells attempt to replicate damaged DNA. Faulty synthesis of DNA, on a damaged template, converts potentially mutagenic damage to altered daughter sequences, which themselves contain normal, undamaged DNA that is not capable of being amended by DNA repair mechanisms.

Responses to DNA damage in mammalian cells

DNA repair

The DNA damaging agent most widely used in repair and mutation studies is UV radiation, which is quick, cheap, and easy to use. What follows is based primarily on the responses of mammalian cells to UV irradiation, but while UV can be used as a model for other types of DNA damaging agent, it is important to remember that cellular responses to other agents may be quantitatively and qualitatively different (e.g. Vos and Hanawalt 1987; Downes et al. 1993).

The main forms of UV-induced DNA damage are the cyclobutane pyrimidine dimer (CB), which is usually a TpT dimer and the (6-4)pyrimidine–pyrimidone photoproduct, usually CpT (Mitchell and Nairn 1989). Both are formed between adjacent pyrimidine bases; CBs, which account for 70–90 per cent of the damage, are formed fairly uniformly through the genome, while (6-4)photoproducts, which cause much greater distortions of the DNA, are formed predominantly between nucleosomes. Both lesions are removed by the same mechanism, namely excision repair. Lesions are recognized by nuclear proteins and repair complexes of at least eight proteins assemble at the damage sites. Damaged DNA is unwound by helicases and the damaged strand excised by repair nucleases. The excised section is then replaced by a repair polymerase, using the opposite undamaged site as a template and the repaired strand is finally sealed by a ligase. This process is error-free, that is, if cells are allowed enough time after UV irradiation to excise all the UV damage, no mutations are formed when the cells subsequently replicate their DNA. In the genetic disease xeroderma pigmentosum, characterized by photosensitivity and a high frequency of skin tumors, excision repair of UV damage is usually defective. The process of excision repair has been extensively described (Friedberg 1985).

Though CB repair and (6-4)photoproduct repair share a common mechanism—a single repair defect reduces both in xerodermas—the processes occur at different rates. CBs are removed relatively slowly and the pattern of their repair varies with cell type and with the species of origin. From human cells, 50–60 per cent of CBs are removed within 24 h (Mitchell et al. 1985; Lehmann et al. 1988) and eventually all are excised (Kantor and Setlow 1981), but from transformed rodent cell lines in culture only 10–20 per cent are removed from the genome as a whole over 24 h (Meyn et al. 1974; Thompson et al. 1989).

This cell-specific difference in CB repair is due to a difference in discrimination by the excision repair system between transcribed and non-transcribed genes (recently reviewed by Downes *et al.* (1993). Both human and rodent cells remove CB from the transcribed strand of active genes with great efficiency; human cells also repair the non-transcribed strand with high efficiency and inactive genes at about half the maximum repair rate. Rodent cells, however, leave inactive genes largely unrepaired as far as CBs are concerned. But the burden of residual CB damage in rodent cells in no way makes them more sensitive to UV irradiation than human cells.

By contrast, (6-4)photoproducts are removed rapidly and entirely from all the DNA within a few hours in all normal mammalian cells. (6-4)Photoproduct repair is not wholly non-discriminatory; in rodent cells, removal occurs at about twice the average rate in active chromatin (Thomas *et al.* 1989).

DNA synthesis arrest

Irradiated cells rapidly cease to synthesize DNA and RNA. Partly, this is because polymerases cease to function when they encounter DNA lesions; but in the case of DNA synthesis, at least, the decline is due to deliberate regulation. Very low levels of UV or other DNA damaging agents, which leave most replicons undamaged, prevent the initiation of replication in all replicons, damaged or not (Painter 1980, 1985). A defect in the ability to inhibit replication occurs in the radiosensitive genetic disease ataxia telangiectasia. It is not yet known whether the rapid decrease in RNA synthesis that follows damage is also a regulated process.

After a few hours, nucleic acid synthesis recovers in damaged cells. In both hamster cells (Mitchell *et al.* 1988) and human cells (Broughton *et al.* 1990) the recovery of DNA synthesis after UV irradiation correlates with (6-4)photoproduct removal. In hamster cells, split-dose experiments indicate that replication in the presence of unrepaired (6-4)photoproducts is the major cause of cytotoxicity (Mitchell *et al.* 1988); evidence from excision-defective mutants implicates the (6-4)photoproduct as a major mutagenic lesion (Zdzienicka *et al.* 1992). It therefore appears that the purpose of replication delay after UV damage is to allow (6-4)photoproducts, but not necessarily CBs, to be removed by error-free excision repair. But how can replication continue in the presence of CBs?

Replication of the damaged template

Replication of mammalian DNA after damage occurs in three phases. In the first, which affects replicons that were already active at the time of damage or where initiation was not suppressed by the damage, the assembly of daughter DNA molecules is greatly retarded by the presence of damage in the parent strand; gaps are formed in the daughter strand at sites where polymerases are unable to proceed. These gaps have usually been detected by alkaline sucrose gradient centrifugation of replication-labelled DNA; we have developed more sensitive assays based on DNA chromatography (Pillidge *et al.* 1986*a*; Musk *et al.* 1990). In electron micrographs, replication forks can be seen to be stalled at damage sites in this case caused by a UV analogue (Armier *et al.* 1988). Immediately after UV irradiation, the spacing of DNA lesions and of arrested replication forks is approximately equal, in rodent and human cells alike (Meyn and Humphrey 1971; Buhl *et al.* 1973).

In the second phase, the gaps in the daughter strand of replicated DNA caused by the stalling of polymerases at lesion sites disappear. This process is usually known by the somewhat misleading name of 'post-replication repair' or PPR; it is important to remember that it is the daughter-strand gaps, not necessarily the lesion in the parental strand, that are repaired by PRR. We will refer to the disappearance of gaps in daughter strands, formed soon after damage, as 'early PRR'. This is to stress the contrast with events in the third phase, which begins several hours after UV damage. Here, replication proceeds unhindered by any CBs that remain in the genome; no lesion-induced gaps can be detected in newly synthesized DNA, even though CBs may still be abundant in the template (Lehmann and Kirk-Bell 1972; Schumacher et al. 1983; Meechan et al. 1986). This third phase occurs even in xeroderma cells which are completely incapable of excision repair (Buhl and Regan 1974). Unhindered replication may also be considered a form of PRR—near-instantaneous PRR, in fact; we will refer to it as 'late PRR'. Its mechanism is not necessarily the same as that of early PRR; but from analysis of the time-dependence of mutation (Stamato et al. 1987) and of sister chromosome exchanges (Fujiwara et al. 1980) we can conclude it must be error-prone.

Much less is known about PRR than about excision repair pathways. Partly this is due to the relative dearth of PRR mutants, but it is also a result of the complexity of the normal replication process in eukaryotic cells. Progress in analysing the mechanism involved in daughter strand assembly has been slow and that in identifying the genes responsible even slower. Human PRR defective mutants, however, occur in the variant form of xeroderma pigmentosum; cells from these patients appear to have normal, error-free DNA excision repair (Watanabe et al. 1985), but are defective in early PRR. The replication defect is important; these individuals are extremely photosensitive and cancer-prone and their cells are hypermutable.

Four general mechanisms have been proposed for PRR. These are:

1. Retreat of the stalled replication fork, followed by reformation of the parental double-stranded DNA and excision repair of the lesion, after which polymerization can proceed. This might conceivably account for some early PRR (but not for late PRR which is certainly excision-independent). It would be symmetric if the selective excision that occurs at stalled RNA polymerases also occurred at stalled DNA polymerases, but selective repair of replicated DNA has not been observed (Waters 1978; Spivak and Hanawalt 1992).

2. Recombination exchange between the gapped daughter strand and the homologous parental strand in the sister chromatid. This is a prominent mechanism in bacteria, but is probably of marginal significance in mammalian cells (Menck and Meneghini 1982; Fornace 1983).

3. Gap filling. A general cellular response to UV irradiation or other forms of damage is to increase the number of replication forks when DNA synthesis recovers (Tatsumi and Strauss 1979; Cleaver et al. 1980; Painter 1980). This corresponds to a decrease in the spacing of replicon origins, as detected by fibre

autoradiography (Griffiths and Ling 1987). Park and Cleaver (1979) have suggested that the activation of new replication origins, downstream of stalled polymerases, would allow new polymerases to approach the lesions from the other direction and so fill most of the gap in the daughter strand, achieving early PRR. This cannot be the defect in xeroderma variant, where replicon spacing decreases normally after UV (Griffiths and Ling 1991). Gap filling must also involve some special activity in its final stage, to permit synthesis over the damage site; in other words, a form of lesion bypass (see below).

4. Bypass of the lesion. In bacteria, one of the normal responses to damage is a decrease in the fidelity of the DNA polymerase, which allows it to bypass a normally non-coding lesion. This bypass normally follows the 'A rule' (reviewed by Strauss (1991)); polymerases tend to choose dATP as the substrate for bypass incorporation. Note that bypassing TpT dimers, the most common UV photoproduct, according to the A rule gives the correct daughter sequence; bypass of other lesions by the A rule or deviations from the rule in bypassing TpT, will be mutagenic. Xeroderma variants are defective in this bypass process; analysis of mutation patterns in variant cells shows that they do not obey the A rule at damage sites (Wang et al. 1991). In early PRR, bypass may operate either as the last step in a gap-filling process or as a delayed response by the original polymerase proceeding in the normal direction. In late PRR, polymerases must bypass lesions as soon as they encounter them.

Selective damage during the post-replication repair period

A number of the single-stranded DNA gaps, formed at arrested replication forks, may be attacked by endonucleases and converted to double-strand DNA breaks. We have measured the occurrence of replication-related double-strand breakage associated with nascent radiolabelled DNA. Data from neutral elution (Wang and Smith 1986) or pulsed field gel electrophoresis studies (S. Squires and A. J. Ryan, unpublished) show that in human cells newly replicated DNA suffers a dose-dependent accumulation of double-strand breaks (DSBs) during the first 12 h after UV treatment (Fig. 4.1). By contrast, bulk DNA is not damaged. Excision repair-defective xeroderma pigmentosum cells (complementation groups A and D) accumulate many more nascent DNA-associated DSBs than wild-type cells and by 24 h there is very considerable destruction of the entire genome. These data suggest that collisions occur between replication forks and DNA lesions which result in stalling and subsequent nuclease-mediated breakage of the template strand. Such fork-associated DSBs are likely to be extremely cytotoxic and there is a good correlation between cell survival and the number of DSBs generated after UV irradiation in the replication-associated DNA (S. Squires, unpublished). The data also confirm that replication-associated DSBs do not arise as a consequence of overlapping excision repair in opposite DNA strands.

In UV-irradiated rodent cells, the situation is different. For example, we find that in Chinese hamster ovary cells DSBs are not generated in nascent regions of the DNA, not even in the excision repair-deficient mutant UV-5, which is the exact

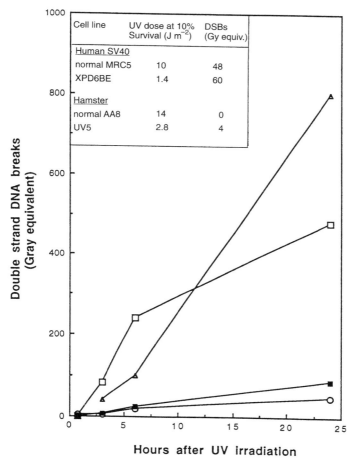

Fig. 4.1. The formation of DNA double-strand breaks (DSBs) in nascent and bulk DNA in human and Chinese hamster cells UV-irradiated with 8 J m^{-2}. Cells in which the bulk DNA had been prelabelled with [^{14}C] thymidine, were UV-irradiated and then incubated with [^{3}H]thymidine for 15 min before and 45 min after UV irradiation. Double-strand breaks were determined at various times after irradiation in the bulk DNA (closed symbols) and in nascent DNA. Human cells (all SV40 transformed fibroblasts): ○, MRC5 (wild-type); ∆, SV20/XPS (xeroderma pigmentosum A); □, XP6BE (xeroderma pigmentosum D). In the inset table data are shown correlating nascent DNA associated DSBs and survival for human cells; for hamster cells there is no correlation.

hamster counterpart of the human xeroderma pigmentosum D mutation (Fleijter *et al.* 1992). In the absence of data on the possible selective repair of replicating DNA in hamster cells, we must conclude that the stalling, which undoubtedly occurs during recovery of daughter strand maturation in hamster cells, does not result in measurable DSBs. A similar lack of nascent DNA breakage after UV was also observed in a transformed Indian muntjac cell line of normal UV sensitivity

(Musk et al. 1990). Indian muntjac and Chinese hamster levels of excision repair are essentially similar, suggesting that many lesions will remain in both for replication tolerance mechanisms to deal with.

Kaufmann (1989) has recently reviewed the PRR literature and suggested that diverse PRR mechanisms may contribute to the recovery of human cells from DNA damage. In this article our aim is to consider the PRR defect in an animal cell mutant, SVM, as one model system that can shed light on the mechanism. Finally, we will consider the effect of caffeine and its analogues on PRR and whether this helps us to understand the modified replication.

SVM, an Indian muntjac cell defective in PRR

SVM is an Indian muntjac fibroblast line, transformed by SV40, which displays considerable chromosome instability and telomere–telomere fusion (K. Sperling

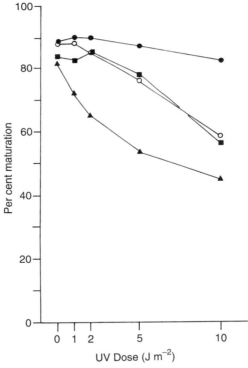

Fig. 4.2. BND-cellulose chromatographic analysis of PRR in (●) normal human fibroblasts (NHD), (○) xeroderma pigmentosum variant fibroblasts (XP30S), (■) Chinese hamster ovary mutant UV-1, and (▲) SVM. UV-irradiated cells were pulse-labelled with [3H]thymidine and then incubated for a 4 h period. The proportion of label associated with mature, entirely double-stranded DNA was separated from that associated with single-stranded regions by BND-cellulose chromatography.

Some insights into the replication of damaged DNA in mammalian cells | 57

and H. Netzel, personal communication). We now know that the phenotype of SVM is complex and that the cell is sensitive to a range of DNA damaging agents. We first discovered that it was unusually sensitive to UV irradiation: Lynda Pillidge and later, Stephen Musk, both research students, were given the task of analysing the UV sensitivity. SVM turned out to be defective not in excision repair, but in PRR. At equal UV dose, SVM's PRR is considerably worse than in either UV-1, the hamster PRR mutant, or in xeroderma variant fibroblasts (Fig. 4.2). A further disparity between SVM and the other mutants is its superinduction of sister chromatid exchanges (SCEs), as well as chromosome aberrations (Pillidge *et al.* 1986*b*). After UV irradiation, there is a steady and dose-dependent increase in the appearance of DSBs in the nascent DNA of SVM but not in that of a wild-type muntjac cell line, as assayed by neutral elution of DNA from polycarbonate filters (Fig. 4.3). The elution under neutral conditions of DNA pulse-labelled shortly after irradiation is an index of the formation of DSBs at sites of inhibition of replication, presumably as a result of endonuclease action at single-strand gaps. By contrast the production of DSBs in bulk DNA is very slight (Musk *et al.* 1990). We know from the work of Mathew Hall, another research student, that in cell-free systems SVM has an exaggerated and destructive nuclease activity, which helps to account for the reduced rejoining of linearized double-stranded plasmids introduced into the cell (Bouffler *et al.* 1990*a*; Ryan *et al.* 1992) and which may help us to understand the UV-induced destruction at or near the replication forks. Caffeine, an agent which potentiates killing of irradiated cells, increases the rate of accumulation of nascent DNA-associated DSBs in both SVM and wild-type cells after UV irradiation (Fig. 4.3); possibly caffeine activates a nuclease function or provides a more suitable substrate for its operation. One consequence of the effect of caffeine

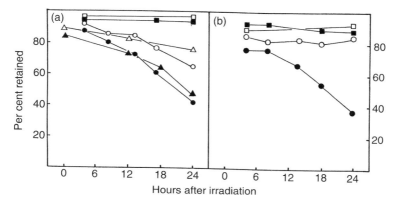

Fig. 4.3. Neutral elution from polycarbonate filters for (a) SVM and (b) DM DNA, from cells labelled for 15 min with [^{3}H]thymidine and chased for various times after irradiation in the presence of 10 μM deoxyribonucleosides. Squares unirradiated; triangles, irradiated with 2 J m^{-2}; circles irradiated with 5 J m^{-2} (SVM) or 10 J m^{-2} (DM). Open symbols, no caffeine; solid symbols, 2 mM caffeine throughout. From Musk *et al.* (1990); reprinted with kind permission of Elsevier Press.

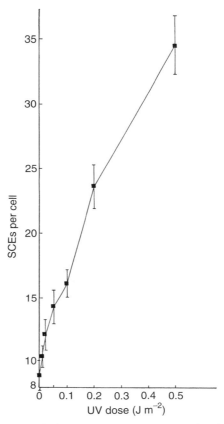

Fig. 4.4. UV dose-dependence of sister chromatid exchange (SCE) induction in the presence of 3 mM caffeine on SVM cells. Cells treated with 10 μM BrdU for 48 h (two cell cycles) after irradiation. Adapted from Musk *et al.* (1988); reprinted with kind permission of the Company of Biologists.

on recently replicated DNA in both SVM and wild-type muntjac cells is an increase in SCE formation, but as Stephen Musk showed this reaches phenomenal levels in the mutant; we estimated that in the presence of caffeine only two CBs per replication cluster may be enough to produce an SCE event (Musk *et al.* 1988; Fig. 4.4).

Isolation of a PRR correcting gene

Indian muntjac cells are readily transfected with genomic DNA and so we set out to isolate a murine gene which conferred UV resistance and PRR ability on SVM. We argued that the DNA from a malignantly transformed mouse cell line with little excision repair, but normal UV resistance, would be a good donor of PRR ability.

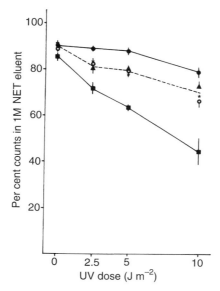

Fig. 4.5. BND-cellulose analysis of PRR in SVM (■), DM wild-type muntjac cells (●), and in genomic transfectants of SVM (▲✪★). Details of the experiment are as in the legend to Fig. 4.2. Counts eluted in 1M NET represent the degree of maturation of the nascent labelled DNA molecules. From Bouffler et al. (1990b); reprinted with kind permission of Plenum Publishing Corporation.

This strategy has allowed Simon Bouffler and Darren Godfrey to isolate, through primary, secondary, and tertiary transfection and, finally, by cosmid rescue, a gene that increased the resistance of SVM to UV irradiation not by improving excision repair but by increasing PRR (Fig. 4.5) (Bouffler et al. 1990b). The gene confers hypermutability on SVM, both per unit dose and per survivor (Fig. 4.6); this is a good indication that an error-prone function is associated with the new PRR process. In its functional state in the transfectant cells, probably under the control of a co-transfected pSV2neo promoter, the PRR gene is presumably regulated differently from its control in the original cell from which it was derived, which is not hypermutable after UV irradiation (Bouffler et al. 1990b).

Further insight into the gene's action in the PRR mechanism has come from the discovery that the UV-resistant SVM transfectants are also resistant to monofunctional alkylating agents such as methylnitrosourea (MNU) (Godfrey et al. 1992). DNA damage introduced by these agents is structurally different from the pyrimidine adducts induced by UV irradiation. Alkyl adducts at the O^6 position of guanine (O^6-alkyl G) are cytotoxic and of particular importance for mutagenicity (Loveless 1969). In mammalian cells repair of O^6-alkyl G is normally achieved by O^6-alkyl G-DNA alkyltransferase, which directly binds the alkyl moiety to one of its cysteine residues, regenerating the unmodified base in an error-free manner (Pegg 1990).

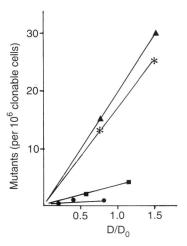

Fig. 4.6. UV-induced mutation to 6-thioguanine resistance in SVM (■), DM (●), genomic transfectant of SVM (▲), and PPR$^+$ transfectant (*), expressed in terms of the D/D_0 value for the given cell line; thus the data are represented in terms of mutants per mean lethal dose. From Bouffler *et al.* (1990*a*); reprinted with kind permission of Plenum Press Corporation.

The MNU resistance of the transfectants is not explained by an increase in the activity of alkyltransferase and O^6-alkyl G adducts are not removed by any alternative mechanism (Godfrey *et al.* 1992). Thus, O^6-alkyl G lesions remain in the DNA of the transfectants as premutagenic and potentially cytotoxic lesions.

A very significant finding, however, is that the transfectants have recovered PRR capacity not only after UV damage, but also after MNU damage. However, the PRR$^+$ transfectants are not corrected for SCE induction by MNU and their O^6-alkyl G residues are not removed (Godfrey *et al.* 1992). This contrasts with the state of *ada* transfectants of SVM, in which introduction of the bacterial alkyltransferase *ada* gene confers repair of O^6-alkyl G, MNU resistance, and protection against MNU-induced SCEs. As with UV irradiation, MNU induces large numbers of mutants per survivor in the PRR$^+$ transfectants, in a dose-dependent manner; replication is again error-prone.

On the basis of this evidence we therefore believe that the MNU and UV-resistant transfectants possess error-prone trans-lesion synthesis through both O^6-alkyl G adducts and UV adducts and that this accounts for the improvement in PRR. One further piece of evidence consistent with this belief is the sensitivity of the transfectants to mitozolomide, a chloroethylating agent. Like MNU, mitozolomide reacts predominantly with oxygen atoms in DNA; if not removed, the initial mitozolomide adduct is converted to a highly toxic interstrand cross-link, which should not permit replication. Removal of the O^6-mitozolomide adducts is normally effected by alkyltransferase and cells (such as the PRR$^+$ transfectants) which lack this enzyme are sensitive to chloroethylating agents, even if they possess a replication bypass mechanism (Godfrey *et al.* 1992).

PRR+ transfectants are also more resistant to 6-thioguanine than SVM

The PRR+ transfectants of SVM represent one example of a growing list of cells which are resistant to O^6-alkyl G, but which possess no alkyltransferase activity or alternative error-free repair mechanism that is able to remove substantial amounts of the damage (see review by Karran and Bignami (1992)). A number of these cells also show cross-resistance to low concentrations of 6-thioguanine (6TG), a deoxyguanine analogue which normally kills cells because of its incorporation into DNA (Le Page 1963; Lee and Sartorelli 1981). 6TG-tolerant Chinese hamster cells differ from their wild-type parents in their ability to replicate on a DNA template heavily substituted with 6TG (Aquilina et al. 1990); excision repair-defective Chinese hamster cells are more sensitive to killing by 6TG than are normal cells (Christie et al. 1984). Together these observations suggest that 6TG residues are recognized as damage by the excision repair complex and mechanism(s) of PRR in the same way as UV-induced photoproducts.

PRR+ transfectants have proved to be much more resistant to the toxic effects of low concentrations of 6TG than the SVM parent (Godfrey et al. 1992). Thus, they are cross-resistant to O^6-alkyl G and to 6TG, a tolerance that is probably achieved by the same mechanism, as has been suggested for other cell strains (Aquilina et al. 1989; Green et al. 1989). The greater sensitivity of SVM to 6TG compared to a wild-type muntjac cell, can be explained by the latter's competent PRR; so, we suggest, can the 6TG tolerance of the PRR+ transfectants. The PRR mechanism, at least in the transfectants, appears best explained by trans-lesion synthesis and we have proposed (Godfrey et al. 1992) that DNA helix distortion, possibly created through the close proximity of a series of persistent O-alkyl residues, can result in polymerase stalling similar to that caused by UV-induced dimers and with the substitution of 6TG for deoxyguanine in the template. A mechanism of PRR, in this case error-prone, could then be used to overcome these replication blocks, providing a common basis for resistance to UV, MNU, and 6TG. Of course, alternative mechanisms for tolerance of O^6-alkyl residues (and 6TG) are possible, as suggested by Goldmacher et al. (1986) and reviewed by Karran and Bignami (1992), though none of these proposed mechanisms suggest how lesions as varied as UV photoproducts, O^6-alkyl adducts, and 6TG can equally be dealt with.

SVM is a dominant mutant

Possibly the most unexpected finding about SVM is its dominant phenotype. We have produced a series of hybrids between SVM and DM (a wild-type, PRR-capable muntjac line) and these have been analysed for resistance to UV, MNU, and the alkylating agent dimethylsulphate (DMS), simply to check the recessive nature of SVM. The hybrids turn out to be as sensitive to these agents as the SVM parent, though as SVM chromosomes segregate from the hybrids, the wild-type phenotype is restored (A. J. Ryan, C. S. Downes, and R. T. Johnson, unpublished). Dominant sensitive phenotypes are rare and not well understood. Two further examples from the literature are the temperature-sensitive Chinese hamster cell cycle mutant *ts42*, which has chromosome breakage at the restrictive temperature (Fainsod *et al.* 1983) and, perhaps even more intriguing, the PRR-defective *mei 41 Drosophila* mutant cell line (Boyd and Setlow 1976; Boyd *et al.* 1976; Gatti 1979; Banga *et al.* 1986). Dominant mutations underlying the SVM and possibly the *ts42* and *mei41* phenotypes, may be regulatory in nature and genes that modify their effect, such as that in the PRR$^+$ transfectants of SVM, may be regulatory suppressors.

The effect of caffeine on PRR

We have already mentioned caffeine briefly as a UV sensitizer. Caffeine and radiobiology have had a strong affinity for one another since the demonstration that caffeine can increase the cytotoxicity of X-ray irradiation (Waldren and Rasko 1978) and UV irradiation (Domon and Rauth 1969). Part of the action of caffeine is that it overrides the normal cell cycle delay that occurs in damaged cells, which would otherwise accumulate in S phase and G_2 (Musk *et al.* 1988; Downes *et al.* 1990; Tolmach 1990). But caffeine and other methylxanthines, notably theophylline, also reduce the survival of UV-irradiated and alkylation-damaged cells by an S phase-dependent mechanism, the inhibition of PRR; this is entirely separate from the action on cell cycle controls (Musk *et al.* 1990).

The action of caffeine on early and late PRR differs. Late PRR is invariably caffeine-sensitive, provided that caffeine is present continuously after UV irradiation. In human cells, this effect is usually masked by the very effective excision repair that removes lesions and thus leaves nothing to prevent DNA polymerase progression, before late PRR can be measured. In xeroderma excision-defective cells, however, lesions remain and late PRR is caffeine-sensitive (Buhl and Regan 1974). If caffeine or theophylline are added several hours after irradiation, there is at first no effect on late PRR (Lehmann and Kirk-Bell 1972) but after a delay of about 3 h, late PRR is inhibited (Doniger and di Paolo 1981).

Curiously, the action of caffeine on early PRR is cell-type specific; typically, it is strong on the early PRR of permanent hamster and mouse cell lines, but not in primary rodent cells (see Sivak *et al.* (1982) for a review). It is also strong in DM, the spontaneously transformed Indian muntjac cell line which we have used as a comparison for SVM (Musk *et al.* 1990), whose overall excision repair activity resembles that of hamster or mouse cells. By contrast, in human non-transformed cell lines the action of caffeine on early PRR cannot be detected by alkaline sucrose gradients (Lehmann *et al.* 1977) and is only slight in transformed lines (Cleaver 1981). More sensitive BND-cellulose chromatographic assays disclose a small, but real, caffeine sensitivity of early PRR even in human diploid fibroblasts (Hatton 1991, C. S. Downes, unpublished).

There are, however, two cases where early PRR in human cells becomes caffeine-sensitive. One is the group of xeroderma variants, where caffeine enormously increases the PRR defect and, hence, UV lethality (Lehmann *et al.* 1975;

Lehmann 1979). The other is an increasingly notorious group of human tumour cells, the melanomas. These are particularly resistant to UV irradiation and have an exaggerated PRR ability, which is caffeine-sensitive (Konishi 1981; Hatton 1991). By contrast, melanoma clones with normal resistance to UV irradiation do not have exaggerated PRR and this is no more sensitive to inhibition by caffeine than a non-melanoma tumour cell (Hatton 1991).

Conversely, there are three cases where rodent-type PRR is caffeine-insensitive. One is the residual PRR in the SVM mutant,which is strikingly caffeine-resistant, though the effect of caffeine on cycle override remains (Musk *et al.* 1988, 1990). Another is the UV-sensitive Q31 mutant of L5178Y mouse lymphoma cells, where caffeine increases UV killing and UV mutagenesis: this has been interpreted as being due to the loss of a caffeine-sensitive, mutagenic PRR mechanism (Sato and Hieda 1980). The third is the UV-1 mutant of CHO cells, which is UV-sensitive, PRR-defective, and hypomutable (Stamato *et al.* 1981) and in which neither UV cytotoxicity or mutagenesis are enhanced by caffeine (Hentosh *et al.* 1990); we have found that the residual PRR capacity of UV-1 is insensitive to caffeine (C. S. Downes, unpublished).

Is it possible to simplify this tangle of data about PRR?

Flag theory: a general model for mammalian PRR

We have been much struck by the coincidence between the distribution of caffeine-sensitive and caffeine-insensitive PRR and selective and (relatively) non-selective CB excision. Transformed rodent and muntjac cells, if they are not UV-sensitive mutants, have caffeine-sensitive early PRR and very limited excision repair; in rodent cells we know and in muntjac we expect by analogy that excision of CBs is concentrated in transcribing genes.

Human cells have largely caffeine-insensitive early PRR and excise CBs throughout their genome.

We propose that the coincidence is in fact due to a cell-dependent variation in the recognition of damage, which affects both excision and early PRR pathways. We suppose that the normal pattern of recognition, except in repair-defective mutants, is this:

1. All (6-4)photoproducts in all cells and all CBs (in human cells) or CBs in transcribing regions (in transformed rodent and muntjac cells) are recognized by an as yet unidentified system which puts a localized signal at or near the lesion. We call this signal a 'flag', but we do not specify what it is. It could be a protein binding to the lesion or a stable modification of the chromatin near the lesion or even a modification of the DNA itself.

2. Excision repair enzymes are then targeted to flagged regions.

3. Where replication forks are not prevented from colliding with lesions (by the combination of replicon initiation inhibition and excision repair), the response of the polymerase to a lesion varies depending on whether the lesion is flagged or

not. If it is flagged, then in early PRR the polymerase pauses and eventually bypasses the lesion (directly or as the last step in gap-filling), in a process which is not sensitive to caffeine and obeys the A rule. If the lesion is not flagged, then the polymerase also pauses but bypasses the lesion in a caffeine-sensitive process. We suspect that this latter process owes its caffeine-sensitivity to its generating substantial regions of single-stranded, unprotected DNA to which caffeine binds; the relative efficiencies of the methylxanthines caffeine, theophylline, and theobromine in inhibiting early PRR correlate with their binding to single-stranded DNA (S. R. R. Musk and C. S. Downes, unpublished). We also suspect, for reasons that will shortly become apparent, that it is inherently very error-prone.

4. Eventually, late PRR takes over; in this, all lesions are rapidly bypassed whether they are initially flagged or not. We suspect that inhibition of this by caffeine is an indirect effect, since it takes several hours to come into effect, as does the indirect effect of caffeine on cell cycle regulation (Downes *et al.* 1991).

Interpreting the various cell responses in terms of this model:

Human cells normally have little caffeine-sensitive repair and extensive excision, because their lesions are mostly flagged.

Xeroderma variants, which have normal excision but caffeine-sensitive early PRR, can place flags normally but their bypass systems do not recognize them; they therefore rely on caffeine-sensitive, mutagenic early PRR.

Melanomas have a normal flagging system and normal excision but an enormous excess of PRR capacity starts operating before lesions are fully flagged or after high fluences of UV operate when the flagging system is utterly saturated; melanoma PRR therefore operates in a caffeine-sensitive mode on unflagged lesions. In some cases at least, this operation can be very error-prone (Hatton 1991).

Transformed rodent cells and DM excise only the limited range of flagged lesions. They remove only the grossly helix-distorting (6-4)photoproducts and CBs from transcribed regions; errors in non-repaired regions may occur as a result of caffeine-sensitive early PRR, but these do not much matter to the cell (and would not show up on normal mutation analysis, since mutations are only detectable in transcribed genes which, in this model, are flagged).

Q31 cells and UV-1 cells appear to have lost the ability to do normal bypass of unflagged lesions; they are therefore UV-sensitive (since polymerases are arrested) but not mutable (because bypass of flagged lesions, in transcribed genes, is normal).

Where does this leave SVM and its derivatives? We suppose that this cell line has lost the ability to bypass unflagged lesions; its residual early PRR is caffeine-insensitive. But it is not hypermutable, suggesting that its early PRR of flagged lesions, in transcribing genes, is normal. The PRR$^+$ transfectant, on the other hand, has acquired a very error-prone early PRR system; we suggest that this operates at random on flagged and unflagged regions.

Further elaborations of the model would probably be premature, given the paucity of data on late PRR, which must be mutagenic and SCE-inducing but for logistic reasons has not been as extensively studied as early PRR. We are also aware that the induction of double-strand DNA breaks at sites of early replication arrest, as discussed above, must have a profound cellular effect and can be enhanced by caffeine (see Roberts (1984) for a review): but the UV-sensitivity and UV repair mechanisms of rodent, muntjac, and human cells do not correlate well with DSB formation. DSB formation cannot currently be explained in terms of the flag model or any other theory of PRR.

Acknowledgements

We thank the Cancer Research Campaign and the Medical Research Council for their continued support and our colleagues past and present for their counsel. It is a great pleasure to contribute to this volume in honour of Henry Harris, whose insights into cell biology have illuminated our scientific lives.

References

Aquilina, G., Zijno, A., Moscufo, N., Dogliotti, E., and Bignami, M. (1989). *Carcinogenesis*, **10**, 1219–23.
Aquilina, G., Giammarioli, A. M., Zijno, A., di Muccio, A., Dogliotti, E., and Bignami, M. (1990). *Cancer Res.*, **50**, 4248–53.
Armier, J., Mezzia, A. M., Leng, M., Fuchs, R. P. P., and Sarasin, A. (1988). *Carcinogenesis*, **9**, 789–95.
Banga, S. S., Shenkar, R., and Boyd, J. B. (1986). *Mutat. Res.*, **163**, 157–65.
Blocher, D., Sigur, D., and Hannan, M. A. (1991). *Int. J. Radiat. Biol.*, **60**, 791–802.
Bouffler, S. D., Jha, B., and Johnson, R. T. (1990a). *Somat. Cell. Mol. Genet.*, **16**, 451–60.
Bouffler, S. D., Godfrey, D., Musk, S. R. R., Raman, M. J., and Johnson, R. T. (1990b). *Somat. Cell Mol. Genet.*, **16**, 507–16.
Boyd, J. B. and Setlow, R. B. (1976). *Genetics*, **84**, 507–26.
Boyd, J. B., Galino, M. D., Nguyen, T. D., and Green, M. M. (1976). *Genetics*, **84**, 485–506.
Broughton, B. C., Lehmann, A. R., Harcourt, S. A., Arlett, C. F., Sarasin, A., Klijer, W. J., et al. (1990). *Mutat. Res.*, **235**, 33–40.
Buhl, S. N. and Regan, J. D. (1974). *Biophys. J.*, **14**, 519–27.
Buhl, S. N., Setlow, R. B., and Regan, J. D. (1973). *Biophys. J.*, **13**, 1266–75.
Christie, N. T., Drake, S., Meyn, R. E., and Nelson, J. A. (1984). *Cancer Res.*, **44**, 3665–71.
Cleaver, J. E. (1981). *Mutat. Res.*, **82**, 159–71.
Cleaver, J. E., Arutyunyan, R. M., Sarkisian, T., Kaufmann, W. K., Greene, A. E., and Coriell, L. (1980). *Carcinogenesis*, **1**, 647–55.
Coquerelle, T. M., Weibezahn, K. F., and Lucke-Huhle, C. (1987). *Int. J. Radiat. Biol.*, **57**, 209–18.
Cornforth, M. N. and Bedford, J. S. (1985). *Science*, **227**, 1589–91.
Domon, M. and Rauth, A. M. (1969). *Radiat. Res.*, **39**, 207–22.
Doniger, J. and di Paulo, H. A. (1981). *Radiat. Res.*, **87**, 565–75.
Downes, C. S., Musk, S. R. R., Watson, J. V., and Johnson, R. T. (1991). *J. Cell Biol.*, **110**, 1855–9.
Downes, C. S., Ryan, A. J., and Johnson, R. T. (1993). *BioEssays*, **15**, 209–16.
Fainsod, A., Voss, R., Spann, P., Goitein, R., and Marcus, M. (1983). *Cytogenet. Cell Genet.*, **35**, 104–9.

Fleijter, W. L., McDaniel, L. D., Johns, D., Friedberg, E. C., and Schulz, R. A. (1992). *Proc. Natl. Acad. Sci. USA*, **89**, 261–5.
Fornace, A. J. (1983). *Nature*, **304**, 552–4.
Friedberg, E. C. (1985). *DNA repair*. Freeman, New York.
Fujiwara, Y., Kano, Y., Tatsumi, M., and Paul, P. (1980). *Mutat. Res.*, **71**, 243–51.
Gatti, M. (1979). *Proc. Natl. Acad. Sci. USA*, **76**, 1377–81.
Gianelli, F., Pawsey, S. A., and Avery, J. A. (1982). *Cell*, **29**, 451–8.
Godfrey, D. B., Bouffler, S. D., Musk, S. R. R., Raman, M. J., and Johnson, R. T. (1992). *Mutat. Res.*, **274**, 225–35.
Goldmacher, V. S., Cuzich, R. A., and Thilly, W. G. (1986). *J. Biol. Chem.*, **261**, 12462–71.
Green, M. H. L., Lowe, J. E., Petit-Frere, C., Karran, P., Hall, J., and Kataoka, H. (1989). *Carcinogenesis*, **10**, 893–8.
Griffiths, T. D. and Ling, S. Y. (1987). *Mutat. Res.*, **184**, 39–46.
Griffiths, T. D. and Ling, S. Y. (1991). *Mutagenesis*, **6**, 247–51.
Hatton, D. H. (1991). Ph.D. thesis, Cambridge University, UK.
Hentosh, P., Collins, A. R. S., Corell, L., Fornace, A. J., Giaccia, A., and Waldren, C. A. (1990). *Cancer Res.*, **50**, 2356–62.
Hittelman, W. N. and Rao, P. N. (1974). *Mutat. Res.*, **23**, 251–8.
Johnson, R. T. and Rao, P. N. (1970). *Nature*, **226**, 717–22.
Johnson, R. T., Collins, A. R. S., and Waldren, C. A. (1982). In *Premature chromosome condensation*, (ed. P. N. Rao, R. T. Johnson, and K. Sperling), pp. 253–308. Academic Press, New York.
Johnson, R. T., Squires, S., Elliott, G. C., and Joysey, V. (1989). *Human Genet.*, **81**, 203–10.
Kantor, G. J. and Setlow, R. B. (1981). *Cancer Res.*, **41**, 819–25.
Karran, P. and Bignami, M. (1992). *Nucleic Acids Res.*, **20**, 2933–40.
Kaufmann, W. K. (1989). *Carcinogenesis*, **10**, 1–11.
Konishi, T. (1981). *Mie Med. J.*, **31**, 15–27.
Lee, S. H. and Sartorelli, A. C. (1981). *Cancer Biochem. Biophys.*, **5**, 189–94.
Lehmann, A. R. (1979). *Nucleic Acids Res.*, **7**, 1901–12.
Lehmann, A. R. and Kirk-Bell, S. (1972). *Eur. J. Biochem.*, **31**, 438–45.
Lehmann, A. R., Kirk-Bell, S., Arlett, C. F., Paterson, M. C., Lohman, P. H. M., de Weerd-Kastelein, E. A., et al. (1975). *Proc. Natl. Acad. Sci. USA*, **72**, 219–23.
Lehmann, A. R., Kirk-Bell, S., Arlett, C. F., Harcourt, S. A., de Weerd-Kastelein, E. A., Keijzer, W., et al. (1977). *Cancer Res.*, **37**, 904–10.
Lehmann, A. R., Arlett, C. F., Broughton, B. R., Harcourt, S. A., Steingrimsdottir, H., Stefanini, M., et al. (1988). *Cancer Res.*, **48**, 6090–6.
Le Page, G. A. (1963). *Cancer Res.*, **23**, 1202–6.
Loveless, A. (1969). *Nature*, **223**, 206–7.
McDaniel, L. D. and Schultz, R. A. (1992). *Proc. Natl Acad. Sci. USA*, **89**, 7968–72.
Meechan, P. J., Carpenter, J. G., and Griffiths, T. D. (1986). *Photochem. Photobiol.*, **43**, 149–56.
Menck, C. F. and Meneghini, R. (1982). *Photochem. Photobiol.*, **35**, 507–13.
Meyn, R. E., and Humphrey, R. M. (1971). *Biophys. J.*, **11**, 295–301.
Meyn, R. E., Vizard, D. L., Hewitt, R. R., and Humphrey, R. M. (1974). *Photochem. Photobiol.*, **20**, 221–6.
Mitchell, D. L. and Nairn, R. S. (1989). *Photochem. Photobiol.*, **49**, 805–20.
Mitchell, D. L., Haipeck, C. A., and Clarkson, J. M. (1985). *Mutat. Res.*, **143**, 109–12.
Mitchell, D. L., Humphrey, R. M., Adair, G. M., Thompson, L. H., and Clarkson, J. M. (1988). *Mutat. Res.*, **193**, 53–63.
Musk, S. R. R., Downes, C. S., and Johnson, R. T. (1988). *J. Cell Sci.*, **90**, 591–9.

Musk, S. R. R., Hatton, D. H., Bouffler, S. D., Margison, G. P., and Johnson, R. T. (1989). *Carcinogenesis*, **10**, 1299–306.
Musk, S. R. R., Pillidge, L., Johnson, R. T., and Downes, C. S. (1990). *Biochim. Biophys. Acta.*, **1052**, 53–62.
Painter, R. B. (1980). *J. Mol. Biol.*, **143**, 289–301.
Painter, R. B. (1985). *Mutat. Res.*, **145**, 63–9.
Park, S. D. and Cleaver, J. E. (1979). *Proc. Natl. Acad. Sci. USA*, **76**, 3927–31.
Pegg, A. (1990). *Mutat. Res.*, **233**, 165–75.
Pillidge, L., Downes, C. S., and Johnson, R. T. (1986a). *Int. J. Radiat. Biol.*, **50**, 119–36.
Pillidge, L., Musk, S. R. R., Johnson, R. T., and Waldren, C. A. (1986b). *Mutat. Res.*, **166**, 265–73.
Roberts, J. J. (1984). In *DNA repair and its inhibition*, (ed. A. R. S. Collins, C. S. Downes, and R. T. Johnson), pp. 193–216. IRL Press, Oxford.
Ryan, A. J., Hall, M., Bouffler, S. D., Evans, A., Coates, J. A., and Johnson, R. T. (1992). *Somat. Cell Mol. Genet.*, **18**, 529–41.
Sato, K. and Hieda, N. (1980). *Mutat. Res.*, **71**, 233–41.
Schumacher, R. I., Menck, C. F., and Meneghini, R. (1983). *Photochem. Photobiol.*, **37**, 605–10.
Sivak, A., Rudento, L., and Teague, L. G. (1982). *Environ. Mutagen.*, **4**, 143–62.
Spivak, G. and Hanawalt, P. C. (1992). *Biochemistry*, **31**, 6799–800.
Stamato, T. D., Hinkle, L., Collins, A. R. S., and Waldren, C. A. (1981). *Somat. Cell Genet.*, **7**, 307–20.
Stamato, T., Weinstein, R., Peters, B., Hu, J., Doherty, B., and Giacca, A. (1987). *Somat. Cell Mol. Genet.*, **13**, 57–66.
Strauss, B. S. (1991). *BioEssays*, **13**, 79–84.
Tatsumi, K. and Strauss, B. S. (1979). *J. Mol. Biol.*, **135**, 435–49.
Thomas, D. C., Okumoto, D. S., Sancar, A., and Bohr, V. A. (1989). *J. Biol. Chem.*, **264**, 18005–10.
Thompson, L. H., Mitchell, D. L., Regan, J. D., Bouffler, S. D., Stewart, S. A., Carrier, W. L., et al. (1989). *Mutagenesis*, **4**, 140–46.
Tolmach, L. J. (1990). *Radiat. Res.*, **123**, 119–37.
Vos, J.-M. and Hanawalt, P. C. (1987). *Cell*, **50**, 289–99.
Waldren, C. A. and Johnson, R. T. (1974). *Proc. Natl Acad. Sci. USA*, **71**, 1137–41.
Waldren, C. A. and Rasko, I. (1978). *Radiat. Res.*, **73**, 95–110.
Wang, T. V. and Smith, K. C. (1986). *Carcinogenesis*, **7**, 389–92.
Wang, Y. C., Maher, V. M., and McCormick, J. J. (1991). *Proc. Natl Acad. Sci. USA*, **88**, 7810–14.
Watanabe, M., Maher, V. M., and McCormick, J. J. (1985). *Mutat. Res.*, **146**, 285–94.
Waters, R. (1978). *J. Mol. Biol.*, **127**, 117–27.
Zdzienicka, M. Z., Venema, J., Mitchell, D. L., van Hoffer, A., van Zeeland, A. A., Vrielag, H., et al. (1992). *Mutat. Res.*, **273**, 73–83.

5. The role of the nucleolus in the transfer of information from nucleus to cytoplasm

Masakazu Hatanaka

Introduction

When the nucleus of a hen erythrocyte is introduced into the cytoplasm of a human or mouse cell in culture, it resumes the synthesis of RNA (Harris 1965, 1967). The reactivated erythrocyte nucleus undergoes great enlargement, but it does not, for at least 2 or 3 days, develop nucleoli. During this period, the heterokaryon, although it may contain active erythrocyte nuclei, does not synthesize any hen-specific surface antigens. But when, later, the erythrocyte nuclei do develop nucleoli, hen-specific antigens reappear on the surface of the heterokaryon and progressively accumulate. These and other nucleolar inactivation experiments suggested that the nucleolus may play a decisive role in the transfer of information from the nucleus to the cytoplasm (Harris *et al*. 1969; Sidebottom and Harris 1969; Deak *et al*. 1972; Harris 1972).

Since the nucleolus was discovered by N. Fontana in 1781, its dynamic morphological changes during the cell cycle (Anastassova-Kristeva 1977) have attracted the attention of cell biologists. However, the very reason for the existence of the nucleolus remains an enigma, except that it is known to be an organelle of ribosome biogenesis (Hadjiolov and Nikolaev 1976; Hadjiolov 1985). The nucleolus is not an isolated floating organelle in the nucleoplasm, but is anchored to skeletal structural elements of interphase chromosomes associated with the nuclear envelope or it is in contact with the nuclear envelope in actively growing eukaryotic cells (Hadjiolov 1985; Bourgeois and Hubert 1988). These chromosomes contain rRNA genes in a so-called nucleolus organizer region (NOR).

The present article focuses on the current understanding of the role of the nucleolus in the transfer of information from nucleus to cytoplasm.

Retroviral mRNA and the nucleolus

The processing of retroviral mRNA is quite different from that of eukaryotic mRNA. The splicing of most cellular mRNAs is completed in the nucleoplasm. In contrast, a certain population of the retroviral mRNAs is transported to the cytoplasm as unspliced RNA, which serves either as mRNA for *gag* and *pol* gene products or as progeny genomes or as partially spliced RNA, which serves as mRNA for the *env* gene. Since retroviruses must use the host cell's splicing machinery in the nucleoplasm, a regulatory system within retroviruses is required to control the balance of spliced and unspliced viral mRNAs. There are three possible mechanisms.

1. The 5' and 3' splice sites may be recognized inefficiently by splicing factors.
2. The *cis*-acting sequences in the viral genome may interact with host proteins to limit the amount of spliced RNA.
3. The viral proteins may be involved in regulating their own mRNA splicing.

In the case of human T-cell leukaemia virus type 1 (HTLV-I) Rex has a crucial role in this regulation. Rex is required for the accumulation of unspliced and partially spliced cytoplasmic mRNAs of viral structural proteins at the expense of the fully spliced *pX* mRNA (Inoue *et al.* 1986, 1987; Hidaka *et al.* 1988), although the mechanism of accumulation of unspliced cytoplasmic viral mRNAs has not been clarified.

Recently, we have reported that Rex is located predominantly in the nucleolus and have identified a highly basic N-terminal sequence (Met-Pro-Lys-Thr-Arg-Arg-Arg-Pro-Arg-Arg-Ser-Gln-Arg-Lys-Arg-Pro-Pro-Thr-Pro) in Rex as a nucleolar targeting signal (NOS) (Siomi *et al.* 1988). It is tempting to hypothesize that the nucleolus is involved in the export of unspliced mRNA. The transport of nucleolar material towards the cytoplasm is suggested by the proximity of the nucleoli to the nuclear envelope (Bourgeois and Hubert 1988).

We have studied the function of the NOS of Rex by using full-length proviral DNA and mutant rex expression plasmids. Partial deletions of the NOS sequence abolished the accumulation of unspliced cytoplasmic mRNA, although the gene products of rex mutants were found in the nucleoplasm (Nosaka *et al.* 1989). These results indicate that the NOS sequence is essential for the function of Rex. Recently, two major nucleolar proteins, nucleolin and No38, have been shown to shuttle constantly between the nucleus and the cytoplasm, suggesting their involvement in the translocation of ribosomal components across the nuclear envelope (Borer *et al.* 1989). Our study on NOS suggests that the nucleolus may serve as a pathway of export of particular mRNAs to avoid the nucleoplasmic splicing machinery or nucleases.

Cellular mRNA and the nucleolus

We have found that Rex extends the half-life of cytoplasmic mRNA of the interleukin-2 receptor (Kanamori *et al.* 1990). There is a correlation between the ability of Rex to localize to the nucleolus and its ability to stabilize the interleukin-2 receptor mRNA (White *et al.* 1991). None of a set of mutants with deletions in the Rex NOS was able to stabilize the mRNA. On the other hand, the substitution mutant sNOS2 could stabilize the mRNA, consistent with its expression in the nucleolus.

It is likely that the change of flow size and export rate of mRNA to the cytoplasm may affect the state of RNA, including processing, splicing, and stability.

HIV and the nucleolus

Human immunodeficiency virus (HIV) has a *rev*-encoded protein Rev, which is known to be similar to Rex in function (Malim *et al.* 1986, 1989*b*; Knight *et al.* 1987; Hidaka *et al.* 1988; Rosen *et al.* 1988; Seiki *et al.* 1988).

We have tested the functional compatibility between Rev and Rex. Each protein recognized the other's *cis*-acting sequence, albeit at reduced levels. Both proteins localize predominantly in the nucleolus. A previously unknown nucleolar targeting signal in Rev was identified, which consists of the sequence 35-RQARRNRRRR WRERQR-50. When the Rev NOS was fused with β-galactosidase, the hybrid protein accumulated in the nucleolus. A deletion mutant that lacks several amino acid residues within the signal failed to function like Rev. These results demonstrate that the NOSs are essential for the functions of Rev and Rex (Kubota *et al.* 1989).

Two chimeric mutant genes derived from *rev* and *rex* were constructed in order to investigate the functions of the two NOSs in the Rev and Rex proteins. A chimeric Rex protein whose NOS region had been substituted with the Rev NOS was located predominantly in the cell nucleolus and functioned like the wild-type protein in the Rex assay system. However, a chimeric Rev containing the Rex NOS abolished Rev function despite its nucleolar localization. This non-functional nucleolus-targeted chimeric protein inhibited the function of both Rex and Rev. In the same experimental conditions, this mutant interfered with the localization of functional Rex in the nucleolus (Kubota *et al.* 1991).

Both NOSs are characterized by an extraordinarily arginine-rich motif. A conserved arginine-rich motif of bacteriophage antiterminators is involved in the ability of the antiterminators to recognize specific RNA hairpin structure (Lazinski *et al.* 1989). Given the analogy of the NOS sequence to this conserved motif, there may be some direct relationship between the NOS and specific RNA recognition. In fact, Rev specifically binds to the Rev-responsive element of HIV-1 mRNA, which forms a stable secondary structure (Zapp and Green 1989; Heaphy *et al.* 1990; Malim *et al.* 1990). Furthermore, the biological activities and *in vitro* Rev-binding abilities of Rev-responsive element mutants were correlated (Malim *et al.* 1990). The Rex protein also recognizes the specific secondary structure of the Rex-responsive element (Derse 1988; Seiki *et al.* 1988; Felber *et al.* 1989; Hanly *et al.* 1989; Toyoshima *et al.* 1990). However, it should be noted that the NOS-exchanged Rex mutant (the product of pH2NOVrex with the Rev NOS) retained the Rex protein function. This suggests that the primary NOS sequence itself has little effect on specific RNA recognition (Olsen *et al.* 1990). If the arginine-rich feature is important in RNA recognition, the NOS may play a positive role in the primary approach of Rex to RNA, after which a secondary specificity event could take place with a second specific region. Therefore, the NOS may serve to approach and attach to RNA before the specific RNA–Rex binding event.

It is known that many mutations of *rev* and *rex* have a dominant-negative phenotype. Our mutant, pH2NOXrev, inhibited a Rex function more strongly than squelching or titration (Kubota *et al.* 1991).

To gain insight into the nature of this inhibitory effect, we examined the subcellular localization of wild-type Rex coexpressed with the pH2NOXrev product, using an anti-Rex antibody. In the presence of pH2NOXrev, intact Rex protein was excluded from cell nucleoli in transfected cells. At the same time, its function was inhibited (Kubota *et al.* 1991).

Another interesting property of Rev was observed recently by using a non-functional mutant of Rev, pH2drev, which was created by deleting seven amino acid residues from the Rev NOS (Kubota *et al.* 1992). The pH2drev protein remains in the cytoplasm, cannot accumulate in the nucleolus, and has no Rev function. However, the mutant strongly inhibits the wild-type Rev function and, surprisingly, blocked the transport of the wild-type Rev into the nucleolus (Kubota *et al.* 1992). The observed interference by dRev with the nuclear/nucleolar localization of Rev further supports the necessity of nuclear/nucleolar localization for Rev function and also suggests that Rev forms multimers *in vivo*.

Rev forms several types of multimers *in vitro* with or without target RNA (Olsen *et al.* 1990; Malim and Cullen 1991; Zapp *et al.* 1991) and several domains (including the NOS region) have been proposed to be required for multimer formation. We have proposed that at least two distinct regions are required for multimer formation—one overlapping the NOS region and the other(s) located outside the NOS region; Rev monomers would be connected via these regions to form multimers (Kubota *et al.* 1992). In dRev, the multimer formation region that overlaps the NOS would be truncated. So, dRev would not be able to form homo-oligomers. However, since the other multimer formation region of dRev would remain intact, dRev would be able to form a hetero-oligomer with wild-type Rev by using the wild-type NOS multimer formation region. Once a Rev molecule was connected to dRev, the Rev NOS would be masked by the attached dRev molecule, which has no NOS. Consequently, such a hetero-oligomer would not be able to migrate into the cell nucleus.

It is also worth mentioning several previous findings that support our hypothesis. Rev mutants with the arginine residues at positions 38 and 39 replaced with other residues have been independently characterized by several groups. These substitution mutants might not be able to form homogenous multimers because of the destruction of the NOS multimer formation region but do form heterogenous multimers with intact Rev. However, such mutants did not act as strong inhibitors of Rev (Malim *et al.* 1989*a*; Hope *et al.* 1990; Olsen *et al.* 1990). We conclude that the phenotypic difference between our dRev and formerly derived NOS mutants is due to their different subcellular localizations. Since our dRev cannot migrate into cell nucleoli, hetero-oligomerization with Rev results in a strong inhibitory effect against Rev. In contrast, other mutants can migrate into nuclei and cannot interfere with the nuclear/nucleolar migration of Rev (Malim *et al.* 1989*a*).

The observed interference by dRev with the nuclear/nucleolar localization of Rev further supports the necessity of nuclear/nucleolar localization for Rev function (Kubota *et al.* 1989; Malim *et al.* 1989*a*) and suggests that Rev forms multimers *in vivo*. Further investigation of the role of Rev localized in nucleoli may give us insights into the nature of post-transcriptional regulation in eukaryotic gene expression.

The nucleolus is believed to be a site of ribosome biogenesis. It may not be limited to this function and may affect post-transcriptional regulation of certain mRNAs. It is possible that it stores some factors that are indispensable for mRNA regulation or provides an extra pathway for unspliced RNA transport to the cytoplasm.

Phosphorylation and the nucleolus

Rev and Rex are phosphorylated proteins localized in the nucleolus; their function may be regulated by their localization and phosphorylation. When an HTLV-1 infected cell line, HUT102, was treated for 2 h by 50 μM of a protein kinase C inhibitor H-7, the phosphorylation of Rex in the H-7-treated cells, decreased dramatically to one-tenth of the control (Adachi *et al.* 1990*a*). The treatment of the cells with H-7 had no effect on the localization of Rex, which remained predominantly in the nucleolus. Therefore, the phosphorylation of Rex is not required for its nucleolar localization and the Rex NOS is solely responsible for its location.

We then examined the effect of H-7 on Rex function by Northern blot analysis of the viral transcripts. HTLV-1 expresses three species of mRNA that are detectable by the *pX* probe. H-7-treated cells contained about one-tenth the amount of unspliced *gag-pol* mRNA compared with that of untreated cells, although the partially spliced *env* mRNA and fully spliced *pX* mRNA showed no differences between the treated and untreated cells. Furthermore, the levels of the cellular transcript of glyceraldehyde 3-phosphate dehydrogenase (GAPDH) did not decrease as a result of H-7 treatment. Therefore, the decrease in the unspliced *gag-pol* mRNA must be specified by the concomitant decrease in phosphorylation of Rex. The total level of viral transcripts changed little, indicating that Tax function was not influenced by H-7 treatment. This was confirmed by a chloramphenicol acetyltransferase (CAT) assay of the transfected HUT-102 cells using the LTR-CAT plasmid as a reporter gene for testing Tax function with or without H-7 treatment. Thus, *gag-pol* mRNA is specifically blocked by H-7 at the level of post-transcriptional accumulation. Although the cells treated with H-7 for 2 h showed a decreased level of unspliced transcript, the amount of Gag proteins in cells did not change dramatically. However, when the H-7-treated cells were cultured with normal medium for 2 h, the drastic decrease in the *de novo* synthesis of a matrix protein of Gag, p19, was observed by Western blot analysis, with no change in an Env product, p20 (Adachi *et al.* 1990*a*).

These results demonstrate that decreased *gag* expression (at both the mRNA and protein levels) is accompanied by deficient phosphorylation of Rex, even though this shows its proper nucleolar location. Therefore, we conclude that the phosphorylationof Rex is required for its function, which may be to partition unspliced, partially spliced, and fully spliced HTLV-1 mRNAs to lead to their balanced expression.

We were therefore interested in investigating in detail the *in vivo* phosphorylation of Rex protein. For this purpose, we isolated ^{32}P-labelled rex from HTLV-1 infected human T-cell lines by immunoprecipitation, performed various biochemical analyses, and determined the phosphorylation sites within the Rex molecule (Adachi *et al.* 1992).

The level of Rex phosphorylation was enhanced by treating the cells with TPA (12-0-tetradecanoylphorbol-13-acetate) alone, a specific activator of cellular protein kinase C (Nishizuka 1986; Pasti et al. 1986). Moreover, the pretreatment of cells with TPA caused down-regulation of cellular protein kinase C, resulting in no augmentation of TPA-dependent Rex phosphorylation after retreatment with TPA. These results closely mirror the previously reported effect of TPA on protein kinase C activation in cultured cells (Chida et al. 1986; Nishizuka 1986; Pasti et al. 1986; Ase et al. 1988; Adachi et al. 1990b).

The *in vivo* phosphorylation sites of Rex were identified as Ser70, Ser177, and Thr 174 by radiosequencing. As the deletion of the first 78 amino acid residues of Rex caused disappearance of TPA-dependent phosphorylation, Ser70 was identified as a TPA-dependent phosphorylation site (Adachi et al. 1992b). This is supported by inspection of its surrounding sequences. The region around Ser70 fits the apparent consensus sequence of the phosphorylation site by protein kinase C, Ser/Thr-X-Arg/Lys (Kishimoto et al. 1985). A wide range of *in vivo* protein kinase C substrate proteins has been reported. However, very few of these proteins are localized predominantly in the nuclei and especially few in the nucleolus (Chan et al. 1986; Nishizuka 1986). Indeed, it has been reported that only low levels of protein kinase C activity are associated with the nuclei (Nishizuka 1986; Cochrane et al. 1989). Therefore, TPA-dependent phosphorylation of Rex may either be directly mediated by protein kinase C in the cytoplasm and the phosphorylated Rex is translocated into the nucleolus or it reflects an additional step of signal transduction, resulting in activation of a nuclear kinase, such as HIV-1 Rev kinase (Cochrane et al. 1989).

Recent studies have revealed that (i) the first 77 N-terminal amino acids of Rex constitute a domain with RNA-binding activity (Grassman et al. 1991), (ii) a region between residues 55 and 132 is required for Rex-mediated *trans*-activation (Hofer et al. 1991) and (iii) an effector domain (residues 66–118) is essential for Rex function (Hope et al. 1991). In this study, we found that the TPA-dependent phosphorylation site, Ser 70 is located within these functional regions. Protein phosphorylation has been recognized as a major post-translational regulatory mechanism and is thought to play an important role in the control of cell growth, differentiation, and tumorigenicity (Hunter and Cooper 1985; Nishizuka 1986). Hence, it is possible that phosphorylation at Ser70 is necessary for the activation of Rex function. We have demonstrated that the protein kinase inhibitor H-7 blocks accumulation of viral unspliced *gag-pol* mRNA corresponding to a decrease in the *in vivo* phosphorylation of Rex (Adachi et al. 1990a). It has been reported that TPA is an inducer of viral gene expression and replication of HTLV-I or HIV-1 in infected human T-cells (Harada et al. 1986; Siekevitz et al. 1987) and enhances Rev phosphorylation *in vivo* (Hauber et al. 1988). Thus, retroviruses such as HTLV-I and HIV-1 that encode their own *trans*-regulatory proteins would be able to adapt if they could also respond to regulatory signals by extracellular stimuli. One way in which this effect could be accomplished would be to make the function of Rex dependent on phosphorylation.

Transcription and the nucleolus

The HIV Tat protein is a *trans*-acting transcriptional activator, which binds to a *cis*-acting element TAR (Muesing *et al.* 1987; Feng and Holland 1988; Berkhout *et al.* 1989). There is also a nucleolar targeting signal in the middle portion of the Tat protein (Endo *et al.* 1989; Siomi *et al.* 1990). We have examined the effects of mutations in the NOS region of Tat on its *trans*-acting activity and cellular localization. Introduction of a stop codon immediately preceding the NOS abolished the activity, while the truncated mutant with the NOS retained some activity (Endo *et al.* 1989). The Tat NOS could be replaced with that of Rev, but not with basic amino acid clusters of adenovirus E1a nor cellular enzyme. The results of immunofluorescence analysis revealed a correlation between the nuclear (especially nucleolar) accumulation and the activities of mutant Tat proteins (Endo *et al.* 1989).

Therefore a correlation exists between nucleolar accumulation of Rev and Tat and their functions. We examined the subcellular localization of these two *trans*-activators within transfected COS-7 cells. Confocal laser scanning microscopy (CLSM) was employed to obtain serial images of tomographic optical slices (microtomographic images) with high resolution of the inner structures of the cells (Miyazaki *et al.* 1992).

In *rev*-transfected cells, fluorescence was observed throughout the nucleoli 66 h after transfection. By contrast, Tat accumulated solely in the perinucleolar region, even though it has an arginine-rich motif similar to that of Rev (Lazinski *et al.* 1989). Nucleoplasmic staining with a speckled pattern was also observed in some *tat*-transfected cells (Miyazaki *et al.* 1992). Both Rev- and Tat-expressing cells were detected in 10–15 per cent of the transfected cells by indirect immunofluorescence; there was no difference in fluorescence intensity between these two cell types. It is therefore unlikely that the topological difference between the viral *trans*-activators in the nuclei results simply from indiscriminate binding of their basic amino acid residues to rRNAs.

The nucleolus is not an isolated floating organelle in the nucleoplasm but is linked to both the chromosomes and the nuclear envelope (Bourgeois and Hubert 1988) and a relationship between the position of the nucleolus near the nuclear envelope and the export of the nucleolar RNA towards the cytoplasm has been suggested (Hadijiolov 1985; Bourgeois and Hubert 1988). Furthermore, two major nucleolar proteins, nucleolin and No38, have been shown to shuttle between nucleus and cytoplasm, suggesting a role for these nucleolar proteins in the nucleocytoplasmic transport of ribosomal components (Borer *et al.* 1989). Rev is thought to operate at the post-transcriptional level. The nucleolar function(s) of nucleoli in post-transcriptional regulation should be investigated further.

Previous studies using transfected cells have demonstrated the nuclear/nucleolar localization of Tat by using conventional microscopy. However, the significance of the nucleolar localization of Tat has not yet been verified. In a study using transfected cells, a *trans*-dominant mutant of Tat was shown to inhibit Tat function by preventing nucleolar migration (Pearson *et al.* 1990). In another study, however, endocytosed rhodamine-labelled Tat was not located in the nucleoli, despite being functional in an assay system (Mann and Frankel 1991), suggesting that nucleolar localization *per se* is not essential for *trans*-activation by Tat. By using CLSM, we demonstrated the specific perinucleolar accumulation of Tat in the transfected COS-7 cells, indicating that Tat need not be destined for inside the nucleoli. However, its specific accumulation in the perinucleolar region suggests that Tat has some relation to the nucleoli. To elucidate the significance of perinucleolar localization of Tat for its function, we examined the subcellular localization of a mutant Tat protein, H2tat-AR (Endo *et al.* 1989), which was made by replacing the highly basic region of Tat with that of Rev (Miyazaki *et al.* 1992). This mutant protein was previously shown to retain 30–40 per cent of the *trans*-acting activity of the wild-type Tat protein (Endo *et al.* 1989). The Tat-Rev fusion protein was less predominantly located in the perinucleolar region than the wild-type Tat protein. This finding suggests the necessity of Tat structure for perinucleolar localization and function.

Tat *trans*-activates viral gene expression at the level of both transcriptional initiation and elongation from the HIV-1 long terminal repeat (Peterlin *et al.* 1986; Kao *et al.* 1987; Muesing *et al.* 1987). It is, however, still under discussion whether Tat increases the expression of viral genes at a post-transcriptional stage. Direct evidence for a post-transcriptional role of Tat was obtained from experiments involving microinjection into *Xenopus* oocytes (Braddock *et al.* 1989). In primate cells permissive for viral replication, however, Tat exerts its effect primarily at the level of transcription (Chin *et al.* 1991). Recently, Carter *et al.* (1991) reported that nuclear poly(A) RNA was found concentrated primarily within several discrete domains which often surround the nucleoli. Their data also suggest that these poly(A) RNA regions are transcription sites. Considering that, as we found, Tat is predominantly located in the perinucleolar region, it is possible that Tat may *trans*-activate viral gene expression in this region. The nucleolus is not separated by any membranous structure from the surrounding nucleoplasm and no unique structure has been identified in the boundary between the nucleolus and the nucleoplasm. The perinucleolar accumulation of Tat could reflect some novel feature of nucleolar architecture in the region where Tat exerts its *trans*-acting effect. Investigation of the relevance of the perinucleolar localization of Tat to its function may help to clarify the mechanism by which Tat *trans*-activates gene expression and should provide a clue for understanding hitherto unknown features of the organization of the nucleolus.

The nucleolus is well known as a site of ribosome biogenesis. However, its other functions are not fully understood as yet and we know very little about how certain proteins are imported into the nucleolus. The HTLV-I Rex protein, which may function in a manner similar to Rev, has been shown to be located in the nucleolus (Siomi *et al.* 1988; Nosaka *et al.* 1989; Kalland *et al.* 1991). The study of HIV-1 Tat, Rev, and HTLV-I Rex will shed light on the functions of the nucleolus.

Information and the nucleolus

Increased flow of genetic information is often observed in the growing cells *in vivo* such as hyperplasia and cancer. For example, we have found that prostatic hyperplasia and brain tumour showed highly increased expression of genes for growth factors (Mori et al. 1990; Takahashi et al. 1990). Gliomas express large amounts of basic fibroblast growth factor (bFGF) and its receptors so that the brain tumour grows in an autocrine fashion and promotes angiogenesis in a paracrine fashion (Takahashi et al. 1991a,b, 1992). The bFGF is a mitogen and a morphogen for a wide range of neuroectoderm- and mesoderm-derived cells as well as a potent angiogenic factor *in vivo* (Folkman and Klagsbrun 1987). It has been suggested that bFGF is involved in autonomous cell growth and tumorigenesis in many kinds of tumour cells. Our Northern blot analyses have suggested that abnormal expression of bFGF mRNA is detected in 94.4 per cent of human glioma tissues. The FGF receptors 1 and 2 are constitutively expressed in the normal brain. We examined the relationship between nucleolar activity and expression of bFGF.

Level of basic FGF and NORs of glioma cells

Positive staining of bFGF was detected in 18 of the 19 gliomas tested, but not in the two normal brain-tissue specimens (Takahashi et al. 1992). Immunoreactivity was chiefly localized in tumour cells. The degree of bFGF expression was classified into four grades according to the number of positively stained glioma cells and the amount of immunoreactivity per cell. The degree of bFGF staining increased in proportion to the malignancy grades of the gliomas.

The number of NOR dots in Grade III and Grade IV gliomas were significantly higher than in normal brain cells (unpaired t-test: $t = 4.43$ and 2.23; $P = 0.021$ and 0.050, respectively). The number of NOR dots in Grade IV gliomas was also higher than in Grade II gliomas (unpaired t-test: $t = 2.56$; $P = 0.023$).

There was a tendency for the number of NORs in glioma cells to increase in proportion to the level of bFGF expression. There were more NOR dots in samples with degrees of bFGF staining graded + and +++ than in those with negative staining (graded –) (unpaired t-test: $t = 2.95$ and 4.34; $P = 0.016$ and 0.023, respectively). The number of NOR dots was also higher in samples with staining graded +++ than in those graded + (unpaired t-test: $t = 4.96$; $P = 0.001$).

Basic FGF staining and NORs of endothelial cells in glioma tissue

Endothelial cells in Grade IV glioma tissue exhibited immunoreactivity toward the anti-bFGF antibody, while those in normal brain or Grade II gliomas did not. The number of NORs in endothelial cells, ranging from 1 to 2.26, was higher in Grade IV gliomas than in normal brain and Grade II gliomas (unpaired t-test: $t = 2.32$ and 2.78; $P = 0.043$ and 0.015, respectively) (Takahashi et al. 1992). In addition, the proportion of cases with incresed vascularity was also greater in high-grade gliomas; increased vascularity was found in one of six Grade II gliomas, two of three Grade III gliomas, and nine of 10 Grade IV gliomas. Almost all of these hypervascular tumours showed high levels of bFGF expression (Takahashi et al. 1992). Most of these cases also showed a higher number of NORs in endothelial cells than did cases without increased vascularity (unpaired t-test: $t = 4.53$; $P = 0.0003$).

Basic FGF and gliomas

A constitutive interaction of cellular growth factors with their cell-surface receptors in involved in autonomous cell growth and tumorigenesis (Sporn and Roberts 1985). Basic FGF has been suggested as one of the autocrine growth factors in a number of tumour cells (Gospodarowicz et al. 1987a,b; Gospodarowicz 1990). Mouse Y-1 adrenal cortical tumour cells and A-204 rhabdomyosarcoma cells can utilize their endogenous bFGF as an autocrine factor, while normal adrenal cortical cells and myoblasts require exogenous bFGF for growth (Gospodarowicz and Handley 1975; Gospodarowicz et al. 1987a,b; Schweigerer et al. 1987). Cells such as NIH3T3 cells and Swiss 3T3 cells can acquire a transformed phenotype after transfection of the bFGF gene (Sasada et al. 1988; Quatro et al. 1989; Yayon and Klagsbrun 1990).

In human gliomas, bFGF RNA is abundantly expressed in more than 90 per cent of cases, while it was not detectable in normal brain (Takahashi et al. 1990). Fibroblast growth factor receptors have also been detected in human gliomas (Takahashi et al. 1991b). In addition, our recent study has shown that anti-bFGF IgGs can inhibit both anchorage-dependent and anchorage-independent growth of U87MG and T98G human glioblastoma cells and their tumorigenesis in nude mice (Takahashi et al. 1991a). Because bFGF lacks a typical signal peptide, the release mechanism is unknown (Gospodarowicz et al. 1987a,b; Gospodarowicz 1990), but U87MG human glioblastoma cells actually release bFGF, which acts extracellularly (Sato et al. 1989). Autonomous cell growth and the genesis of these tumours may result from the acquisition of the ability to produce cellular bFGF and to respond to it (Quatro et al. 1989).

The bFGF peptide was expressed in 18 (94.7 per cent) of 19 gliomas and the expression level of bFGF increased proportionally with the histological malignancy grade. Moreover, gliomas with a high level of bFGF peptide indicated high cell activity, which is reflected by the number of NORs. These results are compatible with the findings that the expression level of bFGF RNA increased in proportion to the histological grade of malignancy (Takahashi et al. 1990).

Basic FGF and tumour vascularity

Histopathologically, malignant gliomas often show pseudopalisading with necrosis and vascular proliferation (Rubinstein 1972). A remarkably increased vascularity is often found on cerebral angiograms (Rubinstein 1972). We hypothesized that glioma-derived bFGF might play a crucial role in tumour angiogenesis as a paracrine factor because bFGF is a potent angiogenic factor *in vivo*. Therefore, we also examined the cell activity of endothelial cells in tumour tissues using the number of NORs and compared it with the degree of bFGF staining and the vascularity on cerebral angiograms. The number of NORs of endothelial cells increased in the cases with a high level of bFGF staining and rich vascularity. Moreover, bFGF staining in endothelial cells was detected in Grade IV gliomas but not in brain cells or Grade II gliomas. Glioma-derived bFGF may promote the growth of endothelial cells and contribute to neovascularization in tumour tissues, which results in rich vascularity in glioma tissues. Basic FGF can also be involved in tumorigenesis due to the increased blood supply to tumour tissues.

Nucleolar organizer regions

Nucleolar organizer regions are loops of DNA that occur in the nucleoli of cells and possess RNA genes (Steitz 1987). They are transcribed by RNA polymerase I and are of vital significance in the ultimate synthesis of protein. Thus, the number and/or configuration of NORs is considered to reflect the activity of cells and nuclei (Crocker and Nar 1987). The number of NORs in human brain tumours has been reported to be correlated well with Ki-67 staining or DNA polymerase staining (Kunishio *et al.* 1989; Plate *et al.* 1990). We used this argyrophilic technique to evaluate the cell activity of glioma cells and endothelial cells in tumour tissues. The range of the NOR number in glioma cells in the present study was consistent with a previous report that the mean number of NORs was 3.04 in glioblastomas (glioma Grade IV), 2.68 for anaplastic astrocytomas (Grade III), and 1.91 for low-grade astrocytomas (Grade II) (Kunishio *et al.* 1989). The number of NORs in endothelial cells of gliomas in the present study ranged from 1 to 2.26, indicating that the endothelial proliferation in malignant glioma tissues was due to reactive hyperplasia as was reported in human liver cells (Crocker and McGovern 1988).

Conclusion

The heterokaryon study pioneered by Harris triggered a new paradigm on the role of the nucleolus in the transfer of information from nucleus to cytoplasm. There is no doubt as to the existence of cross-talk between nucleoplasm and nucleolus, between nucleolus and cytoplasm, and between cytoplasm and nucleoplasm. However, our comprehension of how the nucleolus interacts with the nucleoplasm and cytoplasm at the molecular and topological levels to allow the flow of the genetic information, is sketchy and primitive. It is the time to listen carefully and ponder the intricate intercommunications; this may give rise to another new paradigm in cell biology like that generated by the heterokaryon study (Harris 1965). A bell has rung to open the curtain to the prospect of the future. The drama continues.

References

Adachi, Y., Nosaka, T., and Hatanaka, M. (1990*a*). Protein kinase inhibitor H-7 blocks accumulation of unspliced mRNA of human T-cell leukemia virus type I (HTLV-I). *Biochem. Biophys. Res. Commun.*, **169**, 469–75.
Adachi, Y., Maki, M., Ishii, K., Hatanaka, M., and Murachi, T. (1990*b*). Possible involvement of calpain in down-regulation of protein kinase C. In *The biology and medicine of signal transduction*, (ed. Y. Nishizuka), pp. 478–84. Raven Press, New York.
Adachi, Y., Copeland, T. D., Takahashi, C., Nosaka, T., Ahmed, A., Oroszlan, S., *et al.* (1992). Phosphorylation of the Rex protein of human T-cell leukemia virus type I. *J. Biol. Chem.*, **267**, 21977–81.
Anastassova-Kristeva, M. (1977). The nucleolar cycle in man. *J. Cell Sci.*, **25**, 103–10.

Ase, K., Berry, N., Kikkawa, U., Kishimoto, A., and Nishizuka, Y. (1988). Differential down-regulation of protein kinase C subspecies in KM3 cells. *FEBS Lett.*, **236**, 396–400.

Berkhout, B., Silverman, R. H., and Jeang, K. T. (1989). Tat *trans*-activates the human immunodeficiency virus through a nascent RNA target. *Cell*, **59**, 273–82.

Borer, R. A., Lehner, C. F., Eppenberger, H. M., and Nigg, E. A. (1989). Major nucleolar proteins shuttle between nucleus and cytoplasm. *Cell*, **56**, 379–90.

Bourgeois, C. A. and Hubert, J. (1988). Spatial relationship between the nucleolus and the nuclear envelope: structural aspects and functional significance. *Int. Rev. cytol.*, **111**, 1–52.

Braddock, M., Chambers, A., Wilson, W., Esnouf, M. P., Adams, S. E., Kingsman, A. J., *et al.* (1989). HIV-1 TAT 'activates' presynthesized RNA in the nucleus. *Cell*, **58**, 269–79.

Carter, K. C., Taneja, K. L., and Lawrence, J. B. (1991). Discrete nuclear domains of poly(A) RNA and their relationship to the functional organization of the nucleus. *J. Cell Biol.*, **115**, 1191–202.

Chan, P. K., Aldrich, M., Cook, R. G., and Busch, H. (1986). Amino acid sequence of protein B23 phosphorylation site. *J. Biol. Chem.*, **261**, 1868–72.

Chida, K., Kato, N., and Kuroki, T. (1986). Down regulation of phorbol diester receptors by proteolytic degradation of protein kinase C in a cultured cell line of fetal rat skin keratinocytes. *J. Biol. Chem.*, **261**, 13013–18.

Chin, D. J., Selby, M. J., and Peterlin, B. M. (1991). Human immunodeficiency virus type 1 Tat does not transactivate mature *trans*-acting responsive region RNA species in the nucleus or cytoplasm of primate cells. *J. Virol.*, **65**, 1758–64.

Cochrane, A., Kramer, R., Ruben, S., Levine, J., and Rosen, C. A. (1989). The human immunodeficiency virus *rev* protein is a nuclear phosphoprotein. *Virology*, **171**, 264–6.

Crocker, J. and McGovern, J. (1988). Nucleolar organiser regions in normal, cirrhotic, and carcinomatous livers. *J. Clin. Pathol.*, **41**, 1044–8.

Crocker, J. and Nar, P. (1987). Nucleolar organizer regions in lymphomas. *J. Pathol.*, **151**, 111–18.

Deak, I., Sidebottom, E., and Harris, H. (1972). Further experiments on the role of the nucleolus in the expression of structural genes. *J. Cell Sci.*, **11**, 379–91.

Derse, D. (1988). *Trans*-acting regulation of bovine leukemia virus mRNA processing. *J. Virol.*, **62**, 1115–19.

Endo, S., Kubota, S., Siomi, H., Adachi, A., Oroszlan, S., Maki, M., *et al.* (1989). A region of basic amino-acid cluster in HIV-1 Tat protein is essential for *trans*-acting activity and nucleolar localization. *Virus Genes*, **3**, 99–110.

Felber, B. K., Derse, D., Athanassopoulos, A., Cambell, M., and Pavlakis, G. N. (1989). Cross-activation of the Rex proteins of HTLV-I and BLV and of the Rev protein of HIV-1 and nonreciprocal interactions with their RNA responsive elements. *New Biol.*, **1**, 318–30.

Feng, S. and Holland, E. C. (1988). HIV-1 tat *trans*-activation requires the loop sequence within TAR. *Nature*, **334**, 165–7.

Folkman, J. and Klagsbrun, M. (1987). Angiogenic factors. *Science*, **235**, 442–7.

Gospodarowicz, D. (1990). Fibroblast growth factor. Chemical structure and biological function. *Clin. Orthopaed. Related Res.*, **257**, 231–48.

Gospodarowicz, D. and Handley, H. H. (1975). Stimulation of division of Y1 adrenal cells by a growth factor isolated from bovine pituitary glands. *Endocrinology*, **97**, 102–7.

Gospodarowicz, D., Ferrara, N., Schweigerer, L., and Neufeld, G. (1987*a*). Structural characterization and biological functions of fibroblast growth factor. *Endocrine Rev.*, **8**, 95–114.

Gospodarowicz, D., Neufeld, G., and Schweigerer, L. (1987*b*). Fibroblast growth factor: structural and biological properties. *J. Cell. Physiol. (Suppl.)*, **5**, 15–26.

Grassmann, R., Berchtold, S., Aepinus, C., Ballaun, C., Boehnlein, E., and Fleckenstein, B. (1991). *In vitro* binding of human T-cell leukemia virus *rex* proteins to the *rex*-response element of viral transcripts. *J. Virol.*, **65**, 3721–7.

Hadjiolov, A. A. (1985). *The nucleolus and ribosome biogenesis*, Springer-Verlag, New York.
Hadjiolov, A. A. and Nikolaev, N. (1976). Maturation of ribosomal ribonucleic acids and the biogenesis of ribosomes. *Prog. Biophys. Mol. Biol.*, **31**, 95–144.
Hanly, S. M., Rimsky, L. T., Malim, M. H., Kim, J. H., Hauber, J., and Dodon, M. D. (1989). Comparative analysis of the HTLV-I rex and HIV-1 rev trans-regulator proteins and their RNA response elements. *Genes Dev.*, **3**, 1534–44.
Harada, S., Koyanagi, Y., Nakashima, H., Kobayashi, N., and Yamamoto, N. (1986). Tumor promoter, TPA, enhances replication of HTLV-III/LAV. *Virology*, **154**, 249–58.
Harris, H. (1965). Behaviour of differentiated nuclei in heterokaryons of animal cells from different species. *Nature*, **206**, 583–8.
Harris, H. (1967). The reactivation of the red cell nucleus. *J. Cell Sci.*, **2**, 23–32.
Harris, H. (1972). A new function for the nucleolus. *Aust. J. Exp. Biol. Med. Sci.*, **50**, 827–32.
Harris, H., Sidebottom, E., Grace, d. M., and Bramwell, M. E. (1969). The expression of genetic information: a study with hybrid animal cells. *J. Cell Sci.*, **4**, 499–525.
Hauber, J., Bouvier, M., Malim, M. H., and Cullen, B. R. (1988). Phosphoreglation of the *rev* gene product of human immunodeficiency virus type 1. *J. Virol.*, **62**, 4801–4.
Heaphy, S., Dingwell, C., Ernberg, I., Gait, M. J., Green, S. M., Karn, J., et al. (1990). HIV-1 regulator of virion expression (Rev) protein binds to an RNA stem–loop structure located within the Rev responsive element region. *Cell*, **60**, 685–93.
Hidaka, M., Inoue, J., Yoshida, M., and Seiki, M. (1988). Post-transcriptional regulator (*rex*) of HTLV-I initiates expression of viral structural protein but suppresses expression of regulatory protein. *EMBO J.*, **7**, 519–23.
Hofer, L., Weichselbraun, I., Quick, S., Farrington, G. K., Bohnlein, E., and Hauber, J. (1991). Mutational analysis of the human T-cell leukemia virus type I *trans*-acting *rex* gene product. *J. Virol.*, **65**, 3379–83.
Hope, T. J., McDonald, D., Huang, X., Low, J., and Parslow, T. G. (1990). Mutational analysis of the human immunodeficiency virus type 1 Rev transactivator: essential residues near the amino terminus. *J. Virol.*, **64**, 5360–6.
Hope, T. J., Bond, B. L., McDonald, D., Klein, N. P., and Parslow, T. G. (1991). Effector domains of human immunodeficiency virus type 1 Rev and human T-cell leukemia virus type I Rex are functionally interchangeable and share an essential peptide motif. *J. Virol.*, **65**, 6001–7.
Hunter, T. and Cooper, J. A. (1985). Protein-tyrosine kinases. *Ann. Rev. Biochem.*, **54**, 897–930.
Inoue, J., Seiki, M., and Yoshida, M. (1986). The second pX product $p27^{x-III}$ of HTLV-1 is required for *gag* gene expression. *FEBS Lett.*, **209**, 187–90.
Inoue, J., Yoshida, M., and Seiki, M. (1987). Transcriptional (p40x) and post-transcriptional ($p27^{x-III}$) regulators are required for the expression and replication of human T-cell leukemia virus type I gene. *Proc. Natl. Acad. Sci. USA*, **84**, 3653–7.
Kalland, K. H., Langhoff, E., Bos, H. J., Gottlinger, H., and Haseltine, W. A. (1991). Rex-dependent nucleolar accumulation of HTLV-I mRNAs. *New Biol.*, **3**, 389–97.
Kanamori, H., Suzuki, N., Siomi, H., Nosaka, T., Sato, A., Sabe, H., et al. (1990). HTLV-1 $p27_{rex}$ stabilizes human interleukin-2 receptor α chain mRNA. *EMBO J.*, **9**, 4161–6.
Kao, S. Y., Calman, A. F., Luciw, P. A., and Peterlin, B. M. (1987). Anti-termination of transcription within the long terminal repeat of HIV-1 by *tat* gene product. *Nature*, **330**, 489–93.
Kishimoto, A., Nishiyama, K., Nakanishi, H., Uratsuji, Y., Nomura, H., Takeyama, Y., et al. (1985). Studies on the phosphorylation of myelin basic protein by protein kinase C and adenosine 3′/5′-monophosphate-dependent protein kinase. *J. Biol. Chem.*, **260**, 12492–9.

Knight, D. M., Flomerfelt, F. A., and Ghrayeb, J. (1987). Expression of the art/trs protein of HIV and study of its role in viral envelope synthesis. *Science*, **236**, 837–40.
Kubota, S., Siomi, H., Satoh, T., Endo, S., Maki, M., and Hatanaka, M. (1989). Functional similarity of HIV-1 *rev* and HTLV-1 *rex* proteins: identification of a new nucleolar-targeting signal in *rev* protein. *Biochem. Biophys. Res. Commun.*, **162**, 963–70.
Kubota, S., Nosaka, T., Cullen, B. R., Maki, M., and Hatanaka, M. (1991). Effects of chimeric mutants of human immunodeficiency virus type 1 Rev and human T-cell leukemia virus type I Rex on nucleolar targeting signals. *J. Virol.*, **65**, 2452–6.
Kubota, S., Furuta, R., Maki, M., and Hatanaka, M. (1992). Inhibition of human immunodeficiency virus type 1 Rev function by a Rev mutant which interferes with nuclear/nucleolar localization of Rev. *J. Virol.*, **66**, 2510–13.
Kunishio, K., Matsuhisa, T., Maeshiro, T., Mishima, N., Tsuno, K., and Shigematsu, H. (1989). Cell kinetics of brain tumors using Ag-NOR staining technique (a comparison with DNA polymerase-a immunostaining). *Neuroimmunol. Res.*, **2**, 315–19.
Lazinski, D., Grzadzielska, E., and Das, A. (1989). Sequence-specific recognition of RNA hairpins by bacteriophage antiterminators requires a conserved arginine-rich motif. *Cell*, **59**, 207–18.
Malim, M. H. and Cullen, B. R. (1991). HIV-1 structural gene expression requires the binding of multiple Rev monomers to the viral RRE: implications for HIV-1 latency. *Cell*, **65**, 241–8.
Malim, M. H., Hauber, J., Fenrick, R., and Cullen, B. (1986). Immunodeficiency virus *rev trans*-activator modulates the expression of the viral regulatory genes. *Nature*, **335**, 181–3.
Malim, M. H., Bohnlein, S., Hauber, J., and Cullen, B. (1989*a*). Functional dissection of the HIV-1 Rev *trans*-activator–derivation of a *trans*-dominant repressor of rev function. *Cell*, **58**, 205–14.
Malim, M. H., Hauber, J., Le, S-Y., Maizel, J. V., and Cullen, B. (1989*b*). The HIV-1 *rev trans*-activator acts through a structured target sequence to activate nuclear export of unspliced viral mRNA. *Nature*, **338**, 254–7.
Malim, M. H., Tiley, L. S., McCarn, D. F., Rusche, J. R., Hauber, J., and Cullen, B. R. (1990). HIV-1 structural gene expression requires binding of the Rev *trans*-activator to its RNA target sequence. *Cell*, **60**, 675–83.
Mann, D. A. and Frankel, A. D. (1991). Endocytosis and targeting of exogenous HIV-1 Tat protein. *EMBO J.*, **10**, 1733–9.
Miyazaki, Y., Takamatsu, T., Nosaka, T., Fujita, S., and Hatanaka, M. (1992). Intranuclear topological distribution of HIV-1 trans-activators. *FEBS Lett.*, **305**, 1–5.
Mori, H., Maki, M., Oishi, K., Jaye, M., Igarashi, K., Yoshida, O., *et al.* (1990). Increased expression of genes for basic fibroblast growth factor and transforming growth factor type β_2 in human genign prostatic hyperplasia. *Prostate*, **16**, 71–80.
Muesing, M. A., Smith, D. H., and Capon, D. (1987). Regulation of mRNA accumulation by a human immunodeficiency virus *trans*-activator protein. *Cell*, **48**, 691–701.
Nishizuka, Y. (1986). Studies and perspectives of protein kinase C. *Science*, **233**, 305–12.
Nosaka, T., Siomi, H., Adachi, Y., Ishibashi, M., Kubota, S., Maki, M., *et al.* (1989). Nucleolar targeting signal of human T-cell leukemia virus type I *rex*-encoded protein is essential for cytoplasmic accumulation of unspliced viral mRNA. *Proc. Natl. Acad. Sci. USA*, **86**, 9798–802.
Olsen, H. S., Cochrane, A. W., Dillon, P. J., Nalin, C. M., and Rosen, C. A. (1990). Interaction of the human immunodeficiency virus type I Rev protein with a structured region in env mRNA is dependent on multimer formation mediated through a basic stretch of amino acids. *Genes. Dev.*, **4**, 1357–64.

Pasti, G., Lacal, J. C., Warren, B. S., of Aaronson, S. A., and Blumberg, P. M. (1986). Loss of mouse fibroblast cell response to phorbol esters restored by microinjected protein kinase C. *Nature*, **324**, 375–7.

Pearson, L., Garcia, J., Wu, F., Modesti, N., Nelson, J., and Gaynor, R. (1990). A *trans*-dominant *tat* mutant that inhibits tat-induced gene expression from the human immunodeficiency virus long terminal repeat. *Proc. Natl. Acad. Sci. USA*, **87**, 5079–83.

Peterlin, B. M., Luciw, P. A., Barr, P. J., and Walker, M. D. (1986). Elevated levels of mRNA can account for the *trans*-activation of human immunodeficiency virus. *Proc. Natl. Acad. Sci. USA*, **83**, 9734–8.

Plate, K. H., Ruschoff, J., Behnke, J., and Mennel, H. D. (1990). Proliferative potential of human brain tumours as assessed by nucleolar organizer regions (AgNORs) and Ki67-immunoreactivity. *Acta Neurochirurg.*, **104**, 103–9.

Quatro, N., Talarico, D., Sommer, A., *et al.* (1989). Transformation by basic fibroblast growth factor requires high levels of expression: comparison with transformation by hst/K-fgf. *Oncogene Res.*, **5**, 101–10.

Rosen, C. A., Terwilliger, E., Dayton, A., Sodrosti, J. G., and Haseltin, W. A. (1988). Intragenic *cis*-acting *art* gene-responsive sequences of the human immunodeficiency virus. *Proc. Natl. Acad. Sci. USA*, **85**, 2071–5.

Rubinstein, R. J. (1972). *Tumors of the central nervous system. Atlas of tumor pathology*, Series 2, Fascicle 6, pp. 19–126. Armed Forces Institute of Pathology, Washington, DC.

Sasada, R., Kurokawa, T., Iwane, M., and Igarashi, K. (1988). Transformation of mouse BALB/c 3T3 cells with human basic fibroblast growth factor cDNA. *Mol. Cell. Biol.*, **8**, 588–94.

Sato, Y., Murphy, P. R., Sato, R., and Friesen, H. G. (1989). Fibroblast growth factor release by bovine endothelial cells and human astrocytoma cells in culture is density dependent. *Mol. Cell. Endocrinol.*, **3**, 744–8.

Schweigerer, L., Neufeld, G., Mergia, A., Abraham, J. A., Fiddes, J. C., and Gospodarowicz, D. (1987). Basic fibroblast growth factor in human rhabdomyosarcoma cells: implications for the proliferation and neovascularization of myoblast-derived tumors. *Proc. Natl. Acad. Sci. USA*, **84**, 842–6.

Seiki, M., Inoue, J., Hidaka, M., and Yoshida, M. (1988). Two *cis*-acting elements responsible for posttranscriptional *trans*-regulation of gene expression of human T-cell leukemia virus type I. *Proc. Natl. Acad. Sci. USA*, **85**, 7124–8.

Sidebottom, E. and Harris, H. (1969). The role of the nucleolus in the transfer of RNA from nucleus to cytoplasm. *J. Cell Sci.*, **5**, 351–64.

Siekevitz, M., Josephs, S. F., Dukovich, M., Peffer, N., Wong-Staal, F., and Greene, W. C. (1987). Activator of the HIV-1 LTR by T cell mitogen and the *trans*-activator protein of HTLV-I. *Science*, **238**, 1575–8.

Siomi, H., Shida, H., Nam, S. H., Nosaka, T., Maki, M., and Hatanaka, M. (1988). Sequence requirements for nucleolar localization of human T cell leukemia virus type I pX protein, which regulates viral RNA processing. *Cell*, **55**, 197–209.

Siomi, H., Shida, H., Maki, M., and Hatanaka, M. (1990). Effects of a highly basic region of human immunodeficiency virus Tat protein on nucleolar localization. *J. Virol.*, **64**, 1803–7.

Sporn, M. B. and Roberts, A. B. (1985). Autocrine growth factors and cancer. *Nature*, **313**, 745–7.

Steitz, J. A. (1987). The unexpected structures of eukaryotic genomes. *Molecular biology of the gene*, Vol. 1, (ed. J. Watson, N. H. Hopkins, J. W. Roberts, *et al.*), pp. 621–75. Benjamine/Cummings Publishing, Menopark, CA.

Takahashi, J. A., Mori, H., Fukumoto, M., Igarashi, K., Jaye, M., Oda, Y., *et al.* (1990). Gene expression of fibroblast growth factors in human gliomas and meningiomas:

demonstration of cellular source of basic fibroblast growth factor mRNA and peptide in tumor tissues. *Proc. Natl. Acad. Sci. USA*, **87**, 5710–14.

Takahashi, J. A., Suzui, H., Yasuda, Y., Ito, N., Ohta, M., Jaye, M., *et al.* (1991*a*). Gene expression of fibroblast growth factor receptors in the tissues of human gliomas and meningiomas. *Biochem. Biophys. Res. Commun.*, **177**, 1–7.

Takahashi, J. A., Fukumoto, M., Kozai, Y., Ito, N., Oda, Y., Kikuchi, H., *et al.* (1991*b*). Inhibition of cell growth and tumorigenesis of human glioblastoma cells by a neutralizing antibody against human basic fibroblast growth factor. *FEBS Lett.*, **288**, 65–71.

Takahashi, J. A., Fukumoto, M., Igarashi, K., Oda, Y., Kikuchi, H., Hatanaka, M. (1992). Correlation of basic fibroblast growth factor expression levels with the degree of malignancy and vascularity in human gliomas. *J. Neurosurg.*, **76**, 792–8.

Toyoshima, H., Itoh, M., Inoue, J-I., Seiki, M., Takaku, F., and Yoshida, M. (1990). Secondary structure of the human T-cell leukemia virus type 1 *rex*-responsive element is essential for *rex* regulation of RNA processing and transport of unspliced RNAs. *J. Virol.*, **64**, 2825–32.

White, K. N., Nosaka, T., Kanamori, H., Hatanaka, M., and Honjo, T. (1991). The nucleolar localisation signal of the HTLV-1 protein *p27rex* is important for stabilisation of IL-2 receptor α subunit mRNA by *p27rex*. *Biochem. Biophys. Res. Commun.*, **175**, 98–103.

Yayon, A. and Klagsbrun, M. (1990). Autocrine transformation by chimeric signal peptide-basic fibroblast growth factor: reversal by suramin. *Proc. Natl. Acad. Sci. USA*, **87**, 5346–50.

Zapp, M. L. and Green, M. R. (1989). Sequence-specific RNA binding by the HIV-1 Rev protein. *Nature*, **342**, 714–16.

Zapp, M. L., Hope, T. J., Parslow, T. G., and Green, M. R. (1991). Oligomerization and RNA binding domains of the type 1 human immunodeficiency virus Rev protein: a dual function for an arginine-rich binding motif. *Proc. Natl. Acad. Sci. USA*, **88**, 7734–8.

6. Transcription by immobile RNA polymerases

P. R. Cook

Summary

An unstated assumption in current models for transcriptions is that a polymerase tracks along the template as it synthesizes RNA. However, experiments using 'nucleoids' derived by lysing cells in a non-ionic detergent and 2 M NaCl originally suggested that RNA was synthesized by a polymerase attached to an underlying nucleoskeleton. These experiments were subject to the criticism that the associations seen resulted from the artefactual aggregation of nascent RNA. Further experiments using more physiological conditions have confirmed the existence of an intermediate-filament-like nucleoskeleton and have shown that active polymerases resist electroelution from nuclei, presumably because they are attached to the skeleton. Whether immobilization affects polymerase activity has also been tested directly by attaching to plastic beads a pure enzyme that is widely used for transcription *in vitro*—the RNA polymerase of bacteriophage T7. Although initiation is inhibited, immobilization has no effect on elongation. It is suggested that genes become active by binding to an attached polymerase and then transcripts are generated as the template moves past the fixed enzyme.

Introduction

A highlight of my first year as a graduate student in 1967 were the lectures by Henry Harris in this theatre on the 'Nucleus and Cytoplasm' (Harris 1967). The house was packed, the atmosphere electric, and the delivery word-perfect. The lectures were perfused with a simple message: look at the evidence underlying accepted ideas; if it is unsatisfactory, there is an opportunity to spend 'many hours of simple pleasure' doing experiments to test those ideas.

Fig. 6.1 shows the front cover of the third, paperback, edition of the book that resulted from these lectures (Harris 1974). It illustrates heterokaryons formed by fusion of mouse A9 cells with chick erythrocytes, the subject of my research as a graduate student. The erythrocyte nuclei, initially highly condensed and transcriptionally inert, are swelling, decondensing their chromatin, and becoming transcriptionally active; only when nucleoli appeared within the reactivating nuclei were chick genes expressed. No underlying nucleoskeleton was visible in the light microscope in either kind of nucleus.

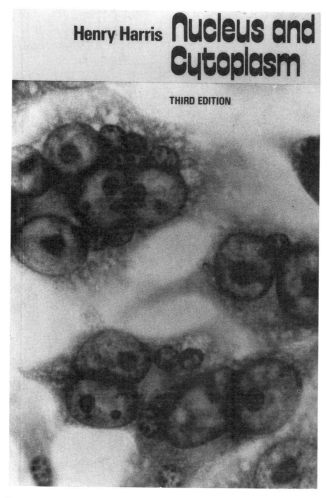

Fig. 6.1. The front cover of the 3rd edition of *Nucleus and cytoplasm*. (With permission of Oxford University Press.)

I worked on hypoxanthine-guanine phosphoribosyl transferase in these heterokaryons, which is encoded by the X-chromosome in mammals. It was only natural, then, that I should wonder what the basis of the inactivation of one of the two X-chromosomes in cells of female mammals might be; eventually I suggested that the linear chromosomes of higher eukaryotes must be organized into loops, and that differences in supercoiling in those loops underpinned differences in X-chromosome activity (Cook 1973, 1974). We then demonstrated that eukaryotic DNA was indeed supercoiled and organized into loops by attachment to a nuclear 'cage' and that supercoiling was lost as chromatin condensed during maturation of chicken erythroblasts into transcriptionally inert erythrocytes (Cook and Brazell 1975,

1976). Inevitably I asked the question: do RNA polymerases work out in the loop or at the base of the loop? I shall describe experiments that go some way to answer this question. But first, what is the evidence for current models for transcription?

The 'textbook' model for transcription

An unstated assumption in current models is that a polymerase tracks along the template as it synthesizes RNA (e.g. Alberts *et al.* 1983; Darnell *et al.* 1986). This assumption follows naturally from the relative size of the polymerase and template; presumably it is the smaller of the two that moves. But despite almost complete acceptance of the 'textbook' model, there seem to be only two kinds of evidence to support it.

The first kind is circumstantial; soluble polymerases work *in vitro* in the absence of any immobilizing skeleton. Why invoke any role for a skeleton, when we can mimic so well what happens *in vivo* without one? However, it is not widely appreciated that most RNA polymerase II in the cell is insoluble (Beebee 1979; Weil *et al.* 1979; Jackson and Cook 1985*b*). Of course, soluble enzymes are found in certain cases, for example in frogs' eggs, but they are inactive stockpiles, awaiting later use. Moreover, soluble polymerases isolated by most biochemists are inefficient and only become active when incorporated into large complexes. Thus, when cell extracts are incubated with appropriate templates, essentially all active RNA polymerases I, II, and III assemble into complexes that can be pelleted by a 5 min spin in a microcentrifuge (Culotta *et al.* 1985). Clearly, even these 'soluble' enzymes quickly form large complexes *before* becoming active. Until a soluble system is developed that initiates correctly at rates approaching those found *in vivo*, this kind of evidence cannot provide definitive proof for a skeleton-free model.

The second kind of evidence is provided by 'Miller' spreads (Miller 1984) and is apparently decisive. These spreads are prepared by dropping nuclei into a solution that is little more than distilled water (sometimes containing the detergent 'Joy'). Removing counter-ions charges chromatin, which expands and bursts the nucleus; individual chromatin fibres and beautiful 'Christmas tree' complexes can then be seen at the edge of the spread chromatin. No skeleton is visible. But, *a priori*, it would seem dangerous to draw general conclusions about structures *in vivo* using such a disruptive procedure and based on visualization of a minority of transcription complexes.

Active polymerases are attached to nucleoid cages

More than 10 years ago Shirley McCready did an analogous experiment to Miller's—she spread HeLa derivatives prepared not by reducing the tonicity, but by *increasing* it with 2 M NaCl (McCready *et al.* 1979). The now naked DNA, initially confined within a residual nucleoid 'cage', spreads to form a skirt that is attached to, and surrounds, the collapsed cage (Fig. 6.2). The DNA is supercoiled. Autoradiography showed that there was **no** nascent RNA in the skirt; **all** remained associated with the cage, which was presumably where it was made (Jackson *et al.*

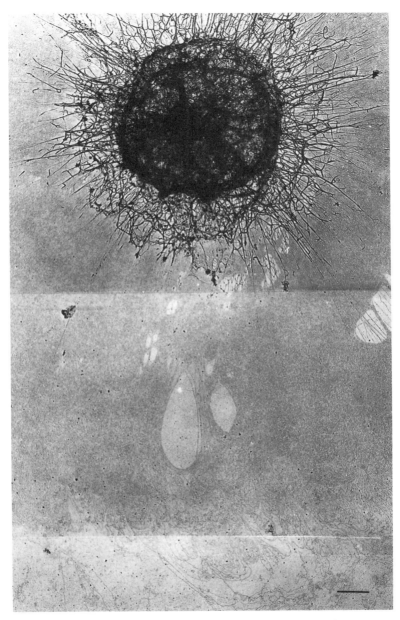

Fig. 6.2. DNA is organized into loops containing supercoils by attachment to a nuclear 'cage'. HeLa cells were treated with Triton (to permeabilize cell and nuclear membranes) and 2 M NaCl (to strip histones from the DNA). The resulting 'nucleoid' has been prepared for electron microscopy using Kleinschmidt's procedure. A tangled mass of supercoiled DNA fibres extend from the 'cage' to the edge of the field. Bar: 10 μm. From Jackson *et al.* (1984).

1981). Dean Jackson also removed most DNA with nucleases and found that transcribed sequences, and especially enhancers, were amongst the minority of sequences that still remained attached (reviewed by Jackson et al. (1984)). This suggests that the cage was the site of transcription and allowed us to rationalize the results we had obtained with nucleoids derived from different cells of the erythrocyte lineage: erythroblasts yielded well-developed cages (associated with supercoiled DNA) and were transcriptionally active; erythrocytes gave no cage (so their DNA was relaxed) and were transcriptionally inert.

Encapsulated cells allow use of a physiological salt concentration

We used unphysiological conditions for these experiments (as does nearly everybody), because chromatin aggregates into an unworkable mess at isotonic salt concentrations. This, coupled to the fact that transcription complexes are very sticky, led to the suspicion that transcript–cage complexes were isolation artefacts (Cook 1988). Therefore we developed a procedure that allowed the use of more physiological conditions (Jackson and Cook 1985a; Jackson et al. 1988). Cells were encapsulated in agarose microbeads of about 50 μm diameter (Fig. 6.3, left). As agarose is permeable to small molecules, cells can be regrown or extracted in 'physiological' buffers containing Triton; then most cytoplasmic proteins and RNA diffuse out to leave encapsulated chromatin surrounded by the cytoskeleton (Fig. 6.3, right). These fragile cell remnants are protected by the agarose coat, but accessible to probes like antibodies and enzymes. Whilst one cannot be certain that any isolate is artefact-free, this type contains intact DNA and essentially **all** the replicative and transcriptional activities of the living cell. As the attachments that I will describe involve polymerases, it seems unlikely that they are generated artefactually when all activity is retained.

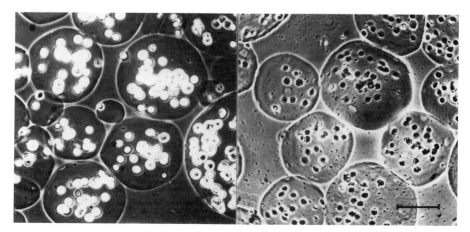

Fig. 6.3. HeLa cells encapsulated in agarose microbeads, before (left) and after (right) lysis with Triton.

Active polymerases are associated with a nucleoskeleton

We used the encapsulated and lysed cells to examine whether polymerases were attached to some large structure in the nucleus. Models involving mobile or immobile (i.e. attached) polymerases can be distinguished by fragmenting the encapsulated chromatin with an endonuclease and then removing any unattached material electrophoretically. If polymerizing complexes are attached to a larger skeletal structure, they should remain in beads: if unattached, they should electroelute from beads with most chromatin (Fig. 6.4). (Note that chromatin containing DNA fragments of 150 kb can escape from beads.) Cutting HeLa chromatin into <10 kb fragments, followed by electroelution of most chromatin, leaves residual clumps of chromatin associated with an intermediate-filament-like skeleton (Fig. 6.5; Jackson and Cook 1988). However, chromatin removal hardly reduces the activity of RNA polymerases I and II (Fig. 6.6) or DNA polymerase α; nascent RNA and DNA also

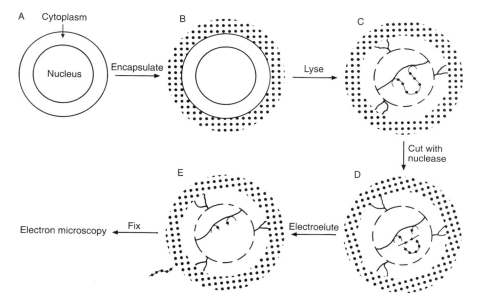

Fig. 6.4. Outline of experimental approach. Cells (A) are encapsulated (B) in agarose microbeads (stippled area), lysed (C), and washed in a 'physiological' buffer. Structures too large to escape through agarose are left in beads and include the cytoskeleton, nuclear lamina (dashed line) and chromatin (looped 'beads on a string') which generally obscures any underlying nucleoskeleton. Chromatin is fragmented (D) by addition of a nuclease (arrows) and small unattached pieces are removed electrophoretically (E). Finally, samples are fixed and viewed in the electron microscope; any underlying nucleoskeleton can now be seen in the relatively empty nucleus. Alternatively, attachments of polymerase can be analysed by comparing polymerizing activities in beads that have been subjected to electrophoresis or stored on ice. If the polymerase is associated with the skeleton, all activity should resist electroelution; if not, most activity should be lost with the electroeluted chromatin.

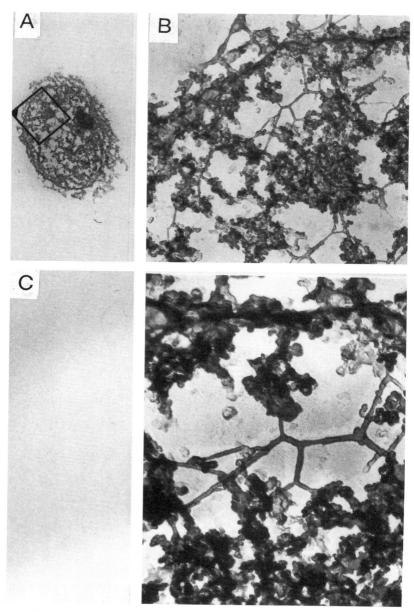

Fig. 6.5. Electron micrographs of thick resinless sections of encapsulated HeLa cells from which 80 per cent of the chromatin had been removed as in Fig. 6.4. (A) Low power, showing a section through a HeLa cell. The surrounding agarose cannot be seen at this magnification. (B) Medium power, showing the region in the square in (A) (the top left-hand corner is filled in for orientation). (C) High power showing residual clumps of chromatin still attached to a nucleoskeleton. The nuclear lamina runs across the top of the field. (From Jackson and Cook 1988.)

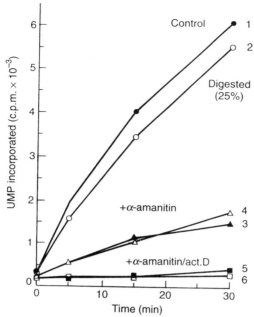

Fig. 6.6. Active RNA polymerases resist electroelution. Cells were encapsulated, treated with or without *Eco*RI, and detached fragments electroeluted as in Fig. 6.4. The time-course of incorporation of [^{32}P]UTP into acid-insoluble material by beads treated in various ways is shown. In some cases cells were treated with actinomycin D before harvesting, in others lysed cells were preincubated with α-amanitin before transcription. Curve 1: control, without inhibitors, digestion with *Eco*RI, or electrophoresis. Curve 2: without inhibitors, but digested and electroeluted (25 per cent of the chromatin remained). Curve 3: without digestion or electrophoresis, but with α-amanitin. Curve 4: with digestion, electroelution (25 per cent of the chromatin remained), and α-amanitin. Curve 5: without digestion or electrophoresis, but with actinomycin D and α-amanitin. Curve 6: with digestion, electrophoresis (25 per cent of the chromatin remained), actinomycin D, and α-amanitin. Despite the removal of 75 per cent of the chromatin, essentially all RNA polymerizing activity, which is mostly RNA polymerase II, is retained in beads (curves 1 and 2). The α-amanitin-resistant and actinomycin D-sensitive activity (i.e. RNA polymerase I) also resisted elution (curves 3 and 4). (From Dickinson *et al.* 1990.)

resisted electroelution, presumably because they are attached to the skeleton (Jackson and Cook 1985*b*, 1986*a*,*b*,*c*; Jackson *et al.* 1988; Dickinson *et al.* 1990).

After removing most chromatin (as in Fig. 6.4), the size of the loops can be deduced from the size of the residual attached fragments and the percentage of chromatin remaining in beads (Jackson *et al.* 1990). Loop sizes ranged from 5 to 200 kb, with an average of 86 kb; the smaller loops were probably the transcriptionally active ones. Loops in nuclei isolated by conventional methods, as well as matrices and scaffolds—which are all prepared in non-isotonic buffers—had smaller loops; many of their attachments of chromatin fibre to the skeleton must be generated artefactually during isolation.

Dean Jackson has recently gone on to map which sequences attach viral minichromosomes to the skeleton in transfected cells (Jackson and Cook, 1993). Non-transcribed minichromosomes in the population eluted from nuclei but transcriptionally active ones did not. Cutting the attached fraction with *Hae*III enabled most resulting ~400 bp fragments to elute and analysis of the residual fragments showed that no single sequence was responsible for attachment; rather each minichromosome was attached at only one or two points through a promoter or somewhere in a transcription unit (i.e. probably through an elongating RNA polymerase II). The latter attachments must change dynamically as the template slides past the attachment site. It is obviously tempting to extrapolate these results to cellular loops and suggest that they, too, are attached only by active polymerases and promoters (Jackson *et al.* 1992).

We have recently visualized sites of transcription by fluorescence microscopy (Jackson, *et al.* 1993). Encapsulated and permeabilized HeLa cells are incubated with Br-UTP to extend nascent RNA chains by ~500 nucleotides; then sites of incorporation are directly immunolabelled using an antibody against Br-RNA.~300 focal sites of incorporation (i.e. RNA synthesis) can be seen in each nucleus; most of these also contain RNA polymerase II and a component of the splicing apparatus detected by anti-Sm antibodies. α-amanitin, an inhibitor of RNA polymerase II, prevents incorporation into these foci so that ~25 discrete foci can be seen more clearly in nucleoli. All these fluorescent foci remain after removing ~90 per cent of the chromatin. As calculations show that each focus contains many transcription units, this suggests that an underlying skeleton must organize groups of transcription units (in both nucleolar and extra-nucleolar regions) into 'factories' where transcripts are both synthesized and processed. We will now visualize these factories by electron microscopy, much as we have done for the analogous replication factories (Hozák, *et al.* 1992).

The use of 'physiological' conditions and recovery of essentially all activity, rather than a minor fraction, make explanations of these results based on artefacts involving aggregated polymerases difficult to sustain. The polymerizing complexes cannot fortuitously have no net charge and so be unable to electroelute as the same results are obtained at a different pH (Jackson *et al.* 1988). If the complex is unattached, it must be so large that the polymerase is effectively attached. But the simplest interpretation is that active polymerases are attached.

The topology of transcription

If active polymerases are attached, presumably they are immobile. How, then, does transcription occur?

Transcription of a double helix poses various topological problems. One concerns templates with ends that are unable to rotate freely, for example if organized into circles or loops (Jackson *et al.* 1981; Liu and Wang, 1987). Another concerns the interlocking of template and transcript that results if the polymerase

1. POLYMERASE TRANSLOCATES AND ROTATES

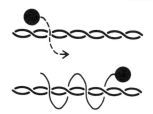

2. POLYMERASE TRANSLOCATES, DNA ROTATES

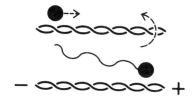

3. DNA TRANSLOCATES, POLYMERASE ROTATES

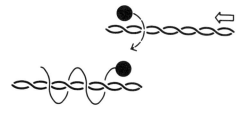

4. DNA TRANSLOCATES AND ROTATES

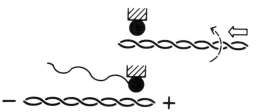

Fig. 6.7. Models for transcriptional elongation involving mobile or static polymerases (black circles) and double-helical templates. The upper figure in each model indicates initial relative positions; subsequent movements are shown by arrows. Lower figures show final positions after generation of transcripts (wavy lines attached to polymerases). + and − indicate the formation of domains of positive and negative supercoiling. In 4, the hatched area immobilizes the polymerase. (From Cook and Gove 1992.)

tracks along a helical strand, as in 'textbook' models. Polymerase and template must move relative to each other, both rotationally around the helix axis and laterally along it, so relative motions can be classified in four ways, depending on which of the two players (polymerase or DNA) performs which of the two movements (rotation or translocation).

The first model in Fig. 6.7 involves a mobile polymerase both rotating about and translocating along a static template. Then the polymerase, plus nascent transcript, must rotate about the template, once for every 10 bp transcribed. This gives a transcript that is intertwined about the template and we have no mechanism for 'untwining' them. This untwining problem seems insuperable, making model 1 unlikely. Model 3 faces the same intractable problem.

This problem is sidestepped if DNA rotates instead of the polymerase. In model 2—the 'twin-supercoiled-domain' model (Liu and Wang 1987)—the enzyme translocates laterally but its rotation is restricted, perhaps by the frictional drag of the

---

Fig. 6.8. Immobilizing T7 RNA polymerase. (A) The structure of the plasmid encoding the hybrid polymerase. The sequence of the linker is shown below: the underlined ACT is codon 392 of *malE* (the gene for the maltose-binding protein), the region in bold encodes the protease Xa recognition sequence (IEGR), and the ATG codes for the first amino acid of the polymerase. (B) Transcription from P_{tac} and subsequent translation leads to the formation of a hybrid protein, with maltose-binding protein and polymerase domains, connected through a peptide linker containing the Xa-cleavage site, IEGR. (C) Cartoon of two hybrid proteins immobilized by attachment via antibodies directed against the maltose-binding moiety (Y-shaped structures) to protein A (circles) covalently attached to plastic (hatched area). The upper hybrid protein has bound template and generated a transcript (wavy line); the lower one is inaccessible to template. Treatment with Xa releases both polymerases. (D) Bound and free RNA polymerases elongate at equal rates. Elongation rates were measured under conditions where initiation was suppressed, both using heparin and by removing excess template. Hybrid protein was bound to beads and transcription initiated by adding ATP, CTP, and GTP, but not UTP. Then initiated complexes with 7–nucleotide-long transcripts are formed, as the first U is incorporated into nascent RNA at position eight. All samples were washed free of excess template, some were incubated for 3 or 30 min at 20 °C (–/+ heparin, –/+ Xa) and some were then rewashed to remove any detached polymerase. Transcriptional elongation was then re-started by addition of $[\alpha^{32}P]UTP$. Equal volume reactions containing labelled transcripts were run on a denaturing gel and an autoradiograph was prepared. Samples were withdrawn at 0.25, 5, and 15 min (indicated by triangles), giving three tracks per reaction. Nucleotide sizes are indicated on the left. At the low UTP concentration used, transcription is inefficient and transcripts stall or terminate prematurely wherever UTP is required. For example, many do so 123 nucleotides downstream from the promoter, where four consecutive uridines are incorporated. Most transcripts synthesized after 3 min pre-incubation in the absence of heparin are shorter than 123 nucleotides (lanes 1–3). Heparin, by preventing reinitiation, suppresses the synthesis of shorter transcripts and stimulates the formation of longer ones (lanes 4–6). 30 min pre-incubation (either with or without factor Xa) has essentially no effect on the length of the resulting transcripts (lanes 7–9 and 10–12); the attached polymerase elongates just as efficiently as the free enzyme. Washing after preincubation with factor Xa removes >80 per cent activity (lanes 16–18; note that band intensities are weaker), showing that treatment with Xa detaches the polymerase. (From Cook and Gove 1992.)

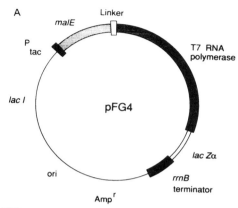

LINKER

malE sequence - CAG <u>ACT</u> AAT TCG AGC TCG GTA CCC GGC
CGG GGA TCC **ATC GAA GGT CGT** - ATG - T7 RNA polymerase

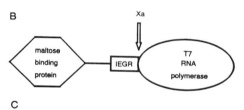

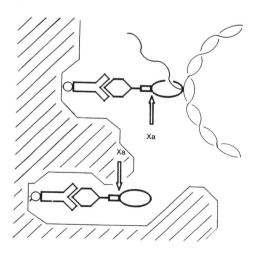

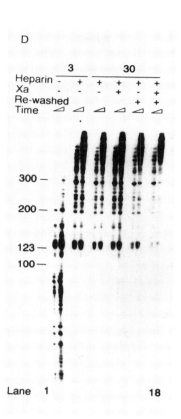

transcript; instead DNA rotates. Polymerase translocation along DNA generates positive supercoiling 'waves' ahead of, and negative supercoiling 'waves' behind, the moving enzyme. The torsional strain associated with these supercoils limits transcription unless removed by topoisomerases. Although there is now considerable support for such twin domains (e.g. Wu *et al.* 1988; Droge and Nordheim 1991), this model faces the problem of preventing the polymerase from rotating whilst allowing it to translocate. Even one accidental rotation—which is especially likely in long transcription units or when the transcript is short and frictional drag limited—would yield an entwined transcript. Heggeler-Bordier *et al.* (1992) have recently suggested that rotation might be restricted if the polymerase deformed the template into an apical loop, so preventing rotation of the loop and associated enzyme about the helical axis. But again, it seems unlikely that this could completely prevent rotation throughout long transcription units. Indeed, it is difficult to imagine any mechanism that would do so without immobilizing the polymerase.

In model 4, threading and untwining problems are completely eliminated because the enzyme is immobilized by attachment to some larger structure (i.e. the nucleoskeleton); instead DNA both translocates and rotates (Jackson *et al.* 1981; Cook 1989). It can be viewed as a special case of the 'twin-domain' model; domains of supercoiling are generated in much the same way and must be removed.

Are immobile polymerases active?

But can an attached polymerase work? Therefore we tested whether immobilization inhibited the activity of one of the most active polymerases known, that of the bacteriophage T7 (Fig. 6.8; Cook and Gove 1992). A bipartite protein consisting of the polymerase connected through a peptide linker with an immobilizing domain was expressed in bacteria. This was attached (via an antibody to the immobilizing domain) to protein A, which was, in turn, covalently linked to plastic beads. The polymerase could be released by cleaving the linker with a specific protease, factor Xa (Fig. 6.8 (*c*)). Comparison of the bound and free forms (i.e. after treatment without or with factor Xa) showed that immobilization reduced the rate of initiation but had little effect on elongation (Fig. 6.8(D)).

Fig. 6.9. A model for transcription. (A) A loop of DNA is shown attached to the skeleton (rod) at two sites. These attachments probably persist whether or not the loop is transcribed or replicated; they are probably adjacent transcription units. A gene out in the loop cannot be transcribed as its promoter (P) is remote from any attached polymerase. (E) marks an upstream activating sequence (e.g. an enhancer). (B) During development, the gene in the loop becomes active by attachment to a transcription complex assembled on the skeleton. The complex contains a polymerase (pol) flanked by two topoisomerases (T), plus a transporter (engine) on a track that leads through 'stations' where the appropriate enzymes for RNA processing, including polyadenylation (p(A)) and splicing (Sp) are concentrated. Initially E attaches at one site (triangle) to become permanently tethered to the complex; this inevitably brings P into close proximity to the polymerase, facilitating its binding. Elements of the complex are drawn spatially separated but they are probably in close contact to allow

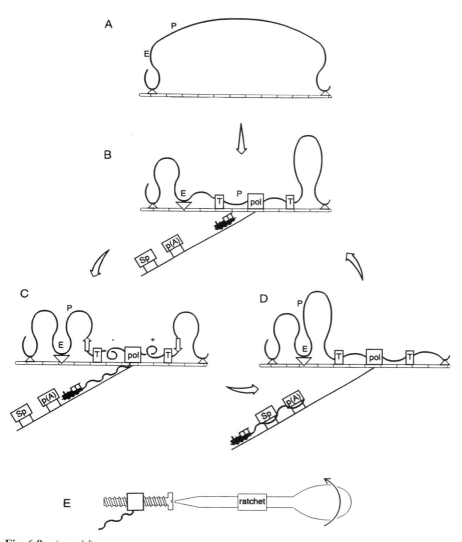

Fig. 6.9 (cont'd)
inter-communication. (C) After initiation, DNA moves (arrows) through the complex as RNA (wavy line) is extruded and attached to the transporter, which has begun to move down the track. The loop on the right shrinks as the loop on the left enlarges. Positive and negative supercoils appear transiently as shown but are removed immediately by topoisomerases. (D) The transcript is complete; it has been spliced and polyadenylated and is being transported to the nuclear pore. The template now detaches from the polymerase and the topoisomerases, but is held at the enhancer so that the promoter can easily rebind to start the whole process again. (E) Transcription is analogues to driving a bolt (DNA) through a nut (polymerase), whilst the ratchet (topoisomerase) in the screwdriver releases torsional strain. The complex is shown below the active transcription unit in (C). Adapted from Cook (1989).

Model for transcription

This leads us to a general model for transcription in which unentangled transcripts can only be made by immobilized enzymes. Bacteria and viral enzymes probably function as dimers, anchoring themselves to one piece of DNA whilst transcribing another, as in bacterial nucleoids. Eukaryotic enzymes adopt a different strategy, becoming immobilized by attachment to a skeleton (Fig. 6.9; Cook 1989). In the special case of the reactivating chick erythrocyte nucleus in the heterokaryon, the initially inert nucleus lacks a skeleton and its associated polymerases; chick genes are remote from polymerases on the mouse skeleton and cannot be transcribed. Only when a skeleton (plus associated transcription machinery) are built in the *chick* nucleus can promoters attach and the template move through the fixed polymerizing site to generate the transcript. Subsequent processing and transport also take place on the skeleton. Nuclear swelling and the appearance of nucleoli are then gross structural correlates of this complicated process.

The skeleton and replication

This essay has concentrated on the role of a nucleoskeleton during transcription. An integrating role for a similar structure during replication is also emerging (reviewed by Cook 1991) and what relationship there is between the two skeletons is obviously of the greatest interest.

Conclusions

These experiments lead us to a very different view of how transcription occurs—the DNA moves rather than the polymerase. People often say that movement is relative, so why should it matter which moves past the other? There are at least two very good reasons. First, I think it important to get the principles governing such a basic process as transcription right. It does not matter to most of us whether or not the earth goes round the sun, but we do like to know which moves. The second reason is more practical. Biochemists find it relatively easy to work with soluble enzymes found in supernatants, discarding pelleted material. But I think more authentic activities can be found in the pellet; we should concentrate on these, instead of throwing them away!

Acknowledgements

The Cancer Research Campaign has continuously supported me and my group over the years; recent work has also been supported by the Wellcome Trust and the Medical Research Council. I have summarized the work of many people, but Dean Jackson's contribution should be specially acknowledged. Henry Harris introduced me to 'many hours of simple pleasure' and has encouraged me throughout my career; I thank him for his help, especially when things were not going well.

References

Alberts, B., Bray, D., Lewis, J., Raff, M., Roberts, K., and Watson, J. D. (1983). *Molecular biology of the cell*. Garland, New York.
Beebee, T. J. C. (1979). A comparison of methods for extracting ribonucleic acid polymerases from rat liver nuclei. *Biochem. J.*, **183**, 43–54.
Cook, P. R. (1973). Hypothesis on differentiation and the inheritance of gene superstructure. *Nature*, **245**, 23–5.
Cook, P. R. (1974). On the inheritance of differentiated traits. *Biol. Rev.*, **49**, 51–84.
Cook, P. R. (1988). The nucleoskeleton: artefact, passive framework or active site? *J. Cell Sci.*, **90**, 1–6.
Cook, P. R. (1989). The nucleoskeleton and the topology of transcription. *Eur. J. Biochem.*, **185**, 487–501.
Cook, P. R. (1991). The nucleoskeleton and the topology of replication. *Cell*, **66**, 627–35.
Cook, P. R. and Brazell, I. A. (1975). Supercoils in human DNA. *J. Cell Sci.*, **19**, 261–79.
Cook, P. R. and Brazell, I. A. (1976). Conformational constraints in nuclear DNA. *J. Cell Sci.*, **22**, 287–302.
Cook, P. R. and Gove, F. (1992). Transcription by an immobilized RNA polymerase from bacteriophage T7 and the topology of transcription. *Nucleic Acids Res.*, **20**, 3591–8.
Culotta, V. C., Wides, R., and Sollner-Webb, B. (1985). Eucaryotic transcription complexes are specifically associated in large sedimentable structures: rapid isolation of polymerase I, II and III transcription factors. *Mol. Cell. Biol.*, **5**, 1582–90.
Darnell, J., Lodish, H., and Baltimore, D. (1986). *Molecular cell biology*. Scientific American Books, New York.
Dickinson, P., Cook, P. R. and Jackson, D. A. (1990). Active RNA polymerase I is fixed within the nucleus of HeLa cells. *EMBO J.*, **49**, 2207–14.
Droge, P. and Nordheim, A. (1991). Transcription-induced conformation change in a topologically closed DNA domain. *Nucleic Acids Res.*, **19**, 2941–6.
Harris, H. (1967). *Nucleus and cytoplasm*. Oxford University Press, Oxford.
Harris, H. (1974). *Nucleus and cytoplasm*, (3rd edition). Oxford University Press, Oxford.
Heggeler-Bordier, B., Wahli, W., Adrian, M., Stasiak, A., and Dubochet, J. (1992). The apical localization of transcribing RNA polymerases on supercoiled DNA prevents their rotation around the template. *EMBO J.*, **11**, 667–72.
Hozak, P., Hassan, A. B., Jackson, D. A. and Cook P. R. (1993). Visualization of replication factories attached to a nucleoskeleton. *Cell*, **73**, 361–73.
Jackson, D. A. and Cook, P. R. (1985a). A general method for preparing chromatin containing intact DNA. *EMBO J.*, **4**, 913–8.
Jackson, D. A. and Cook, P. R. (1985b). Transcription occurs at a nucleoskeleton. *EMBO J.*, **4**, 919–25.
Jackson, D. A. and Cook, P. R. (1986a). Replication occurs at a nucleoskeleton. *EMBO J.*, **5**, 1403–10.
Jackson, D. A. and Cook, P. R. (1986b). A cell-cycle dependent DNA polymerase activity that replicates intact DNA in chromatin. *J. Mol. Biol.*, **192**, 65–76.
Jackson, D. A. and Cook, P. R. (1986c). Different populations of DNA polymerase α in HeLa cells. *J. Mol. Biol.*, **192**, 77–86.
Jackson, D. A. and Cook, P. R. (1988). Visualization of a filamentous nucleoskeleton with a 23 nm axial repeat. *EMBO J.*, **7**, 3667–77.
Jackson, D. A. and Cook, P. R. (1993). Transcriptionally active minichromosomes are attached transiently in nuclei through transcription units. *J. Cell Sci.*, **105**, 1143–50.

Jackson, D. A., McCready, S. J., and Cook, P. R. (1981). RNA is synthesized at the nuclear cage. *Nature*, **292**, 552–5.

Jackson, D. A., McCready, S. J., and Cook, P. R. (1984). Replication and transcription depend on attachment of DNA to the nuclear cage. *J. Cell Sci. Suppl.*, **1**, 59–79.

Jackson, D. A., Yuan, J., and Cook, P. R. (1988). A gentle method for preparing cyto- and nucleo-skeletons and associated chromatin. *J. Cell Sci.*, **90**, 365–78.

Jackson, D. A., Dickinson, P., and Cook, P. R. (1990). The size of chromatin loops in HeLa cells. *EMBO J.*, **9**, 567–71.

Jackson, D. A., Dolle, A., Robertson, G., and Cook, P. R. (1992). The attachments of chromatin loops to the nucleoskeleton. *Cell Biol. Int. Rep.*, **16**, 687–96.

Jackson, D. A., Hassan, A. B., Errington, R. J. and Cook P. R. (1993). Visualization of focal sites of transcription within human nuclei. *EMBO J.*, **12**, 1059–65.

Liu, L. F. and Wang, J. C. (1987). Supercoiling of the DNA template during transcription. *Proc. Natl. Acad. Sci. USA*, **84**, 7024–7.

McCready, S. J., Akrigg, A., and Cook, P. R. (1979). Electron microscopy of intact nuclear DNA from human cells. *J. Cell Sci.*, **39**, 53–62.

Miller, O. L. (1984). Some ultrastructural aspects of genetic activity in eukaryotes. *J. Cell Sci. Suppl.*, **1**, 81–93.

Weil, P. A., Luse, D. S., Segall, J., and Roeder, R. G. (1979). Selective and accurate initiation of transcription at the Ad2 major late promoter in a soluble system dependent on purified RNA polymerase II and DNA. *Cell*, **18**, 469–84.

Wu, H.-Y., Shyy, S., Wang, J. C. and Liu, L. F. (1988). Transcription generates positively and negatively supercoiled domains in the template. *Cell*, **53**, 433–40.

7. Lateral mobility of membrane proteins—a journey from heterokaryons to laser tweezers

Michael Edidin

Introduction

Soon after virus-induced heterokaryons were described (Harris and Watkins 1965) it was observed by several laboratories that species-specific surface markers of the parent cells were randomly distributed over the heterokaryon surface (Watkins and Grace 1967; Harris *et al.* 1969; Gordon and Cohn 1970). All of these observations were made hours after the fusion of the parent cells, and were made with fairly low spatial resolution. Larry Frye, then a graduate student in my laboratory, decided to refine the experiments by resolving the time required for randomization of surface markers from parent cells. He also improved the spatial resolution of the distribution by using fluorescent antibodies to the class I MHC (major histocompatibility complex) molecules of the mouse cell parent and antibodies to undefined, but membrane-integral, antigens of the human cell parent of the heterokaryons. With this system, Frye was able to show that randomization of surface marker antigens began within minutes of fusing mouse and human cells with virus (Frye and Edidin 1970). Controls, including fixation of newly formed heterokaryons, showed clearly that redistribution was due to diffusion and not to cell metabolism (Edidin and Wei 1977). Together with studies of lipid motion in the bilayer, these experiments set us on the road to viewing cell membranes, especially cell surface membranes, as ordered fluids whose components are free to move in the plane of the membrane. The experiments that showed diffusional motions of lipids and proteins are summarized in an early review on the subject (Edidin 1974); the model embodying them, together with data on the solubility of membrane proteins is the well-known 'fluid mosaic' model of Singer and Nicolson (1972).

Frye's experiments detected lateral diffusion of proteins because he changed the time and spatial resolution of his measurements from that previously used. In the years since his work, we have found that lateral diffusion in cell membranes can be quantified by a number of techniques that yield information on different spatial and temporal scales. The results obtained using these techniques show clearly that one cannot picture cell surface membranes as fluid lipid continua. Membranes are patchy, containing metastable domains enriched for particular molecules or classes of molecules (Edidin 1992; Sheetz 1993). The data also show that membrane proteins do not diffuse as if they are in dilute solution; they interact with their neighbours and with molecules adjacent to either side of the bilayer. No membrane protein seems to be in 'splendid isolation'. Indeed, one model for their behaviour aptly terms them 'the milling crowd' (Eisenger et al. 1986).

From qualitative to quantitative descriptions of lateral diffusion

The rapid redistribution of membrane markers, visualized with fluorescent antibodies, dramatically demonstrates the fact of lateral diffusion, but the geometry of this redistribution allows only a rough estimate of diffusion coefficents. Despite this, the method has also been used to show the lateral diffusion of erythrocyte proteins (Fowler and Branton 1977; Koppel and Sheetz 1981) and of molecules in plant protoplast membranes (Mastrangelo and Mitra 1981). Another form of the method showed that proteins of the apical surface of polarized epithelial cells redistributed when an epithelial layer was dissociated into single cells (Pisam and Ripoche 1976; Ziomek et al. 1980). However, none of these experiments could do more than set upper or lower limits to the diffusion coefficients, D, of membrane proteins. It was of considerable interest to know D for lipids and proteins in the cell membrane, since mobile molecules moving in a fluid lipid phase offered the possibility of coupling chemical reactions by lateral diffusion and the possibility of rapid functional reorganization of cell surfaces.

The first measurements of D for a membrane protein were the determinations, by Poo and Cone (1974) and by Liebman and Entine (1974), of D for visual rhodopsin in the discs of vertebrate rods. The measured value is the benchmark for all later work: $D \approx 0.3$–0.5 $\mu m^2 s^{-1}$. This value could be calculated from the viscosity of membrane lipids, the diffusion of lipid probes in synthetic bilayer, and the dimensions of the rhodopsin molecule.

The technique used for rhodopsin involved bleaching a small region of the rod, a stack of disc membranes, using a light flash directed through a mask of known size and shape. The absorption spectrum of bleached rhodopsin differs from that of unbleached rhodopsin and so diffusion could be followed using dim light to detect the return of unbleached rhodopsin to the region defined by the light flash. This limited the method. However, the experimental approach of imposing a defined geometry on the membrane of interest proved to be the key to a general quantitative method for measuring lateral diffusion (Axelrod et al. 1976; Edidin and Zagyansky 1976; Edidin et al. 1976; Jacobson et al. 1976). This method, fluorescence photobleaching and recovery, or FPR (also know as fluorescence recovery after photobleaching, FRAP) defines the geometry by focusing laser light onto a small area of a cell surface labelled with a fluorescent label. Using fluorescent labels allows the method to be generalized to almost any molecule in any membrane and greatly increases the sensitivity of the method. The general approach is shown in the top part of Fig. 7. 1, The bottom part of the figure shows typical experimental results from the method. (In descending order the curves show: no diffusion, diffusion and complete recovery of fluorescence, and diffusion with partial recovery of fluorescence.)

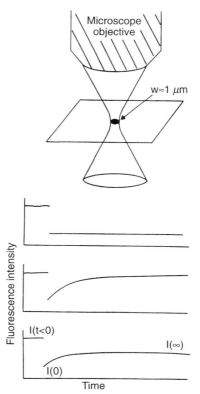

Fig. 7.1. Fluorescence photobleaching and recovery (FPR), a general method for measuring lateral diffusion of membrane proteins. (Top) An attenuated laser beam is focused on to a small spot on the cell surface. Fluorescence excited by the focused laser light is collected by the objective. A fraction of the fluorophores may be bleached by briefly removing the attenuator from the light path. Recovery of fluorescence is monitored using the attenuated beam. (Bottom) Three typical FPR curves. The first shows no recovery of fluorescence. The second shows complete recovery of fluorescence. The third shows partial recovery of fluorescence.

Knowing the area illuminated by the laser beam, and the time for recovery of fluorescence, it was relatively easy to calculate D for the label. Following the publication of the FPR method there was an explosion of determinations of D for a wide variety of membrane proteins and lipids (reviewed in Edidin (1991)). The results were surprising. D for most membrane proteins was at least 50-fold smaller than that predicted and that measured for visual rhodopsin, ~0.01 $\mu m^2/s$. This did not seem to be an artefact of the method. A number of controls appeared to rule out artefacts of photodamage. D for fluorescent rhodopsin, measured by FPR (Wey et al. 1982) was the same as D measured by the original flash photolysis methods. D estimated from heterokaryons and from other methods of measuring lateral diffusion agreed to within an order of magnitude with D from FPR. It appeared that molecules in a cell membrane were constrained in their lateral diffusion by one or more factors of cell architecture and cell metabolism.

FPR measurements also characterize a second parameter of the labelled molecules, R, the fraction of all molecules free to diffuse in the time, tens of seconds, of a single measurement. Here too, there was a discrepancy between results on diffusion of lipids and proteins in artificial lipid bilayers, results on visual rhodopsin, and results on most cell surface proteins. Recovery of fluorescence approached 100 per cent in the first two instances; all labelled molecules were mobile. In contrast, only a fraction of any labelled molecule was free to diffuse in most cell membranes. This fraction ranged between 25–30 and 70–80 per cent. Some of the immobile fraction proved to be due to technical artefacts. For example, in mobile fractions of molecules labelled by fluorescent Fab fragments, Fab increase if aggregates are removed from the Fab solutions before labelling cells (Bierer *et al.*, 1987). However, once the technical problems were recognized and resolved, it became evident that a substantial fraction of Fab-labelled membrane proteins, fluorescent lipid analogues, and glycolipid-anchored proteins, molecules like Thy-1, is immobile in an FPR experiment.

The differences between D and R for molecules of cell surface membranes and for molecules in artificial bilayers and in disc membranes imply constraints to lateral diffusion in the cell surface. Logically such constraints could involve interactions of membrane molecules with one another, with molecules of the cell cytoplasm (the membrane- and cyto-skeleton), or with molecules of the extracellular matrix and the medium bathing the surface. Examples of all possible interactions have been found. Indeed, it appears that the D and R observed for a given protein on a given cell type reflect several different constraints to lateral diffusion. The importance of any one of these constraints—cytoplasmic, extracellular, or lateral, will depend upon the biology of the cell and the protein investigated.

The first clear evidence that a cytoskeleton affected D was obtained by Sheetz and co-workers (1980) who used FPR to compare D and R of lipid analogues and membrane proteins of normal and spectrin-deficient (sphaerocytic) mouse erythrocytes. D of lipid analogues was similar in the two sorts of erythrocytes. D of labelled proteins was about 100-fold higher in spectrin-deficient erythrocytes than in normal ones. Cherry and co-workers (1976) had shown that the *rotational* diffusion of the erythrocyte membrane protein, band 3, was close to that calculated for rotation of this protein in a lipid bilayer with a viscosity of about 1 P. The two data sets were interpreted as showing that the lateral mobility of band 3 and other proteins in erythrocyte membranes was limited by a matrix of cytoskeletal proteins, principally spectrin. Lateral diffusion over an area larger than the unit area of the spectrin matrix (which has sides 100–200 nm long) could only occur if the matrix was labile, creating defects through which molecules could move. On the other hand, rotational diffusion of a membrane protein could readily occur within the area of a single unit of the matrix. Membrane proteins tethered to the proteins of the matrix would neither rotate nor diffuse laterally. This is consistent with the observation of an immobile fraction of label when measuring lateral diffusion and with the finding that the anisotropy whose change measures rotational diffusion did not decay to 0. Similar conclusions are reached from the demonstration by Golan and Veatch (1980) that chemical removal of the membrane skeleton from normal erythrocyte ghosts increases D and R of labelled band 3.

As has been pointed out by Peters (1988) it is lateral diffusion in the small area, 0.01 μm^2, that is the unit of the spectrin matrix (reviewed by Shen (1985)) and not lateral diffusion in the 1–2 μm^2 area measured by FPR, that is important in coupling reactions in membranes.

The data on constraints to lateral diffusion in red cells ought to apply to other cells and several kinds of experiments have been done to show this. None of these experiments is conclusive, indeed several seem to show that the membrane or cytoskeleton is not an important constraint to lateral diffusion.

Webb and co-workers (Tank et al. 1982; Wu et al. 1982) induced blebs in cultured cells that lifted the plasma membrane away from the underlying cytoplasm. The blebs appeared to be free of cytoskeleton. They did not stain for actin, and small particles trapped in the blebs were in Brownian motion. D of membrane proteins was 100 times higher in the blebs than on the surface of intact cells and R increased to 1; all the labelled molecules were mobile. In one instance, namely acetylcholine receptors on adult mouse muscle fibres, no diffusion could be detected in the native membranes ($R = 0$, $D < 10^{-12}$ cm^2/s, but D approached the viscosity limit and R increased to 1 in blebs.

The same experiments also suggest that lateral interactions between membrane molecules constrain their diffusion. D of fluorescent lipid analogues and of a fluorescent stearoyldextran is 10 times higher in blebs than in native membranes; however, both of these probes are anchored to the membrane by acyl chains that only extend into the outer leaflet of the bilayer; they cannot interact directly with proteins of the cytoplasm. The concentration of membrane proteins in the blebs has never been measured, but it is likely to be lower than in the native membrane. In that case at least some of the effect of blebbing on D could be due to dilution of membrane proteins to a point where they do not interfere with one another's lateral diffusion.

Other FPR experiments pointing towards a role for cyto- or membrane skeleton in restricting D and R correlate localization of spectrin with localization of membrane ATPase in morphologically polarized cells. The Na$^+$K$^+$ ATPase of differentiated MDCK cells is localized to the basal/lateral surface and forms a complex with the underlying spectrin (Nelson et al. 1990). Jesaitis and Yguerabide (1986) found that only 50 per cent of the Na$^+$K$^+$ ATPase was mobile in such cells. Other basal/lateral molecules are also immobilized when MDCK cells differentiate (Salas et al. 1988). Similarly, the Na$^+$K$^+$ ATPase of cultured chicken photoreceptors localizes with the spectrin of those cells and this localization, as the cells differentiate, correlates with a 50 per cent reduction of R measured by FPR (Madreperla et al. 1989). Note that in both instances, the main effect is on R not D.

If the cytoskeleton or membrane skeleton is an important constraint on the lateral diffusion of membrane proteins, then we expect that truncation of the cytoplasmic portions of these proteins ought to affect D or R. This is not the case for four different proteins, class I MHC molecules in L-cells (Edidin and Zuniga 1984), epidermal growth factor receptor (Livneh et al. 1986) a viral glycoprotein (Scullion et al. 1987) and the high-affinity IgE receptor ($Fc_\epsilon RI$) (Mao et al. 1991). Three of these are so-called type I proteins, whose bulk lies outside the bilayer in the extracellular medium. Only the multichain IgE receptor has components, the β and γ chains, whose bulk lies within the bilayer.

There is one instance in which truncation of the cytoplasmic domains of type I membrane proteins, class II MHC molecules, does change D, by 10-fold, when cytoplasmic domains are removed from both chains (Wade et al. 1989). In that instance we can also associate truncation with reduced signalling via class II molecules. D appears to reflect the extent of physical coupling between class II molecules and the proteins of the signalling cascade.

Even D of fully truncated class II MHC molecules, molecules with no cytoplasmic domains, is about 10-fold smaller than D for similar molecules reconstituted into synthetic lipid bilayers. Like the results with lipid analogues in blebs, this implies that there must be other constraints on lateral mobility for this type of protein, especially for proteins whose bulk lies outside the bilayer.

The role of the extracellular matrix in constraining D has not been thoroughly investigated. Some early experiments suggested that membrane proteins were immobile where they were in contact with fibronectin. Zhang and co-workers (1991) pointed out that the extracellular matrix is often coarsely patched on cells and so only proteins beneath the patches would be affected by the extracellular matrix. On the other hand, recent work from the same laboratory (Lee et al. 1993), using a particle-tracking method (described in detail below) implies that proteoglycans and other molecules of the cell glycocalyx can interact with the extracellular portions of membrane proteins to impede their diffusion.

Cell membranes are crowded; membrane molecules effectively move in two dimensions, not three, and this limits both their lateral and their rotational diffusion (Grassberger et al. 1986). These limits crowd molecules, so that they have to take longer paths through a given area than they would if it were not occupied by other proteins. The crowding and the restricted rotation also enhance weak interactions between molecules in the plane of the membrane and there is evidence that such interactions are important in constraining lateral diffusion (Sheetz 1993). We have shown (Wier and Edidin 1988) that D of class I MHC molecules increases as sites for N-linked glycosylation are removed by directed mutation. When all three sites are removed, D approaches the limit imposed by the viscosity of bilayer lipids. We suggested that the effect was due to excluded volume effects, rather than to specific interactions between oligosaccharides and other surface molecules. Branched oligosaccharides are large, approximately 20 Å end to end (Wu et al. 1991); hence in solution they sweep through volumes of $\sim 10^4$ Å^3.

Effects of glycosylation on D are also seen if cells are treated with tunicamycin (M. Wier and M. Edidin, unpublished) and D measured by FPR. However, other experiments showed no effect of under-glycosylation on D of viral glycoproteins (Scullion et al. 1987).

The experiments with mutant glycoproteins reflect diffusion of these proteins in an otherwise normally glycosylated cell surface. Tunicamycin treatment should change glycosylation of most membrane proteins. We tried to achieve in another way by comparing D of a single species of membrane protein, again a class I MHC molecule, expressed in lectin-resistant CHO cell mutants, cells whose glycosylation is altered, potentially resulting in shorter or smaller than normal N-linked oligosaccharides (Barbour and Edidin 1992). To our surprise we found that D was little affected by global changes in surface glycosylation.

Our first observations on the strong effect of glycosylation on D were made in L-cells, not in CHO cells. We believe that the difference in the results between the two cell types reflects differences in the extent of their cytoskeleton and the extent to which membrane glycoproteins are coupled to the cytoskeleton. In one cell type, L-cells, constraints on D are dominated by interactions of molecules in the plane of the membrane. In the other cell type, CHO cells, constraints are dominated by interactions of membrane proteins with the cytoskeleton. This implies that one cannot speak of a single locus for constraints on D. Rather, D as measured by FPR reflects a mixture of the effects on D of the cytoskeleton and proteins of the cytoplasm, by bilayer viscosity, by other membrane proteins, and by the extracellular matrix or glycocalyx. The extent to which any one of these is important depends upon the biology of the cells examined.

FPR and spatial heterogeneity of membranes

Thus far we have dealt with questions about D, the area explored by a molecule per unit time. These questions are posed in terms of the entire membrane area; they neglect issues of local inhomogeneities of membrane composition and physical properties. Such issues are certainly important; there are many examples of differentiated patches, domains, in a single membrane. FPR and other probes of cell surfaces show clearly that smaller domains are also found in membranes (for reviews see Edidin (1990, 1992). Proteins seem to be confined to these domains by barriers, rather than by tethering, a situation similar to that described earlier for erythrocytes. A summary of our FPR data will serve to show how a model of membrane domains is developed. It will also show further limits to FPR and serve to introduce newer methods to characterize lateral mobility in membranes, methods of higher resolution than FPR.

The geometry of an FPR experiment (Fig. 7.1) clearly predicts that for a given value of D, the time for recovery of fluorescence increases as the area of the laser spot is increased; if a is area, then $t_{\frac{1}{2}} \propto a/D$. D calculated from $t_{\frac{1}{2}}$ should be independent of the area of the laser spot. R, the mobile fraction of molecules, also is not expected to change with a, except when a is a significant fraction of the total membrane area. In a typical experiment $a \approx 1-2$ μm^2, while the membrane area is hundreds of μm^2.

In fact, when we measured D and R of surface glycoproteins on human fibroblasts as a function of the laser spot area, we found that both D and R changed with spot area (Yechiel and Edidin 1989). Over a 15-fold range of spot areas from ~0.3 μm^2 to 15 μm^2, D increased with increasing spot size, while R decreased. Careful measurements of the spot area, and control experiments with some fluorescent lipid analogues and with fluorescent solutions, made it clear that the effects were not optical or technical artefacts. We extended the approach to a number of other cultured cell types and found the same effects. They were most readily interpreted as reflecting the partition of what appeared to be a continuous membrane bilayer into domains whose boundaries were impermeable to diffusing glycoproteins. Larger laser spots would bleach one or more complete domains and these could not contribute to the recovery of fluorescence. The scale of the domains could be estimated from the areas of the laser spots used. It appeared that they are 1–2 μm in diameter. We believed that the increase in D with increasing spot area reflected increased sampling of proteins in relatively protein-poor domains, in which random walks were little impeded by molecular crowding.

The estimated size of the domains suggested that they might be formed by a membrane- or cyto-skeleton. Though each mesh of a spectrin matrix has sides measuring ~0.1 μm, an imperfect matrix would allow percolation of proteins over longer distances and could explain the estimates of 1–2 μm for the linear dimensions of a domain. We tested this idea by comparing the diffusion behaviour of a conventional class I MHC molecule, H-2L^d, with the behaviour of a lipid-anchored form of the molecule, Qa2 (Edidin and Stroynowski 1991). The two molecules are more than 80 per cent homologous and differ mainly in their anchorage to the bilayer. The lipid-anchored form of MHC class I molecules was not confined to membrane domains: R did not change as the spot area changed. The conventional form of the class I molecule, anchored to the bilayer by a transmembrane helix, did appear to be organized into domains.

D as a function of laser spot area increased for both types of class I molecule. This meant that our initial interpretation of changes in D was wrong. If Qa2 molecules were not collected in membrane domains, then our experiments could not be detecting the differences in D for molecules in and out of domains. We asked if the difference could be due to the time resolution of the FPR experiment at different spot areas. Recall that $t_{\frac{1}{2}} \propto a/D$. The time resolution of our FPR instrument is ~50 ms and we need about five samples to measure the slope of the recovery curve. Hence, for a small spot, say 0.3 μm^2, we cannot resolve D values above ~1×10^8 cm^2/s and we will underestimate D in the range of 1–10^{-9} cm^2/s. On the other hand, using a spot of 15 μm^2 we will easily detect large D, but molecules with D of ~10^{-11} cm^2/s will not contribute to a recovery curve taken for 30 s, the usual time for one of our experiments. We tested this idea by measuring D for solutions containing mixtures of large (fluorescein–IgM) and small (free fluorescein) molecules. The results are shown in Table 7.1. The value of D given in the literature for free fluorescein in water is 5×10^{-6} cm^2/s. That for IgM in water is about 2×10^{-7} cm^2/s. It can be seen that even the largest spot used gives a D of fluorescein or IgM of about one-tenth that measured by other methods. However, the important point to note is that the apparent D changes as the proportions of fluorescein and fluorescein–IgM

Table 7.1 Diffusion of mixtures of fluorescein and fluorescein–IgM in water

	Objective/measured spot radius					
	10 ×/2.2 μm		3 ×/10 μm		1 ×/30 μm	
	D	%R	D	%R	D	%R
Fluorescein	6×10^{-8}	96	2×10^{-7}	100	6×10^{-7}	100
IgM	5×10^{-9}	90	5×10^{-9}	100	1×10^{-8}	50–70
IgM:fluorescein 1:1	1×10^{-8}	90	$3–4 \times 10^{-8}$	100	2×10^{-7}	70
IgM:fluorescein 3:1	9×10^{-9}	90	8×10^{-9}	100	1×10^{-7}	50–80
IgM:fluorescein 1:3	2×10^{-8}	95	5×10^{-8}	100	3×10^{-7}	80

change and as the spot size changes. This implies that FPR is detecting heterogeneity of D within a population of membrane glycoproteins, but that it cannot resolve D for a population into its components.

Beyond FPR—the behaviour of small numbers of molecules in a fluid bilayer

The FPR data summarized in the last section suggest a great complexity and heterogeneity of membrane organization and of molecular behaviour in membranes. Since FPR samples populations, it cannot further resolve this complexity. Advances in understanding membrane organization at the level of small numbers of molecules in small areas have come only from newer techniques which use video microscopy and computer analysis of images. The techniques, single-particle

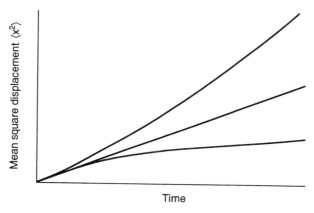

Fig. 7.2. Idealized plots of the area explored over time, $\langle x^2 \rangle$, by a marker particle bound to membrane proteins or lipids. If the labelled molecules are moving in a random walk (i.e. diffusing), then $\langle x^2 \rangle$ versus time is linear (centre). If the particle is being moved by membrane flow or other directed forces, then $\langle x^2 \rangle$ versus time is concave up (top). If the diffusing particle meets barriers as it moves, then $\langle x^2 \rangle$ versus time is concave down (bottom).

tracking and the laser optical trap, have confirmed and extended the picture of membranes painted from the FPR data.

Gold beads, or brightly fluorescent low density lipoprotein (LDL) particles, can be used to track Brownian motion of membrane proteins and lipids (Barak and Webb 1981, 1982; Gelles *et al.* 1988). A single membrane protein, the LDL receptor, is visualized by fluorescent LDL (Barak and Webb 1981, 9182; Ghosh and Webb 1990; Anderson *et al.* 1991). Fluorescent viruses (Anderson *et al.* 1991) and antibody- or lectin-coated gold particles have a lower resolution than this but each particle probably detects 5–30 molecules of membrane protein (Anderson *et al.* 1991), a small number relative to the hundreds of molecules whose mobility is measured in an FPR experiment; hence, even these markers offer greater spatial resolution as well as greater temporal resolution of mobility than FPR.

Whatever label is used, whether detected by fluorescence microscopy or by interference contrast microscopy, its track can be described overall in terms of the area, $<x^2>$, that it covers per unit time, t. The overall shape of a plot of $<x^2>$ versus t can report unrestricted diffusion, or flow, or free diffusion in a bounded region, a domain (Sheetz *et al.* 1989) (Fig. 7.2) Details of the records of particle tracks are also of interest and show that a given particle may have several different modes of lateral transport. Fig. 7.3(A) shows the area swept out by a 30 nm gold bead coated with antibodies to a GPI-anchored class I molecule, Qa2. The particle moved about 1 μm along in each axis in the 20 s of observation, but did so wandering about a given point for a time before moving away. This is the picture of a random walk (see Fig. 1.4 in Berg (1983) for a simulated random walk). However, if the time history of the particle's movement along each axis is read (Fig. 3(B) and (C)), it is clear that for about one of the 20 seconds, in the interval from 6 to 8 s, the particle was motionless. Thus, even the tracking method gives evidence that a small cluster of molecules moving as a group may alternate periods of free diffusion with periods of anchoring. Published data on the behaviour of gold-labelled molecules also show this effect, as well as changes from random walk to directed flow (Sheetz *et al.* 1989).

Results with fluorescent LDL showed the same phenomena and also resolved percolation and other particle movements consistent with the presence of membrane domains on the micrometre scale, suggested by FPR experiments (Ghosh and Webb 1990). This percolation behaviour is even seen for mutant LDL receptors lacking cytoplasmic tails. If the cytoskeleton plays a role in restricting the lateral diffusion of these particles it must be mediated through other transmembrane proteins.

It is harder to resolve domains when tracking gold beads (Lee *et al.* 1991) but some evidence for their existence has been obtained by this method (de Brabander *et al.* 1991).

Fig. 7.3. Random walk of a GPI-anchored molecule, Qa2, labelled with a 30 nm gold bead. The top panel shows the overall two-dimensional walk in a period of 20 s. The pattern is typical of diffusion. The bottom panels show the particle's displacements along the y and x axes as a function of time. These panels show that the particle's walk includes some periods when it is anchored and others when it is free to diffuse.

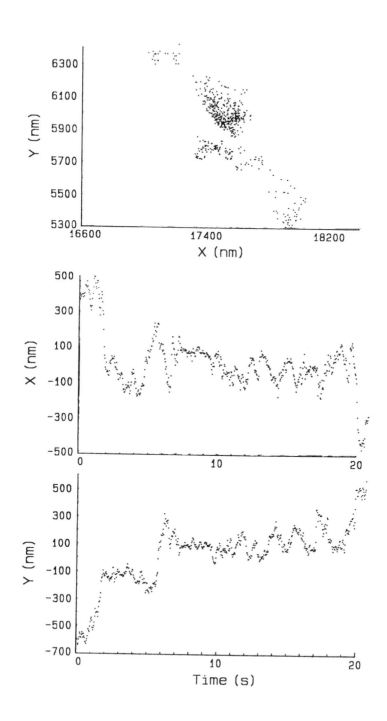

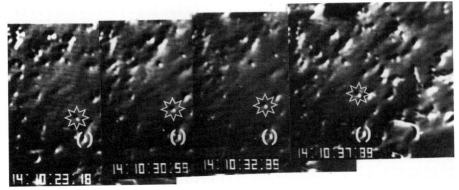

Fig. 7.4. Determining the barrier-free path of MHC molecules labelled with an antibody-coated gold bead. The bead of interest, marked by a star, was trapped in a laser optical trap. Then the stage was translated in one direction. The particle is displaced relative to an immobile particle, marked by ( ).

Antibody-coated gold beads have been used to probe for membrane domains, not by following their random walks, but by using a laser optical trap (Ashkin *et al.* 1987; Kuo and Sheetz 1992) to move beads along the cell surface. The distance dragged before a bead was pulled from the trap, the barrier-free path, again proved to be about 1 μm at room temperature (Fig. 7.4). The BFP of lipid-anchored molecules was three times that of homologous molecules anchored by a transmembrane peptide and cytoplasmic tail (Edidin *et al.* 1991). The BFP of each type of molecule increased 3-fold when the temperature was raised to 34 °C.

The experiment with laser tweezers gives a different view of membrane domains from that given either by FPR or by particle tracking. This view suggests that the domains are dynamic and reflect the statistics of barriers, possibly formed by the membrane skeleton. Thus they are consistent with the view that membranes are patchy and heterogeneous. Indeed they revive the model made to explain the effect of the erythrocyte membrane skeleton on lateral diffusion (Sheetz *et al.* 1980).

Closing remarks

We began by describing what was then a new technique, virus-induced cell fusion, for creating heterogeneities in otherwise uniform-appearing membranes. This technique allowed resolution of large-scale molecular mobility in plasma and changed our view of how membranes could be constructed and how they function. Since then other techniques have been devised, to quantify and differentiate forms of molecular lateral mobility. They have resulted in a refined and consistent picture of membranes, but they also continue to yield surprising results.

References

Anderson, C. A., Georgiou, G. N., Morrison, I. E. G., Stevenson, G. V. W., and Cherry, R. J. (1991). *J. Cell Sci.* **101**, 415–25.
Ashkin, A., Dziedzic, J. M., and Yamane, T. (1987). *Nature*, **330**, 769–71.
Axelrod, D., Koppel, D. E., Schlessinger, J., Elson, E. L., and Webb, W. W. (1976). *Biophys. J.*, **16**, 1055–69.
Barak, L. S. and Webb, W. W. (1981). *J. Cell Biol.*, **90**, 595–604.
Barak, L. S. and Webb, W. W. (1982). *J. Cell Biol.*, **95**, 846–52.
Barbour, S. and Edidin, M. (1992). *J. Cell. Physiol.*, **150**, 526–33.
Berg, H. (1983). *Random walks in biology*. Princeton University Press, Princeton, NJ.
Bierer, B., Herrmann, S. H., Brown, C. S., Burakoff, S. J., and Golan, D. (1987). *J. Cell Biol.*, **105**, 1147–52.
Cherry, R. J., Burkli, A., Busslinger, G., Schneider, G., and Parrish, G. R. (1976). *Nature*, **263**, 389–93.
de Brabander, M., Nuydens, R., Ishihara, A., Holifield, B., Jacobson, K., and Geerts, H. (1991). *J. Cell Biol.*, **112**, 111–24.
Edidin, M. (1974). *Annu. Rev. Biophys. Bioeng.*, **3**, 179–201.
Edidin, M. (1990). *Curr. Top. Membr. Transport*, **36**, 81–96.
Edidin, M. (1991). In *The Structure of biological membranes*, (ed. P. Yeagle), pp. 539–72. CRC Press, Boca Raton, FL.
Edidin, M. (1992). *Trends Cell Biol.*, **2**, (In press.)
Edidin, M. and Stroynowski, I. (1991). *J. Cell Biol.*, **112**, 1143–50.
Edidin, M. and Wei, T. (1977). *J. Cell Biol.*, **75**, 475–82.
Edidin, M. and Zagyansky, Y. (1976). *Biochim. Biophys. Acta*, **433**, 209–14.
Edidin, M. and Zuniga, M. (1984). *J. Cell Biol.*, **99**, 2333–5.
Edidin, M., Zagyansky, Y., and Lardner, T. J. (1976). *Science*, **191**, 466–8.
Edidin, M., Kuo, S. C., and Sheetz, M. P. (1991). *Science*, **254**, 1379–82.
Eisenger, J., Flores, J., and Peterson, W. P. (1986). *Biophys. J.*, **49**, 987–1001.
Fowler, V. and Branton, D. (1977). *Nature*, **268**, 23–6.
Frye, L. D. and Edidin, M. (1970). *J. Cell Sci.*, **7**, 319–35.
Gelles, J., Schnapp, B., and Sheetz, M. P. (1988). *Nature*, **331**, 450–53.
Ghosh, R. N. and Webb, W. W. (1990). *Biophys. J.*, **57**, 286a.
Golan, D. E. and Veatch, W. (1980). *Proc. Natl Acad. Sci. USA*, **77**, 2537–41.
Gordon, S. and Cohn, Z. (1970). *J. Exp. Med.*, **131**, 981–1003.
Grassberger, B., Minton, A. P., De Lisi, C., and Metzger, H. (1986). *Proc. Natl Acad. Sci. USA*, **83**, 6258–0000.
Harris, H. and Watkins, J. F. (1965). *Nature*, **205**, 640–6.
Harris, H., Sidebottom, E., Grace, D. M., and Bramwell, M. E. (1969). *J. Cell Sci.*, **4**, 499–525.
Jacobson, K., Wu, E.-S., and Poste, G. (1976). *Biochim. Biophys. Acta*, **433**, 215–22.
Jesaitis, A. J. and Yguerabide, J. (1986). *J. Cell Biol.*, **102**, 1256–63.
Koppel, D. and Sheetz, M. P. (1981). *Nature*, **293**, 159–61.
Koppel, D., Sheetz, M. P., and Schindler, M. (1981). *Proc. Natl Acad. Sci. USA*, **78**, 3576–80.
Kuo, S. C. and Sheetz, M. P. (1992). *Trends Cell Biol.*, **2**, 116–7.
Lee, G., Ishihara, A., and Jacobson, K. (1991). *Proc. Natl Acad. Sci. USA*, **88**, 6274–8.
Lee, G., Zhang, F., Ishihara, A., McNeil, C. L., and Jacobson, K. (1993). *J. Cell Biol.* (In press.)
Liebman, P. A. and Entine, G. (1974). *Science*, **185**, 457–9.
Livneh, E., Benveniste, M., Prywes, R., Felder, S., Kam, Z., and Schlessinger, J. (1986). *J. Cell Biol.*, **103**, 327–31.

Madreperla, S., Edidin, M., and Adler, R. (1989). *J. Cell Biol.*, 109, 1483–93.
Mao, S.-Y., Varin-Blank, N., Edidin, M., and Metzger, H. (1991). *J. Immunol.*, **146**, 958–66.
Mastrangelo, I. A. and Mitra, J. (1981). *J. Hered.*, **72**, 81–6.
Nelson, W. J., Shore, E. M., and Wang, A. Z. (1990). *J. Cell Biol.*, **110**, 349–57.
Peters, R. (1988). *FEBS Lett.*, **234**, 1–19.
Pisam, M. and Ripoche, P. (1976). *J. Cell Biol.*, **71**, 907–20.
Poo, M.-M. and Cone, R. A. (1974). *Nature*, **247**, 438–41.
Salas, P. J. I., Vega-Salas, D. E., Hochman, J., Rodriguez-Boulan, E., and Edidin, M. (1988). *J. Cell Biol.*, **107**, 2363–76.
Scullion, B. F., Hou, Y., Puddingeton, L., Rose, J. K., and Jacobson, K. (1987). *J. Cell Biol.*, **105**, 69–75.
Sheetz, M. P. (1993). *Annu. Rev. Biophys. Biomol. Struct.*, **22**, 417–31.
Sheetz, M. P., Schindler, M., and Koppel, D. E. (1980). *Nature*, **285**, 510–12.
Sheetz, M. P., Turney, S., Qian, H., and Elson, E. L. (1989). *Nature*, **340**, 284–8.
Shen, B. W. (1985). In *Red blood cell membranes* (ed. P. Agre and J. C. Parker), pp. 261–97. Marcel Dekker, New York.
Singer, S. J. and Nicolson, G. (1972). *Science*, **175**, 720–31.
Tank, D. W., Wu, E.-S., and Webb, W. W. (1982). *J. Cell Biol.*, **92**, 207–12.
Wade, W. F., Freed, J. H., and Edidin, M. (1989). *J. Cell Biol.*, **109**, 3325–31.
Watkins, J. F. and Grace, D. M. (1967). *J. Cell Sci.*, **2**, 193–204.
Wey, C.-L. Cone, R. A., and Edidin, M. (1982). *Biophys. J.*, **33**, 225–32.
Wier, M. and Edidin, M. (1988). *Science*, **242**, 412–14.
Wu, E.-S., Tank, D. W., and Webb, W. W. (1982). *Proc. Natl Acad. Sci. USA*, **79**, 4962–6.
Wu, P., Rice, K. G., Brand, L., and Lee, Y. C. (1991). *Proc. Natl Acad. Sci. USA*, **88**, 9355–9.
Yechiel, M. and Edidin, M. (1989). *J. Cell Biol.*, **105**, 755–60.
Zhang, F., Crise, B., Hou, Y., Rose, J., Bothwell, A., and Jacobson, K. (1991). *J. Cell Biol.*, **115**, 75–84.
Ziomek, C. A., Schulman, S., and Edidin, M. (1980). *J. Cell Biol.*, **86**, 849–57.

8. Intracellular transport of secretory proteins within intact transient heterokaryons and homokaryons

Conny Valtersson, Masahiro Mizuno, Anne H. Dutton, and S. J. Singer

Introduction

The overall intracellular pathway taken by secretory proteins inside eukaryotic cells, from their synthesis to their exteriorization, was laid out some time ago (Palade 1975). Initially, proteins to be secreted are synthesized on ribosomes that become attached to the cytoplasmic surfaces of the membranes of the endoplasmic reticulum (ER), and as the proteins are translated they are translocated across the membrane into the lumen of the ER. Following this, the secretory protein (suitably modified) is transferred from the ER to the *cis*-faces of the Golgi apparatus (GA). After passage through successive Golgi saccules and the *trans*-Golgi network, the secretory protein is deposited in granules which ultimately fuse with the plasma membrane to release the protein to the exterior of the cell.

The part of this complex pathway that we address in this paper is the transfer from the ER to the GA. This transfer is thought to be mediated by transition vesicles that bud off transitional elements of the ER, and then fuse with the *cis*-face of a stack of GA saccules (see Palade 1975; Merisko et al. 1986; Lodish et al. 1987). The molecular components involved in the ER to GA transfer are under intense investigation by biochemical and genetic methods (for reviews, see Balch (1989) and Pryer et al. (1992)). In addition to these molecular features, however, there are ultrastructural and mechanochemical aspects of the transfer process that need to be addressed. In interphase eukaryotic cells, the ER is generally a highly convoluted, membrane-bounded organelle that is widely dispersed throughout the cytoplasm, whereas the GA is most often a compact organelle that is confined near and to one side of the cell nucleus. These grossly different intracellular distributions of the ER and GA raise certain questions about the ER to GA transfer. Do transition vesicles bud off the ER at random sites anywhere inside a cell, or are they formed only at confined and perhaps specialized sites in the ER? After formation, do the transition vesicles find their way to the *cis*-faces of the stacks of GA by a random diffusional process, as is thought to characterize vesicular transfer between

Golgi saccules (Rothman *et al.* 1984), or is the transfer a directed process, mediated, for example, by tracking along one or more cytoskeletal elements?

In order to address some of these questions, we have carried out a series of experiments involving cell fusion hybrids, the first results of which have been published (Valtersson *et al.* 1990). These studies are therefore one of a wide array of different kinds of investigations, many of which have been reported in this symposium, that have taken advantage of the unique possibilities provided by the cell fusion methodology pioneered by Dr Henry Harris. Our approach was as follows. Heterokaryons were formed in culture between two different types of human cells secreting different proteins. Several hours before the cells were fused, they were treated with cycloheximide (or puromycin) to arrest protein synthesis and empty the cells of their previously synthesized secretory proteins (Keller *et al.* 1986). Shortly after cell fusion was complete, the cycloheximide was washed out and protein synthesis was reinitiated in the intact heterokaryons. There is a period after cell fusion (~90 min) during which the two GAs contributed by the two cell moieties of a heterokaryon remain spatially separate and physically distinct. After that time the two GAs overlap and presumably merge. Our experiments were carried out during this time window when the two GAs of a heterokaryon remained distinguishable. The *in situ* intracellular transfers of the newly synthesized secretory proteins were then followed by immunofluorescence microscopy at different times after reinitiation of protein synthesis. If ER to Golgi transfer occurs by vesiculation at random sites throughout the cell followed by free diffusion of the transition vesicles to the GA, it seemed likely that a secretory protein might appear simultaneously in both GAs of a transient heterokaryon. If, on the other hand, the vesicular transfer is not random, but is either spatially confined or mechanically directed, or both, we considered it possible that a secretory protein might appear first in the homologous GA, that is, the one of the two GAs of the heterokaryon that was contributed by the parental cell for which that secretory protein was specific; only later might that protein appear in the other, heterologous, GA of the heterokaryon.

Our first reported experiments (Valtersson *et al.* 1990) involved the binary fusion of cells of the HepG2 human hepatoma line, secreting serum albumin, with cells of the WI38 human fibroblastic line, secreting procollagen I. We found that in such heterokaryons, newly synthesized serum albumin always appeared first in the GA contributed by the HepG2 cell, while newly synthesized procollagen I always appeared first in the GA contributed by the WI38 cell, suggesting that ER to Golgi transfer was not a set of stochastic events, but was instead restricted, either spatially, or mechanically, or both.

We have extended these initial studies in several directions, in order to determine their generality, and also to probe further into the mechanism of the transfer process. The results with two more heterokaryon systems are presented herein, using parental cells and secretory proteins additional to the ones initially studied. Furthermore, we also examined a homokaryon cell fusion system, in which a cell expressing a transfected secretory protein ordinarily foreign to it was fused with a cell of the same type but which had not been transfected. Finally, in a preliminary set of experiments, we have explored whether microtubules might be involved in a directed vesicular transfer between the ER and the GA.

Materials and methods

The cell culture techniques and experimental methods, except for additional features mentioned in this section or in the text, were carried out as described in the previous study (Valtersson et al. 1990). One feature that was inadvertently omitted in that study that is also relevant to the present investigation is that the medium in which the WI38 human fibroblasts was cultured contained 0.25 mM ascorbic acid to promote procollagen I synthesis and secretion. The HeLa cells that were stably transfected to produce retinal binding protein (RBP) (tHeLa) were generously given by Dr Per A. Peterson, as was the rabbit antiserum to RBP. The tHeLa cells were grown in Dulbecco's modified Eagle's medium/10 per cent fetal calf serum containing 160 μg/ml xanthine, 16 μg/ml adenine, 19 μg/ml mycophenolic acid, 11 μg/ml hypoxanthine, 0.146 μg/ml aminopterin, 3.15 μg/ml thymidine, 240 μg/ml glutamine, and 20 μg/ml streptomycin. The polyclonal rabbit antibodies specific for Glu-tubulin (αT12; Kreis 1987) were obtained through the kindness of Dr Thomas E. Kreis. The fluorescent beads used to mark the untransfected HeLa cells were 0.74 μm diameter Fluoresbrite Carboxylate Microspheres (Catalogue no. 7766, Polysciences, Warrington, PA). The HeLa cells were incubated with the beads for 8–12 h to permit their uptake, and the free non-phagocytosed beads were then carefully washed away. After trypsinization, the loaded cells were then co-plated with the tHeLa cells, prior to further treatments.

Results

Heterokaryon systems

HepG2–HeLa cell fusions: steady state In these experiments, serum albumin-secreting HepG2 cells were fused with HeLa cells instead of the WI38 cells used in our initial studies (Valtersson et al. 1990). WI38 cells are fibroblastic, with well-organized stress fibres in the cytoplasm which might interfere with any diffusion of vesicles in a heterokaryon. By contrast, HeLa cells have a convoluted and less rigid cytoskeleton. As controls for the experiments described in the next section, HepG2 and HeLa cells in the steady state were fused, and the heterokaryons were doubly labelled for fluorescence microscopy with antibodies to serum albumin and with wheat germ agglutinin (WGA) (Fig. 8.1). WGA serves as a marker for the GA. Observations of the cells with time after adding the fusogen polyethylene glycol (PEG) showed that by 30 min, 100 per cent of the fusion hybrids had formed (Fig. 8.2). At this time, serum albumin labelling was found concentrated in the GA of the HepG2 half of each heterokaryon (Fig. 8.1(a), filled arrow) and absent from the GA of the HeLa half (Fig. 8.1(a), open arrow), as expected. (WGA labelling of the GA in the HeLa cell moieties (Fig. 8.1(b), (e), and (h); open arrows) was

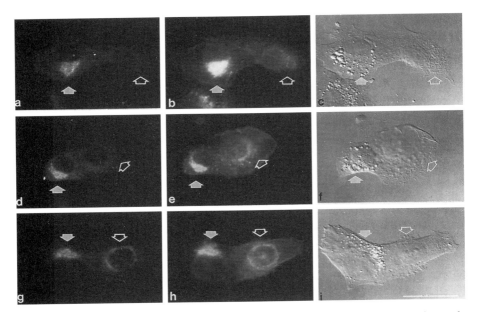

Fig. 8.1. Heterokaryons made by the fusion of HepG2 and HeLa cells which were in steady states, examined at 30 min ((a), (b), and (c)), 40 min ((d), (e), and (f)), and 50 min ((g), (h), and (i)) after fusion was initiated. The heterokaryons were immunolabelled for serum albumin ((a), (d), and (g)), and double labelled with WGA ((b), (e), and (h)) to localize the two Golgi apparatuses in the heterokaryon (filled arrows indicate the HepG2 Golgi, open arrows the HeLa Golgi). The respective Nomarski images of the heterokaryons are also shown ((c), (f), and (i)). Serum albumin labelling in the HeLa Golgi does not become prominent until 50 min after fusion (g). Bar in (i) represents 50 μm.

always less intense than for the GA in the HepG2 moieties (Fig. 8.1(b), (e), and (h); filled arrows). It was not until 50 min after PEG addition that a substantial number of heterokaryons showed serum albumin labelling of both GAs (Fig. 8.1(g)). In another set of experiments, the HepG2 cells were treated with cycloheximide for 10 min prior to fusion with the HeLa cells in order to arrest serum albumin synthesis without emptying the GA of that protein. Fusion was then carried out with PEG containing cycloheximide, and cycloheximide was present in the rest of the experiment. The fluorescence microscopic results of these experiments (not shown) were indistinguishable from the first set without the cycloheximide treatment (Fig. 8.1). A large number of such observations are collected in Fig. 8.2. They indicate that there was a time lapse of about 26 min after fusion before serum albumin appeared in the GA of the HeLa moiety of the heterokaryons, whether or not cycloheximide was present. As detailed in the Discussion section, we infer from these results that the serum albumin that eventually appeared in the GA of the HeLa

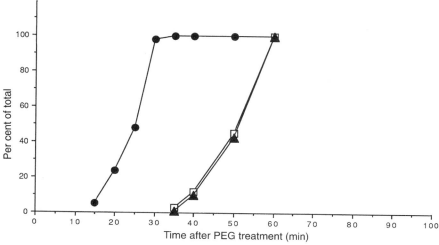

Fig. 8.2. Time-course of HepG2–HeLa cell fusions (●) after treatment with polyethylene glycol (PEG), and of the appearance of serum albumin in the Golgi apparatus of the HeLa cell moiety of a heterokaryon, either without cycloheximide treatment (□), or with cycloheximide treatment begun 10 min before cell fusion (▲).

moiety was largely derived by direct transfer from the GA of the HepG2 moiety of the same heterokaryon.

HepG2–HeLa cell fusions: protein synthesis reinitiated In these experiments, the HepG2 cells (along with the co-plated HeLa cells) were treated for 2.5 h with cycloheximide to empty them of their serum albumin prior to cell fusion. After fusion in the presence of PEG and cycloheximide was complete (~30 min, Fig. 8.2), the heterokaryons were rapidly washed free of the reagents at 37 °C, and protein synthesis was reinitiated. This is the zero time for this series of experiments. Serum albumin immunofluorescence within individual heterokaryons was followed with time, and double labelling with WGA was used to locate the two GAs in each heterokaryon. At zero time (Fig. 8.3(a)), immunolabelling for serum albumin was largely absent from the heterokaryons, depleted by the prior prolonged cycloheximide treatment. By 15 min, the GA of the HepG2 half of some of the heterokaryons showed significant serum albumin labelling (Fig. 8.3(g), filled arrow), but the GA of the HeLa half of the heterokaryon (Fig. 8.3(h), open arrow) was still devoid of serum albumin (Fig. 8.3(g), open arrow). By 30 min, both GAs of most of the heterokaryons were labelled for serum albumin (Fig. 8.3(j) and (k)). The results from 350 such individual observations in five independent HepG2–HeLa fusion experiments are plotted in Fig. 8.4. The figure shows that serum albumin appears in the homologous GA about 10 min before it does in the

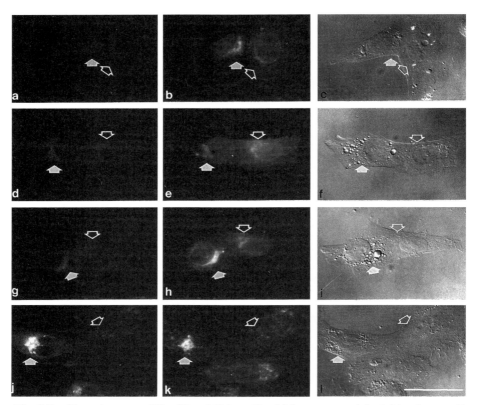

Fig. 8.3. Heterokaryons of HepG2 and HeLa cells which were treated for 2.5 h prior to cell fusion with cycloheximide to empty the cells of their secretory proteins; following fusion, the cycloheximide was washed out and protein synthesis was reinitiated at zero time ((a), (b), and (c)). At 5 min ((d), (e), and (f)), 15 min ((g), (h), and (i)), and 30 min ((j), (k) and (l)) later, immunolabelling for serum albumin ((a), (d), (g), and (j)) and double labelling for WGA ((b), (e), (h), and (k)) was carried out. The respective Nomarski images are shown ((c), (f), (i), and (l)). Filled arrows indicate the HepG2 Golgi apparatus, open arrows the HeLa Golgi apparatus. Serum albumin appears in the HepG2 Golgi (g), but not in the HeLa Golgi, by 15 min after protein synthesis is resumed. By 30 min, however, serum albumin labelling is prominent in both Golgis (j). Bar in (l) represents 50 μm.

heterologous GA in these heterokaryons. These data are virtually indistinguishable from those for serum albumin appearance in the previously studied HepG2–WI38 heterokaryons (Valtersson *et al.* 1990).

tHeLa–WI38 cell fusions: protein synthesis reinitiated Another heterokaryon system that we studied utilized WI38 cells fused with HeLa cells that had been stably transfected with the cDNA for the retinol-binding protein (RBP). These

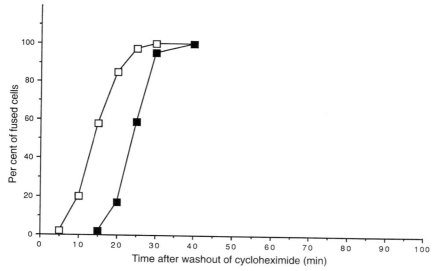

Fig. 8.4. Time-course of appearance of serum albumin in the two Golgi apparatuses in HepG2–HeLa cell heterokaryons after reinitiation of protein synthesis; summary of experiments such as those shown in Fig. 8.3. The appearance of serum albumin in the HeLa Golgi (■) is later than that in the HepG2 Golgi (□).

transfected HeLa cells (tHeLa) secreted RBP, while the WI38 cells secreted procollagen I. Treatment of these cells for 2.5 h with cycloheximide prior to cell fusion completely emptied them of their respective secretory proteins, as demonstrated with heterokaryons from which the cycloheximide had just been washed out (Fig. 8.5(a)–(c)). By 35 min after protein synthesis had been resumed, immunolabelling for RBP was detected in the GA of the tHeLa moiety of many heterokaryons (Fig. 8.5(d), filled arrow), but not in the GA of the WI38 moiety (Fig. 8.5(d), open arrow). Conversely, at this time procollagen I was present in the GA of the WI38 moiety (Fig. 8.5(e), open arrow); but not in the GA of the tHeLa moiety (Fig. 8.5(e), filled arrow) of most heterokaryons. By 45 min, however, RBP was found in both GAs (Fig. 8.5(g)) as was procollagen I (Fig. 8.5(h)), in many heterokaryons. Observations of about 130 such heterokaryons from six independent experiments are summarized in Fig. 8.6. This figure shows that RBP appeared first in the GA of the tHeLa moiety, and only about 10 min later in the GA of the WI38 moiety, whereas procollagen I (which was slower to appear than RBP) first occupied the GA of the WI38 moiety and only about 15 min later the GA of the tHeLa moiety. In these heterokaryons, therefore, a newly synthesized secretory protein first appears in the homologous GA, and only later in the heterologous GA, in both directions in the same heterokaryon.

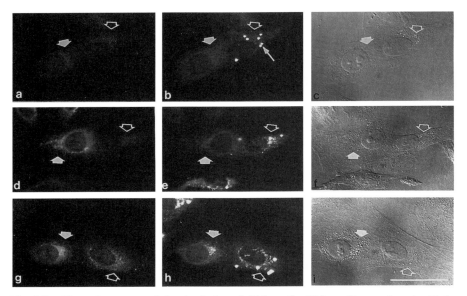

Fig. 8.5. Heterokaryons made by the fusion of tHeLa cells (HeLa cells transfected with the cDNA for RBP) and WI38 cells which were treated for 2.5 h prior to fusion with cycloheximide to empty the cells of their secretory proteins; following fusion, the cycloheximide was washed out and protein synthesis was reinitiated at zero time ((a), (b), and (c)). Double immunolabelling for RBP ((a), (d), and (g)) and for procollagen I ((b), (e), and (h)) was carried out 35 min ((d), (e), and (f)) and 45 min ((g), (h), and (i)) later. The respective Nomarski images are shown ((c), (f), and (i)). The filled arrows indicate the Golgi apparatus of the tHeLa moiety of a heterokaryon, the open arrows that of the WI38 moiety. Fluorescent beads taken up by the WI38 cells prior to cell fusion are indicated by the long arrow in (b). At 35 min after reinitiation of protein synthesis, RBP labelling (d) is apparent in the Golgi of the tHeLa moiety but not in that of the WI38. Procollagen I labelling at this time (e) shows the reverse distribution. By 45 min, RBP labelling (g) and procollagen I labelling (h) are seen in both Golgis. Bar in (i) represent 50 µm.

Homokaryon system

tHeLa–HeLa cell fusions: protein synthesis reinitiated In the studies that we have so far described, heterokaryons were made by the fusion of two disparate human cell types. It seemed worthwhile to ask what would happen if we fused two essentially identical cells to make a homokaryon, but only one of which produced a secretory protein. The availability of the tHeLa cell line permitted us to achieve this by making cell fusions with the parental untransfected HeLa cells. However, whereas the half-cell moieties of the heterokaryons could usually be distinguished morphologically, the homokaryon system presented a problem. This was solved by having the untransfected HeLa cells take up fluorescent beads prior to co-plating and cell fusion. After incorporation, these beads tended to become clustered near

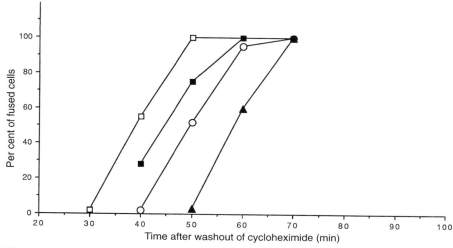

Fig. 8.6. Summary plot of experiments such as those shown in Fig. 8.5, showing the time-course of appearance of RBP in the Golgi apparatus of the tHeLa (□) and of the WI38 (○) moiety of tHeLa–WI38 heterokaryons, as well as of procollagen I in the WI38 (■) and tHeLa (▲) Golgis. For each secretory protein, its appearance in the homologous Golgi precedes that in the heterologous Golgi.

the GA of the HeLa cell, and in homokaryons the beads remained associated with the GA in the half-cell moiety that had originally ingested them. The beads served two purposes. First, they allowed tHeLa–HeLa homokaryons to be instantly recognized and discriminated from any tHeLa–tHeLa and HeLa–HeLa homokaryons in the same field. Second, they distinguished the HeLa half-cell moiety from that of the tHeLa in their homokaryons.

Treatment of the tHeLa cells for 2.5 h with cycloheximide prior to cell fusion emptied them of RBP, as demonstrated with tHeLa–HeLa homokaryons immediately upon removal of the cycloheximide (Fig. 8.7(a)). By 30 min after protein synthesis had been reinitiated, RBP began to appear in the GA, but equally so in both GAs of the homokaryon (Fig. 8.7(d)). At later times, equivalent labelling of both GAs for RBP continued to be observed (Fig. 8.7(h) and (k)). About 200 such tHeLa–HeLa homokaryons were examined in six independent experiments, and the results collected in Fig. 8.8 show that there was no significant difference in the rates of appearance of RBP in the homologous as compared with the heterologous GA, a finding that was clearly different from that obtained for the heterokaryons studied.

Microtubules in HepG2 cells

In our initial report (Valtersson *et al.* 1990), we suggested that a selective transfer of a secretory protein to the homologous GA in a heterokaryon might involve the directed tracking of transition vesicles from the ER to the GA, and that micro-

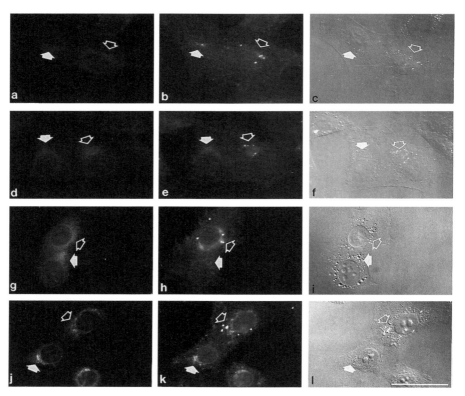

Fig. 8.7. Homokaryons made by the fusion of tHeLa cells (secreting transfected RBP) and untransfected HeLa cells which were treated for 2.5 h prior to fusion with cycloheximide to empty the cells of their secretory proteins; following fusion, the cycloheximide was washed out and protein synthesis was reinitiated at zero time ((a), (b), and (c)). Immunolabelling for RBP ((a), (d), (g), and (j)) and double labelling with WGA ((b), (e), (h), and (k)) was carried out 30 min ((d), (e), and (f)), 40 min ((g), (h), and (i)), and 50 min ((j), (k), and (l)) later. The respective Nomarski images are shown ((c), (f), (i), and (l)). Fluorescent beads (as in Fig. 8.5) were taken up by the untransfected HeLa cells prior to fusion and were generally located near the HeLa cell nucleus in the homokaryons. The filled arrows indicate the Golgi apparatus of the tHeLa moiety, and the open arrows that of the HeLa moiety, of the homokaryons. By 30 min after protein synthesis was reinitiated, RBP labelling (d) was beginning to appear in the Golgi apparatus, but equally so in both Golgis of the homokaryon. This pattern continued at 40 min (g) and 50 min (j). Bar in (l) represents 50 μm.

tubules could mediate such tracking. In particular (see Discussion) the stable microtubules in a cell (rather than the much more abundant dynamic microtubules, which undergo rapid depolymerization and repolymerization) might serve as the tracking elements involved. As part of a more extended study to be reported elsewhere, we have therefore examined by double labelling immunofluorescence microscopy the

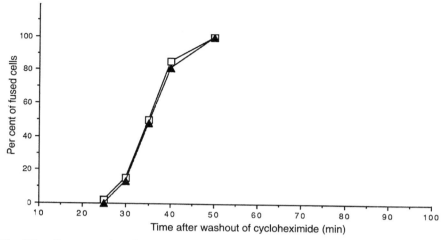

Fig. 8.8. Summary plot of experiments such as those of Fig. 8.7, showing the time-course of appearance of RBP in the Golgi apparatus of the tHeLa moiety (□) and that of the HeLa moiety (▲) of the homokaryons. No significant difference in the rates of appearance in the two Golgis was observed. Contrast these results with those of Fig. 8.6 for heterokaryons.

distribution of total microtubules (Fig. 8.9(a)) compared with serum albumin (Fig. 8.9(b)) in isolated HepG2 cells in their steady state, as well as the distribution of stable microtubules (Fig. 8.9(c), immunolabelled with anti-Glu-tubulin antibodies, see Discussion) compared with the GA (Fig. 8.9(d); as labelled with WGA) in the same cells. It is evident that the stable microtubules (Fig. 8.9(c)) represent only a small fraction of the total microtubules (Fig. 8.9(a)) at any given time, and furthermore, whereas the dynamic microtubule population is spread throughout the cell, the stable microtubules are largely localized to the region around the cell nucleus in proximity to the GA. The stable microtubules therefore have the properties and the location to serve as mediators of ER to GA traffic (see below).

Discussion

In this and our previous paper (Valtersson *et al.* 1990) we have reported ER to GA transfer studies with three different cell fusion heterokaryons formed between two types of human cell, and several different secretory proteins processed by these cells. These heterokaryons involved either HepG2–WI38 fusions (examining the secretory proteins serum albumin or RBP of the HepG2 cell moiety, and procollagen I of the WI38 moiety), or HepG2–HeLa fusions (following the serum albumin of the HepG2 moiety), or tHeLa–WI38 fusions (examining the RBP of the tHeLa moiety and procollagen I of the WI38 moiety). In every instance, a newly synthesized secretory protein first appeared in the homologous GA, that is, the GA of the cell

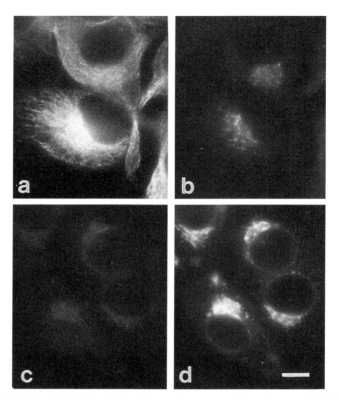

Fig. 8.9. Individual HepG2 cells in the steady state subjected to two double labelling experiments. In the pair (a) and (b)), immunolabelling for total tubulin (a) and for serum albumin (b) is shown. In another pair, (c) and (d), immunolabelling for Glu-tubulin (c) and labelling for WGA (d) are shown. Whereas total tubulin is spread throughout the HepG2 cell, the much less abundant Glu-tubulin (c) is largely confined to the region near the nucleus and the Golgi apparatus, as delineated by WGA (d). Bar in (d) represents 10 μm.

moiety contributing the mRNA for the particular protein. Only significantly (~10–15 min) later did that protein appear in the heterologous GA of the heterokaryon.

In contrast, in the cell fusion homokaryon formed between transfected HeLa cells secreting RBP and untransfected HeLa cells, newly synthesized RBP appeared simultaneously in both GAs of the homokaryon.

We believe that these results rule out a completely stochastic model for ER to GA transfer, that is, a model where a transition vesicle buds off the ER at random anywhere in the cell, and then freely diffuses to and fuses with the *cis*-face of a stack of Golgi saccules. Instead, these results imply that some more spatially restricted or mechanically directed process is involved in the ER to GA transfer.

In order to justify these conclusions, certain trivial explanations of these results must first be eliminated, namely, that the time lag is attributable to a shorter diffusion path for transfer from the ER to the homologous GA in the same half-cell as compared with the path to the heterologous GA in the other half-cell, or that cytoskeletal elements in one or both half-cells obstruct the transfer from the homologous ER to the heterologous GA. The critical point is that very rapidly (within minutes) after cell fusion, the ERs of the two half-cells both appear to permeate the entire heterokaryon. This conclusion is derived from the fact that invariably the immunolabelling of a secretory protein appearing very early after synthesis, and only in the ER of a half-cell moiety of a heterokaryon, is found distributed throughout the heterokaryon (see for example the labelling for serum albumin in Fig. 8.3(d)). On the other hand, although the two ERs in a heterokaryon are spatially intermingled, they apparently do not rapidly fuse together (see below). The rapid physical (but not chemical) mixing of the two ERs in a heterokaryon means that shortly after fusion each ER has essentially the same proximity to each of the two GAs, and therefore the time lag between homologous and heterologous ER to GA transfer cannot be attributed to different diffusional distances for vesicular transfer.

Another factor to consider is that the delayed appearance of a secretory protein in the heterologous GA of a heterokaryon might conceivably arise by any of several routes: (i) from homologous ER first to heterologous ER and only then to heterologous GA: (ii) from homologous ER directly to heterologous GA; or (iii) only from homologous GA to heterologous GA. In particular, only the last route is consistent with the results of the steady state cell fusions presented in Figs. 8.1 and 8.2. In these experiments, the presence of cycloheximide 10 min before cell fusion, and its continued presence thereafter, was intended to curtail new serum albumin synthesis and deposition in the ER, thus substantially decreasing the extent of any possible subsequent transfers of serum albumin to the heterologous GA via routes (i) or (ii) above. However, such brief exposure to cycloheximide (in contrast to the 2.5 h treatment for the experiments recorded in Figs. 8.3 and 8.4) did not significantly deplete the homologous GA of its serum albumin content and therefore permitted a possible transfer via route (iii). Therefore, in view of the fact that such brief cycloheximide treatment had no effect on the rate of appearance of serum albumin in the heterologous GA (Fig. 8.2), we conclude that the serum albumin in the heterologous GA arrived mostly via route (iii), i.e. by direct transfer from the homologous GA. That such inter-Golgi transfer in heterokaryons can occur in the time-frame of these experiments has previously been shown by Rothman *et al.* (1984) in other cell systems.

It must also be realized that in order for the appearance of a newly synthesized secretory protein in the heterologous GA to be delayed relative to that in the homologous GA in our heterokaryon experiments, two other salient points are implied:

1. The mRNA for a secretory protein must somehow remain associated with its homologous ER for some considerable time after cell fusion, because if the mRNA rapidly became associated with the heterologous ER, the secretory protein would then appear simultaneously in both GAs.

2. The two ERs cannot rapidly fuse and chemically mix together, because likewise the secretory protein would then appear simultaneously in both GAs.

This brings us to consider the homokaryon experiments. We originally intended these experiments to serve as a control for the heterokaryon experiments, and were surprised that they yielded a different result, namely that the newly synthesized RBP, associated with the tHeLa moiety of the homokaryon made by fusion with HeLa, appeared simultaneously in both GAs (Figs. 8.7 and 8.8). Although there are many possible explanations for these results, and we have no evidence to distinguish between them experimentally, we favour the possibility that in the homokaryon, in contrast to the heterokaryons, there is a rapid fusion and chemical mixing of the ERs of the two half-cell moieties, obliterating the distinction between the homologous and heterologous ERs shortly after cell fusion takes place. This in turn implies that the ERs of two different cell types are somehow distinctive so that in a heterokaryon the fusion and chemical mixing of the two ERs is retarded or inhibited. This possibility of chemical mixing could be investigated experimentally by immunoelectron microscopy of transient heterokaryons if one had available antigenic markers that distinguished between the ERs of two different cell types.

Returning to the heterokaryon results, the Discussion so far leads us to conclude that the preferential appearance of a newly synthesized secretory protein in the homologous GA compared with the heterologous GA in a heterokaryon is due to some kind of restrictions that favour the transfer from an ER to its homologous GA. The delayed appearance in the heterologous GA may in fact only occur by direct transfer from the homologous GA (see above). As an example of such a restriction on ER to GA transfer, vesicle budding from the ER may occur only at sites where so-called 'transitional elements' of the ER are situated proximally to the *cis*-face of a stack of Golgi saccules. This is the inference to be derived from electron micrographs such as that of Massalki and Leedale (1969, Fig. 19), of the protozoan, *Trebonema vulgare*, which shows a region of the outer nuclear membrane (part of the ER) located directly opposite to the *cis*-face of a Golgi stack, apparently caught in the process of budding off a number of transition vesicles, whereas immediately adjacent portions of the same nuclear membrane that were not apposed to the Golgi stack were quiescent. Such a precisely demarcated spatial relationship between ER vesicle formation and the *cis*-face of a Golgi stack could result from the mediation of some localized cytoskeletal elements in the formation and transfer of the ER-derived transition vesicles to the *cis*-face of the GA.

There are several lines of indirect evidence that microtubules might mediate such vesicle formation from the ER and its directed transfer to the GA. Microtubules and elements of the ER have closely similar distributions in lamellar extensions of cells (Terasaki *et al.* 1986) suggesting an association of the two structures. Furthermore, elements of the ER appear to attach to microtubules and elongate by locomotion along microtubule tracks (Dabora and Sheetz 1988). On the other hand, most microtubules in interphase cells are in rapid dynamic flux, not simply exchanging subunits, but completely depolymerizing and repolymerizing in a matter of minutes (see Schulze and Kirschner 1987). Microtubules that were in such rapid dynamic flux therefore could not alone serve as tracking elements in our heterokaryon experiments, since they would not persist long enough after cell fusion to account for the selective transfer from the ER to its homologous GA. However, in addition to the major fraction of microtubules in interphase cells that is dynamically unstable, there is a minor fraction that is much more stable (Kreis 1987; Webster *et al.* 1987), persisting in cells for times of the order of hours. Such stable microtubules in many cell types can be distinguished from the unstable ones by their content of α-tubulin subunits with the amino acid Glu at their C-termini. The unstable microtubules have α-tubulin subunits ending in Tyr residues (Gundersen *et al.* 1984). Specific antibodies directed against the Glu-terminal α-subunits can therefore be used to distinguish the stable microtubules *in situ* in a cell (Gundersen *et al.* 1984; Kreis 1987; Wehland and Weber 1987; Skoufias *et al.* 1990). The immunofluorescence results shown in Fig. 8.9 demonstrate that a minor fraction of stable microtubules are present in isolated HepG2 cells, and that in contrast to the bulk of the microtubules (the dynamically unstable ones) which are spread throughout the cell (Fig. 8.9(a)), the stable microtubules are to a considerable extent localized around the nucleus and in proximity to the Golgi (Fig. 8.9(c) and (d)). The latter result parallels similar findings in other cell types (Kreis 1987; Skoufias *et al.* 1990).

More directly, in immunoelectron microscopic experiments to be reported elsewhere (M. Mizuno and S. J. Singer, unpublished) we have doubly labelled HepG2 cells with antibodies to Glu-tubulin and to serum albumin, and have demonstrated at high resolution an association of stable microtubules with transitional elements that contain serum albumin, located between the ER and the GA.

Our evidence is therefore consistent with the following proposals. The traffic of secretory and other proteins from the ER to the GA is not a random, stochastic process, but may instead be localized to specific sites in the ER that are located in proximity to the *cis*-faces of stacks of Golgi saccules; tubular extensions and vesiculation of the ER may be mediated by ER attachment to, and tracking on, a subset of microtubules that are dynamically stable, leading to the delivery of the transition vesicles to the *cis*-face of a Golgi stack. We suggest that this structural relationship between local regions of ER and the Golgi, mediated by stable microtubules, largely persists after fusion with another cell to produce a heterokaryon, and accounts for the preferential transfer of a newly synthesized secretory protein from its ER to the homologous GA. The later appearance of the secretory protein in the heterologous GA may be the result of transfer only from the homologous GA. In the case of homokaryons, fusion of the ERs of the two half-cells may occur, thus obliterating the physical and chemical distinction between homologous and heterologous ERs, leading to the simultaneous occupancy of both GAs.

In a more general vein, it seems clear that transient heterokaryons, such as we have utilized in studying the ER to Golgi transfer process, might also prove useful in examining many other interesting problems in cell biology. When two cells fuse, what happens to a variety of similar elements from the two half-cell moieties? For example, we have already referred to the question, when do the two initially distinct ERs of a heterokaryon or a homokaryon undergo fusion and chemical intermixing? What about other organelles, such as mitochondria? Do they remain physically distinct within the heterokaryon, until biosynthetic incorporation of proteins from the intermixed cytoplasm makes them indistinguishable? Or do they undergo fusion before then? If transient heterokaryons are subjected to a temperature of 4 °C, or treated with nocodazole to depolymerize their microtubules and thus cause the dispersal of the elements of their two GAs, upon reversing these treatments and re-collecting the Golgi elements, do these elements intermix? Do intermediate filaments derived from the two half-cell moieties remain distinct, or are they sufficiently dynamic that their subunits intermix upon cell fusion? Questions like these can be investigated immunocytochemically if one has antigenic markers that discriminate between the same element derived from the two cells used for cell fusion.

References

Balch, W. E. (1989). *J. Biol. Chem.*, **264**, 16965–8.
Dabora, S. L. and Sheetz, M. P. (1988). *Cell*, **54**, 27–35.
Gundersen, G. G., Kalnoski, M. H., and Bulinski, J. C. (1984). *Cell*, **38**, 779–89.
Keller, G. A., Glass, C., Louvard, D., Steinberg, D., and Singer, S. J. (1986). *J. Histochem. Cytochem.*, **34**, 1223–30.
Kreis, T. (1987). *EMBO J.*, **6**, 2597–606.
Lodish, H. F., Kong, N., Hirani, S., and Rasmussen, J. (1987). *J. Cell Biol.*, **104**, 221–30.
Massalski, A. and Leedale, G. F. (1969). *Br. Phycol. J.* **4**, 159–80.
Merisko, E. M., Fletcher, M., and Palade, G. E. (1986). *Pancreas*, **1**, 95–109.
Palade, G. E. (1975), *Science*, **189**, 347–58.
Pryer, N. K., Wuestehube, L. J., and Schekman, R. (1992). *Annu. Rev. Biochem.*, **61**, 471–516.
Rothman, J. E., Urbani, L. I., and Brands, R. (1984). *J. Cell Biol.*, **99**, 260–71.
Schulze, E. and Kirschner, M. (1988). *Nature*, **334**, 356–9.
Skoufias, D. A., Burgess, T. L., and Wilson, L. (1990). *J. Cell Biol.*, **111**, 1929–37.
Terasaki, M., Chen, L. B., and Fujiwara, K. (1986). *J. Cell Biol.*, **103**, 1557–68.
Valtersson, C., Dutton, A. H., and Singer, S. J. (1990). *Proc. Natl Acad. Sci. USA*, **87**, 8175–9.
Webster, D. R., Gundersen, G. G., Bulinski, J. C., and Borisy, G. G. (1987). *Proc. Natl Acad. Sci. USA*, **84**, 9040–4.
Wehland, J. and Weber, K. (1987). *J. Cell Sci.*, **88**, 185–203.

II

Gene mapping

9. Radiation hybrid mapping: an idea whose time has finally arrived

David R. Cox

In 1975, Stephen Goss and Henry Harris published a short paper in the journal *Nature* that promised to revolutionize the field of human gene mapping. Entitled 'New method for mapping genes in human chromosomes', this paper described a systematic approach for determining the linear order of human genes and estimating the distances between them. Previous workers had used rodent–human somatic cell hybrids to map genes to specific human chromosomes or to specific chromosomal segments in those regions of the genome where human translocation chromosomes were available. However, such maps were very crude in comparison with the genetic maps available for many other organisms. By exposing diploid human cells to a lethal dose of ionizing radiation followed by fusion to rodent cells to obtain viable hybrid clones segregating human chromosomal fragments, Goss and Harris were able to measure the frequency with which pairs of human genes were cotransferred after irradiation, thereby creating maps of much higher resolution than was possible previously. The novel aspect of this work was not the use of X-rays to generate human chromosomal fragments for mapping purposes, as this had been described by others two years previously (Burgerhout *et al.* 1973). Rather, the new insight provided by Goss and Harris was that by performing a statistical analysis of marker segregation in a large number of hybrids it was possible to determine marker order and distances. Statistical analysis of the cosegregation of markers to generate maps was certainly not a new idea in the field of prokaryotic genetics. However, it was a very new idea in the field of mammalian somatic cell genetics. Virtually all mapping studies that employed rodent–human somatic cell hybrids prior to the work of Goss and Harris used an approach which involved extensive characterization of the human chromosomal complement present in each hybrid cell line, thereby allowing that cell line to be used as a well-defined mapping reagent. In contrast, the Goss–Harris method did not require any prior knowledge of the precise human chromosomal fragments present in each hybrid in order to generate a high resolution map.

Despite its promise, the mapping approach described by Goss and Harris in 1975 did not revolutionize the field of human gene mapping. In part, this was due to the fact that the method relied heavily on the presence of a selectable marker for each human chromosomal region to be mapped, and few such selectable markers were available at that time. However, a more serious difficulty was that the method relied on the fact that human chromosomal fragments not physically linked to a selectable marker would be rapidly eliminated from the hybrid cell lines, and this often turned out not to be the case. Recognizing this to be a serious problem, Goss and Harris published a landmark paper in 1977 in the *Journal of Cell Science* in which they took advantage of the retention of 'non-selected' human chromosomal fragments in hybrids formed by fusing irradiated diploid human cells with mouse cells to generate a statistical map based on the cosegregation of eight loci on human chromosome 1. In contrast to the *Nature* paper, which was widely recognized by the somatic cell genetics community, this later paper, which provides a truly practical and general method for constructing detailed maps of human chromosomes, was largely ignored. Statistical analysis of the cosegregation of human markers in rodent–human somatic cell hybrids was an idea whose time had not yet come.

Despite the failure of the scientific community to take advantage of the statistical mapping approach developed by Goss and Harris, their 1975 *Nature* paper formed the basis for a series of studies carried out by other workers in the 1980s and 1990s in which specific fragments of human chromosomes carrying selectable markers and generated by exposure to X-rays were isolated in viable rodent–human somatic cell hybrids. In one of the first of these studies, Cirullo *et al.* (1983) transferred a human chromosomal fragment containing the genes encoding human leucyl- and asparaginyl-tRNA synthetases into Chinese hamster cell mutants. Unlike the original Goss and Harris experiments, in which irradiated diploid human cells were fused to rodent cells, in these later studies, a hamster–human hybrid cell line retaining only a single human chromosome was fused to mutant Chinese hamster cells following exposure to a lethal dose of X-rays. This approach was a very powerful way of isolating a specific human chromosomal fragment which could then be used to obtain DNA probes and/or genes. However, as in the original Goss and Harris studies, the method was restricted to those segments of the genome which contained selectable markers.

In 1986, my laboratory began a series of experiments aimed at generalizing the approach pioneered by Cirullo *et al.* (1983) to regions of the human genome which did not contain known selectable markers. The rationale for our approach was based on two observations. First, although human chromosomes are rapidly lost in the initial stages of hamster–human hybrid formation, many established hybrids retain human chromosomes in the absence of selection. Second, intraspecific somatic cell hybrids formed by fusing two Chinese hamster cell lines usually undergo only limited chromosomal segregation (Kao *et al.* 1969). Thus, we reasoned that a high proportion of hybrids formed by fusing lethally irradiated hamster–human hybrid cells containing a single human chromosome with unirradiated hamster cells would retain human chromosomal fragments even in the absence of selection for human fragments. Much to our delight, this proved to be case, providing a general method for isolating any desired human chromosomal fragment. I remember my excitement when describing our initial results to Hunt Willard and Peter Goodfellow at the *International Congress of Human Genetics* in Berlin in 1986. They expressed an appropriate degree of scepticism. Undaunted, in collaboration with my colleague Rick Myers, we proceeded to use our 'non-selective' approach to isolate a small region of human chromosome 4 containing the Huntington disease gene, enabling us efficiently to obtain additional DNA probes from this specific region of the genome (Cox *et al.* 1989; Pritchard *et al.* 1989). Following up our initial observations, Benham *et al.* (1989) reported similar success generating non-selected fragments of the human X-chromosome. Subsequently we and others have demonstrated the generality of the approach using a variety of different human chromosomes and rodent cell lines (Cox *et al.* 1990; Goodfellow *et al.* 1990; Richard *et al.* 1991; Warrington *et al.* 1991; Zoghbi *et al.* 1991; Ceccherini *et al.* 1992; Frazer *et al.* 1992).

In our studies, we routinely expose each hamster–human hybrid containing a single human chromosomes to 6000–8000 rad of X-rays, which results in chromosomal fragments with an average size of 5 Mbp. Following irradiation, the human and hamster chromosomal fragments rapidly rejoin with one another, resulting in complex chromosomal rearrangements: Nevertheless, it is very unusual to observe internal DNA rearrangements within individual human chromosomal fragments (Cox *et al.* 1989, 1990). This is in contrast to what is observed when human chromosomal fragments are generated by using the technique of chromosome-mediated gene transfer (Goodfellow and Pritchard 1988). At the dose of X-rays which we employ in our studies, each viable radiation hybrid usually retains between 20 and 60 per cent of the human fragments from any particular chromosome with an average retention frequency of about 25 per cent for any given marker. Although fragments containing centromeric heterochromatin are generally retained at a somewhat higher frequency than fragments from other chromosomal regions, the probability of retention of any given fragment is remarkably uniform, irrespective of whether that fragment contains a human centromere or a human telomere. This surprising result can be explained by the observation that human chromosomal fragments are often inserted into the middle of hamster chromosomes in these 'radiation hybrids', with functional hamster centromeres and telomeres resulting in relatively stable mitotic segregation of most human fragments. Given that the average size of a human chromosomal fragment following a dose of 6000–8000 rad of X-rays is approximately 5 Mb, and given that an average of 25 per cent of all markers are retained in any given radiation hybrid, it follows that most hybrids retain multiple, non-contiguous human chromosomal fragments. The number of such fragments depends on the size of the human chromosome. Radiation hybrids generated with 6000–8000 rad of X-rays for a 40 Mb chromosome like chromosome 21 result in an average retention of 2 human fragments per radiation hybrid while hybrids generated for a 200 Mb chromosome like chromosome 4 result in an average retention of 10 human fragments per hybrid. Although higher doses of X-rays appear to result in a lower retention of human chromosomal material in radiation hybrids, the higher dose also yields a much smaller average fragment size. Thus, the net effect of increasing the X-ray dose is to produce numerous small chromosomal fragments in each radiation hybrid (Goodfellow *et al.* 1990).

Despite the presence of several unlinked human chromosomal fragments in most radiation hybrids, the fact that such hybrids retaining a chromosomal region of interest are likely to be enriched for that segment of chromosome relative to other regions makes such hybrids very useful tools for obtaining region-specific DNA probes. By using techniques that result in the PCR amplification of human-specific inter-Alu sequences (Nelson et al. 1989; Cotter et al. 1990) to analyse radiation hybrids, it is a fairly simple matter to identify a hybrid with only a small amount of extraneous human chromosomal DNA in addition to the region of interest, and to isolate region-specific probes from such a hybrid rapidly and efficiently (Zoghbi et al. 1991).

Although the presence of multiple unlinked chromosomal fragments in each radiation hybrid does not detract from the utility of these hybrids for isolating probes, the presence of multiple unlinked fragments from different regions of the chromosome precludes the use of any single hybrid as a tool for mapping without some additional characterization of the human material retained. In those cases where a map of ordered markers spanning a chromosome already exists, one can simply score a particular radiation hybrid for the presence of markers to determine which segments of the chromosome have been retained in the hybrid. While this may seem to be a straightforward solution, more detailed consideration of this problem indicates that for a radiation hybrid generated by our standard conditions, one must score the hybrid DNA with ordered probes spaced at an average of approximately 1 Mb in order to have a 95 per cent probability of accurately determining all of the human fragments present in that cell line. At present, very few if any maps spanning entire human chromosomes at this level of resolution are available. Thus, by employing the standard somatic cell genetic approach, which requires detailed characterization of the human material present in any given radiation hybrid, one would come to the conclusion expressed by Benham et al. (1989) that radiation hybrids 'are not particularly well-suited to high resolution mapping because of the occurrence of multiple human fragments within a recipient cell line'. However, we reasoned that if a statistical approach similar to that described by Goss and Harris for 'non-selected' human fragments (Goss and Harris 1977) were applied to our radiation hybrids, these cell lines should be ideal reagents for generating high resolution maps of human chromosomes. Indeed, this proved to be the case.

Although the statistical model described by Goss and Harris (1977) for the analysis of non-selected fragments of human chromosome 1 was designed for the case in which irradiated diploid human cells were fused with mouse cells, it was an easy matter to simplify their model for the haploid case corresponding to irradiation of a hamster–human hybrid containing a single human chromosome. Following the lead of Goss and Harris, we assumed that X-rays cause random breaks along human chromosomes, and that each human chromosome fragment is retained independently of any other human fragment in a radiation hybrid. Thus, by using the frequency of breakage between any two markers as a measure of distance, it was possible to describe the probability that these two markers were 'significantly linked' in terms of a Lod score, as is conventional in meiotic linkage analysis. By estimating the frequency of breakage between markers in approximately 100 independent radiation hybrids, it has been possible to determine the order as well as the distance between otherwise unordered markers at high resolution (Cox et al. 1990). Both non-parametric as well as maximum likelihood statistical methods have been used to analyse radiation hybrid data, yielding very similar results for any single data set (Boehnke *et al.* 1991; Lawrence *et al.* 1991; Boehnke 1992). At present, there appears to be no single preferred method of analysis for all radiation hybrid data sets. While I have had stimulating interactions with many colleagues over the past few years concerning the analysis of radiation hybrid data, I am particularly indebted to Aravinda Chakravarti for his early insights into this problem, and more recently to Michael Boehnke, who in collaboration with Kenneth Lange has written a comprehensive computer program, RHMAP, for the analysis of radiation hybrid data (Boehnke *et al.* 1991).

Radiation hybrid mapping does not depend on the availability of DNA markers that detect polymorphic variation, unlike meiotic mapping. Furthermore, each radiation hybrid is informative for every marker, which is seldom if ever the case in meiotic mapping. Any human DNA marker that can be distinguished from rodent DNA sequences can be mapped using the radiation hybrid approach. Both Southern blot analysis and the analysis of products of locus-specific PCR assays have bee used to score radiation hybrids for the presence of human DNA sequences (Cox *et al.* 1990; Richard *et al.* 1991). PCR analysis appears to be the method of choice, however, given the small amounts of DNA required for each PCR assay and the ease with which PCR products can be analysed on ethidium bromide stained gels following electrophoresis. Although false negative PCR assays could in theory result in inaccurate maps, we presently circumvent this potential difficulty by carrying out each assay in duplicate. We find that in approximately 4 per cent of cases, the duplicate disagree. Such inconsistent data points are not used for map construction and are considered incomplete data.

The resolution of a radiation hybrid map generated using a particular set of radiation hybrids depends on the X-ray dose used to generate the hybrids. The ideal X-ray dose depends on the particular problem at hand. For instance, one would use a higher dose of X-rays to generate hybrids if the goal were to map a series of 20 markers from a 10 Mb chromosomal region than one would for 20 markers spanning the entire length of a 200 Mb chromosome. Hybrids generated using a high X-ray dose are not useful for ordering distant markers with respect to one another, as such markers will segregate independently in these hybrids, resulting in no useful information for determining marker order. Similarly, closely placed markers that display no breaks between them in a series of radiation hybrids generated with a low dose of X-rays must be grouped together as a single locus, since the hybrids provide no information regarding the relative order of such markers.

Surprisingly, the frequency of X-ray breakage appears to be linearly related to physical distance, without evidence for hot-spots or cold-spots of X-ray breakage along the chromosome. A useful rule of thumb appears to be that 1 per cent breakage between two markers corresponds to a physical distance of approximately 50 kbp when radiation hybrids are generated using a 6000–8000 rad X-ray dose (Cox *et al.* 1990; Richard *et al.* 1991; Warrington *et al.* 1991). In contrast to the results reported by Goss and Harris (1977) for their map of human chromosome 1, we have not observed evidence for compression of radiation map distances in

Giemsa dark bands versus Giemsa light bands. Based on our experience as well as that of other workers, a set of approximately 100 radiation hybrids generated with 6000–8000 rad of X-rays, in conjunction with 20 previously unordered markers derived from a 10 Mb chromosomal region, can be expected to give a framework map of markers ordered at greater than 1000 : 1 odds with an average distance between framework markers of 1 Mbp.

Things have changed a lot since 1977 when Goss and Harris first reported their map of eight human chromosome 1 markers based on the statistical analysis of the segregation of 'non-selected' X-ray induced human chromosomal fragments in rodent human–somatic cell hybrids. The ability to score hybrids for the presence of human genes using PCR and DNA-based probes, as opposed to histochemical staining for human-specific isozymes following starch gel electrophoresis, has dramatically increased the speed with which mapping data can be generated. In the past two years, members of our Human Genome Mapping Center have generated a radiation hybrid map of over 650 chromosome 4-specific PCR based markers spanning the length of the chromosome and ordered at an average resolution of approximately 500 kbp. The reality of such large scale throughput in conjunction with the availability of large numbers of PCR-based unique human markers from throughout the genome, combined with improved methods for statistical analysis of radiation hybrid data, now make construction of a radiation map of the entire genome using a single set of 100 X-ray induced hybrids, an idea first proposed by Goss and Harris in their 1977 paper, a realistic possibility. But will it be possible to generate high resolution radiation hybrid maps as first proposed by Goss and Harris by fusing irradiated diploid human cells with rodent cells, as opposed to using the approach our group has developed which involves fusing an irradiated human –hamster hybrid cell containing a single human chromosome to hamster cells? We have preliminary results which are very encouraging in this regard. We recently generated a series of radiation hybrids by irradiating diploid human lymphoblastoid cells with 8000 rad of x-rays followed by fusion with unirradiated Chinese hamster cells, and have scored these hybrids for the retention of 10 human chromosome 4 DNA sequences covering the length of the chromosome. The average retention frequency for these chromosome 4 markers was 18 per cent. These findings are very similar to the marker retention results reported by Goss and Harris (1977), as well as the results we obtained with our set of radiation hybrids formed by fusing an irradiated hamster–human hybrid containing a single copy of human chromosome 4 with unirradiated hamster cells. We are presently scoring our 'total genome' radiation hybrids with additional markers scattered across the genome to establish that the results we have obtained with chromosome 4 markers are representative for each of the other human chromosomes. If this proves to be the case, we plan to score these 100 hybrids with 5000 to 10 000 PCR-based markers distributed throughout the genome over the next two years. It seems that the construction of maps of human chromosomes based on the cosegregation of human markers in rodent– human somatic cell hybrids is an idea whose time has finally arrived. I am pleased that Henry Harris is able to witness this moment and I hope he is enjoying it as much as I am.

References

Benham, F., Hart, K., Colla, J., Borrow, M., Francavilla, M., and Goodfellow, P. N. (1989). *Genomics*, **4**, 509–17.
Boehnke, M. (1992). *Cytogenet. Cell Genet.*, **59**, 74–6.
Boehnke, M., Lange, K., and Cox, D. R. (1991). *Am. J. Human Genet.*, **49**, 1174–88.
Burgerhout, W. G., Van Somern, H., and Bootsma, D. (1973). *Humangenetik*, **20**, 159–62.
Ceccherini, I., Romeo, G., Lawrence, S., Breuning, M., Harris, P. G., Himmelbauer, H., et al. (1992). *Proc. Natl Acad. Sci. USA*, **89**, 104–8.
Cirullo, R. E., Dana, S., and Wasmuth, J. J. (1983). *Mol. Cell. Biol.*, **3**, 892–902.
Cox, D. R., Pritchard, C. A., Uglum, E., Casher, D., Kobori, J., and Myers, R. M. (1989). *Genomics*, **4**, 397–407.
Cox, D. R., Burmeister, M., Price, R. E., Kim, S., and Myers, R. M. (1990). *Science*, **250**, 245–50.
Frazer, K., Boehnke, M., Budarf, M., Wolff, R. K., Emanuel, B. S., Myers, R. M., and Cox, D. R. (1992). *Genomics*, **14**, 574–84.
Goodfellow, P. N. and Pritchard, C. A. (1988). *Cancer Surv.*, **7**, 251–65.
Goodfellow, P. J., Povey, S., Nevanlinna, H. A., and Goodfellow, P. N. (1990). *Somat. Cell Mol. Genet.*, **16**, 163–71.
Goss, S. J. and Harris, H. (1975). *Nature*, **255**, 680–84.
Goss, S. J. and Harris, H. (1977). *J. Cell Sci.*, **25**, 39–57.
Kao, F.-T., Johnson, R. T., and Puck, T. T. (1969). *Science*, **164**, 312–14.
Lawrence, S., Morton, N. E., and Cox, D. R. (1991). *Proc. Natl Acad. Sci. USA*, **88**, 7477–80.
Nelson, D. L., Ledbetter, S. A., Corbo, L., Victoria, M. F., Ramirez-Solis, R., Webster, T. D., et al. (1989). *Proc. Natl Acad. Sci. USA*, **86**, 6686–90.
Pritchard, C. A., Casher, D., Uglum, E., Cox, D. R., and Myers, R. M. (1989). *Genomics*, **4**, 408–18.
Richard, C. W., Withers, D. A., Meeker, T. C., Maurer, S., Evans, G., Myers, R. M., and Cox, D. R. (1991). *Am. J. Human Genet.*, **49**, 1189–96.
Warrington, J. A., Hall, L. V., Hinton, L. M., Miller, J. N., Wasmuth, J. J., and Lovett, M. (1991). *Genomics*, **11**, 701–8.
Zoghbi, H. Y., McCall, A. E., and LeBorgne-Demarquoy, F. (1991). *Genomics*, **9**, 713–20.

10. The cloning of tumour suppressor genes

David E. Housman

The identification of a class of genes whose action is required to regulate cell growth and whose absence leads to unregulated cell proliferation can be attributed to two lines of experimental investigation. The term tumour suppressor gene is most clearly associated with the pioneering somatic cell genetic work of Dr Harris and his collaborators and contemporaries in the late 1970s. Terms such as anti-oncogene were first associated with the work of Knudson (1971) and his collaborators who carried out a statistical analysis of the age of onset of childhood cancers and suggested that a specific small number of rate-limiting events was involved in the aetiology of childhood cancers and that hereditary predisposition could reduce the number of such steps to a single rate-limiting event. These two intellectual pathways in fact crossed and merged in the late 1980s and early 1990s. The demonstration that a series of genes exhibiting precisely the characteristics of these negative regulators of cell growth played a key role in the aetiology of a series of both childhood and adult cancers has led to a dramatic shift in our perception of the pathway to tumorigenesis. The number of tumour suppressor genes which have thus far been isolated and characterized remains small relative to the number of oncogenes identified to date. It seems apparent, however, that the balance of genes in these two classes is likely to shift significantly during the next few years as the lessons of the tumour suppressor gene searches to date are understood and appreciated and as the technology required to carry out these searches becomes more powerful. In this paper, I will review two searches for tumour suppressor genes in which my laboratory has played a significant role, the Wilms' tumour suppressor gene, WT1, and the human chromosome 9 tumour suppressor gene for cutaneous malignant melanoma. Through this discussion, I will attempt to lay out the framework within which the identification of the remainder of the tumour suppressor genes which play an important role in human malignancy is likely to be carried out.

Tumour suppressor genes may be defined in the broadest sense as the group of genes for which gene inactivation contributes to the unregulated proliferation of cells. Two basic lines of evidence led to this concept. First, somatic cell hybridization experiments demonstrated that fusion between tumour cells and normal cells led to hybrids which had a phenotype more like the normal cell parent than the tumour cell parent (Harris *et al.* 1969). As chromosome segregation proceeded in such hybrids, the re-emergence of the transformed phenotype could often be observed

(Sager 1985). In a number of specific instances, loss of growth control could be correlated with the absence of a specific chromosome derived from the normal parental cell (Sager 1985). These results led to the inference that the normal cell contributed an active copy of a gene (the tumour suppressor gene), which regulated the growth of the hybrid cell and hence suppressed its potential tumorigenicity (Harris et al. 1969). The second line of evidence supporting the existence of tumour suppressor genes came from analysis of specific childhood cancers, with particular attention to the age of onset and the number of tumours per individual. In an analysis of retinoblastoma, Knudson (1971) was able to demonstrate that children with a hereditary predisposition to this eye tumour showed an early age of onset and a distribution in the number of tumours per individual which was mathematically most compatible with a single rate-limiting hit (Knudson 1971). Children with sporadic retinoblastoma, on the other hand, showed a later age of onset and only a single tumour per individual, consistent with two rate-limiting hits. A parallel analysis with similar conclusions was carried out by Knudson and Strong (1972) for Wilms' tumour. These analyses were subsequently interpreted to indicate that the rate-limiting events in the formation of a retinoblastoma or a Wilms' tumour are the loss of function of both copies of a key tumour suppressor gene. In hereditary cases, one copy of the tumour suppressor gene was presumably inactivated by a germ-line mutation. The loss of the second copy by a somatic event (mutation, non-disjunction, or mitotic crossing over) would be the single rate-limiting event in hereditary tumours. Sporadic tumours would, according to this analysis, be due to the occurrence of a somatic event (point mutation or deletion) inactivating one of the two homologues of a tumour suppressor gene followed by a second somatic event inactivating the second copy of the gene, thus giving rise to a cell in which the tumour suppressor gene was functionally inactive. It is important to note that the Knudson (1971) hit-kinetics analysis applied primarily to childhood cancers, while the somatic cell genetic technique provided support for the relevance of tumour suppressor genes in adult cancers as well. While both lines of evidence provided direct support for the existence of tumour suppressor genes, precise identification of the relevant genes has involved a series of complementary experimental strategies.

Locating and isolating tumour suppressor genes

The identification of tumour suppressor genes has been a formidable task. The isolation and characterization of positive oncogenes, a class of genes which plays an important role in regulating cell proliferation, proceeded by a direct route. Because oncogenes act by providing cells with the ability to circumvent normal growth regulation by the introduction of a new function into the cell, they can be assayed directly, via DNA or viral transformation. In contrast, the genetic strategies which have led to the isolation of tumour suppressor genes have (with one exception, p53) taken a less direct strategy utilizing gene mapping techniques. Several genetic strategies have contributed to the isolation of tumour suppressor genes. First, as noted above, the inactivation of a tumour suppressor gene may involve loss of the normal copy of the gene via a chromosomal event, non-disjunction, or mitotic crossing over. The occurrence of such events can be assayed by the loss of heterozygosity of polymorphic DNA markers on the chromosome carrying the tumour suppressor gene. In each case where inactivation of a tumour suppressor gene has been associated with a specific tumour type, it has been observed that the loss of heterozygosity is more frequent for the chromosome carrying the tumour suppressor gene than for other chromosomes in the tumour. Loss of heterozygosity is therefore considered to be a significant clue in the search for a tumour suppressor gene. However, loss of heterozygosity studies do have limitations and pitfalls. While in most childhood cancers, the karyotype may remain near normal, in adult cancers karyotypic instability in the tumour cells can be extreme. In one study of metastatic melanomas, as many as 25 per cent of the DNA markers assayed had lost heterozygosity (Dracopoli et al. 1987). While it is possible that all of the karyotypic changes which occur in these tumours relate directly to meaningful losses in gene function in the tumour cells, this is by no means certain. The significance of loss of heterozygosity in the context of such extreme karyotypic instability must thus be considered carefully. A further limitation of loss of heterozygosity studies is the likelihood that they can give at best only an approximation of the position of the relevant tumour suppressor gene. Loss of heterozygosity due to chromosomal non-disjunction can provide support for the localization of a tumour suppressor gene to a specific chromosome, but provides no information as to where on that chromosome the tumour suppressor gene may lie. Mitotic recombination or other mechanisms involving partial chromosome loss can be more informative. The location of the chromosome segment which most frequently demonstrates loss of heterozygosity can thus be a clue to the location of the tumour suppressor gene; however, such data do not necessarily provide a precise localization of the gene involved. Another approach which can potentially give information on the position of a tumour suppressor gene is the analysis of families that show inherited predisposition for a given cancer or set of cancers using genetic linkage techniques. While this approach has been particularly valuable in relating the contribution of genes isolated by other techniques to familial cancer, the precision of linkage analysis is also not high in locating specific genes unless very large numbers of genetically homogeneous families are available.

To date, the most powerful discriminator of the position of a tumour suppressor gene has been the identification and analysis of internal chromosome deletions. Chromosomal deletions which have contributed to the identification of tumour suppressor genes occur in two contexts. First, deletions of a chromosome segment which includes a tumour suppressor gene can and do occur during gametogenesis. A sperm or egg carrying such a deletion can thus contribute an incomplete set of genetic instructions to the embryo, resulting in hemizygosity for all the genes within the deleted chromosomal segment including the tumour suppressor gene. A child carrying such a deletion falls into the 'one-hit' category of Knudson (1971) and is thus likely to have multiple tumours of independent origin. Children with multiple Wilms' tumours or multiple retinoblastoma tumours associated with hemizygosity of specific chromosomal segments in all tissues have been identified in the clinical literature for more than a decade (Riccardi et al. 1978; Yunis and Ramsey 1978). In addition to the development of tumours, these children often exhibit multiple congenital malformations leading to ready recognition in the clinic. The germ-line karyotypic abnormalities in these children are quite specific, in contrast to the more disorderly karyotypic picture observed even in the tumours themselves. Analysis of the deletions from a series of such children has helped to define the location of genes for both retinoblastoma and Wilms' tumour. However, while this approach for localization of tumour suppressor genes provides higher definition than methods discussed previously, the deletions observed are quite large and can usually be measured in megabases. To localize a tumour suppressor gene further, much smaller deletions are required. Such deletions can, and do, occur in sporadic tumours at a measurable frequency for both retinoblastoma and Wilms' tumour. Of particular value are tumours in which both copies of the tumour suppressor gene have been lost by deletion, a situation referred to as homozygous deletion. For both retinoblastoma and Wilms' tumour, the frequency of tumours carrying homozygous deletion appears to be between 1 in 50 and 1 in 100. Tumours of this type were particularly valuable in pinpointing the location of the retinoblastoma (Dryja et al. 1986) and Wilms' tumour (Lewis et al. 1988) genes to within a few hundred kilobases as well as identifying the location of the DCC tumour suppressor gene which contributes to the aetiology of adult colon cancer (Fearon et al. 1990). Once a candidate transcription unit has been identified within the region of homozygous deletion, the definitive identification of the tumour suppressor gene can still be a challenging task. Two strategies have been taken to date. First, analysis of a series of tumours which have undergone homozygous deletion can limit the smallest region of overlap to the candidate transcription unit. Second, inactivating point mutations, small deletions, or rearrangements can be demonstrated within the coding region of the gene itself for both sporadic tumours and germ-line mutations in hereditary cases.

The development of a physical and genetic map of the short arm of chromosome 11 played a crucial role in the identification and localization of some of the genes contributing to the WAGR phenotypes. Somatic cell hybrids have made a particularly crucial contribution in the molecular dissection of 11p.

The J1 hybrid series

One of the most important sets of reagents developed for chromosome 11 mapping has been derived from the hybrid cell line J1. J1, a human–hamster somatic cell hybrid which stably maintains chromosome 11 as its only human DNA, has served as the parent in generating a series of cell lines which retain overlapping segments of chromosome 11. The J1 hybrid cell line had its origin in 1971 when Puck et al. (1971) generated a series of somatic cell hybrids by fusing human amniocytes with an auxotrophic Chinese hamster cell line (CHO-K1). J1, a subclone of the original fusion, was found to maintain chromosome 11 stably despite years of passage in the apparent absence of selection (Kao et al. 1976). The unusual stability of chromosome 11 in J1 is due to complementation of a mutation present in the CHO-K1 fusion

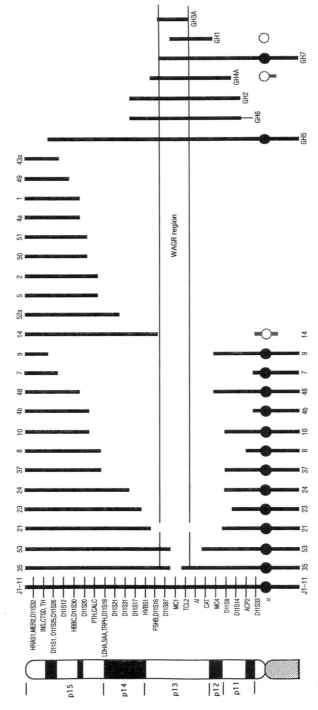

Fig. 10.1. Schematic representation of J1 series of somatic cell hybrids.

partner by a gene which is tightly linked to the *INS* and *HRAS1* genes on chromosome 11. J1 deletion segregants were isolated by mutagenesis and negative selection for *MIC1*, a cell surface glycoprotein which can be targeted using monoclonal antibodies (Kao et al. 1976). All J1 deletion derivatives retained at least the distal portion of 11p and are nested contiguous deletions which span the *MIC1* gene whose product was originally selected against. A deletion map of 11p was constructed based solely upon the pattern of marker segregation among an expanded panel of J1 deletion derivatives (Glaser 1988). This map (Fig. 10.1) was consistent both internally and with independently obtained mapping data for 11p. It subdivides 11p into over 20 segments, ordering 35 breakpoints and 36 genetic markers into intervals of average size < 2000 kb. The J1 hybrid mapping panel has contributed significantly to the development of a fine structure map of 11p. First, it provides a means to localize rapidly any gene or genetic marker to a defined interval on 11p. Second, it has been used to assess the extent of deletions and or translocations associated with WAGR or familial aniridia patients' chromosomes 11. Third, individual clones within the mapping panel have been used to generate radiation-reduced hybrids as a means of isolating specific, subchromosomal fragments of 11p. Finally, genomic libraries of individual J1 series hybrids have produced many new 11p specific markers including several which map to the WAGR region (Glaser 1988).

Cell lines constructed by a modification of the irradiation-reduction procedure of Goss and Harris (1975) have proved to be particularly valuable in the isolation of the WT1 tumour suppressor gene. Using cell line J1-11 (a J1 deletion derivative retaining the short term 4 chromosome 11 as donor and CHO-K1 cells as recipient, hybrid clones retaining the p13 segment of chromosome 11 were selected for their expression of the *MIC1* surface antigen (Glaser et al. 1990). Clones that retained large fragments of 11p were identified by their co-expression of both the *MIC1* and *MER2* cell surface antigens and were eliminated from the panel. A total of seven clones was isolated in this manner. They all carried intact segments of 11p that spanned *MIC1*, ranging in size from 3200 kb to over 50 000 kb. One of these clones, Goss Harris no. 3A (GH3A), stably maintains a single 3200 kb segment of 11p13 that envelopes the WAGR region. The reduced amount of 11p DNA carried by each of these hybrids makes them especially useful for both probe isolation and pulsed field gel electrophoresis mapping in the immediate vicinity of the WT1 gene.

Several human candidate clones were identified by virtue of their ability to hybridize to human Alu 1 repetitive DNA (Glaser et al. 1988; Call et al. 1990). Of particular importance in screening these candidates clones was a sporadic Wilms' tumour, Wit 13, identified with homozygous deletion of genetic material in band 11p13 (Lewis et al. 1988). The region of homozygous deletion in this sporadic tumour proved to be within the boundaries for the WT gene locus previously identified from WAGR patients. Assuming that this tumour had acquired its transformed phenotype by deleting both copies of the 11p13 WT gene, then candidate genes must fall within the smallest region of overlap (SRO) of the deletions, estimated to be ~350 kbp (Rose et al. 1990). One clone, J8-3p4, which fell within this SRO (see Fig. 10.1) was further characterized because of the interesting hybridization pattern it gave when a single copy probe was used in Northern blotting experiments. This clone represents a gene which spans ~50 kbp and whose transcription proceeds in a centromeric to telomeric direction (Call et al. 1990).

Several cDNAs were isolated using single copy probes derived from J8-3p4. One of these, WT33, showed several sequence characteristics of a transcription factor. (The gene which WT33 represents is referred to as WT1). It contains four zinc fingers of the Cys-His variety at the C-terminus, as well as a proline/glutamine-rich N-terminus (Call et al. 1990; Gessler et al. 1990). Such proline/glutamine-rich regions are found in other DNA binding proteins, including Krüppel, and the CTF-NF-1 family of closely related polypeptides (Mitchell and Tjian 1989). Three of the four zinc finger domains of WT1 show 51–63 per cent homology to the three zinc fingers of the early growth response proteins EGR-1 and EGR-2 (Call et al. 1990; Gessler et al. 1990). These proteins are induced when serum-starved fibroblasts are stimulated by the addition of fresh serum to enter G_1 (Joseph et al. 1988). The WT1 gene product binds DNA sequence similar to the EGR recognition site, as recently demonstrated by Rauscher et al. (1990). The EGR proteins are thought to be important nuclear intermediates in the signal transduction pathway for growth response. The homology and conservation in DNA binding between WT1 and the EGRs suggested to us that WT1 may be involved in the growth proliferation response of nephroblasts, and that this regulation is uncoupled in nephroblastomas. Using immunofluorescence, we have shown that WT1 localizes to the nucleus after introduction into COS-1 cells, consistent with its predicted role as a transcription factor (Pelletier et al. 1991a).

The zinc finger amino acid sequence of WT1 is also closely related to the yeast gene product, MIG1. The similarity is particularly pronounced in the fingertip of the loops, which is thought to be the region important for DNA binding (Nardelli et al. 1991). MIG1 has been demonstrated to regulate transcription levels of genes involved in sugar metabolism (Nehlin and Ronne 1990). The MIG1 recognition site is similar to the EGR and WT1 binding sites. The similarities between WT1 and MIG1 suggest that WT1, like MIG1, may function as a repressor. This would be consistent with its proposed role in development and tumorigenesis.

Northern blot analysis demonstrated that the WT1 cDNA recognized a ~3 kb mRNA whose expression was highest in kidney and spleen (Call et al. 1990; Gessler et al. 1990). Four alternatively spliced mRNA species are synthesized due to the presence of two alternative splice choices in the WT1 pre-mRNA (Haber et al. 1990, 1991). Unlike RBl, WT1 is expressed in a tissue-specific manner (Call et al. 1990; Gessler et al. 1990; Pritchard-Jones et al. 1991). WT1 RNA expression is very abundant in human fetal kidney, but not detectable in adult kidney (Haber et al. 1990). In addition, WT1 mRNA levels were examined in several tumour cell lines and found to be present in K562 (erythroleukaemia), CEM (acute lymphocytic leukaemia) (Call et al. 1990), and 293 cells (adenovirus transformed human embryonic kidney cell line) (Pritchard-Jones et al. 1991).

Some Wilms' tumours contain deletions that encompass the WT1 gene along with adjacent genomic DNA sequences (Call et al., 1990; Gessler et al. 1990). However, several tumours were identified in which the region of homozygous deletion specifically involved the WT1 transcription unit. These tumours show deletions extending into the upstream exons (Ton et al. 1991) or into the downstream exons of WT1 (Cowell et al. 1991) (see Fig. 10.1). Small internal deletions which directly alter mRNA structure have also been identified within WT1 in Wilms' tumours (Haber et al. 1990; Huff et al. 1991; Pelletier et al. 1991b). These findings support the identification of this transcription unit and its gene product as the 11p13 Wilms' tumour gene.

The number of tumour suppressor genes which have been identified to date by molecular cloning represent, in my view, a small fraction of the tumour suppressor genes yet to be discovered. For a number of cases discussed above significant genetic clues have led to the identification of small numbers of tumours with homozygous deletions pinpointing the location of the tumour suppressor gene. For many more cases the genetic clues are more difficult to piece together. One example of such a tumour suppressor gene is the gene originally described by Dr Harris (1969) and his co-workers on mouse chromosome 4. In their original studies, this gene was shown to be a potent tumour suppressor for mouse melanoma. Despite elegant cell physiological characterizations, the molecular isolation of this gene has not been possible to date. More recent work on human melanoma strongly supports the location of a tumour suppressor gene for melanoma on chromosome 9, in the region syntenic to the mouse chromosome 4 tumour suppressor gene for melanoma. In the case of the human gene, melanoma cell lines with homozygous deletions in the central regions of 9p have now been identified opening the pathway to isolation of this tumour suppressor gene via a route analogous to that used for the WT1 gene (Fountain et al. 1992).

The implications of these studies for the identification of additional tumour suppressor genes seem clear. As lines of evidence from somatic cell genetic and human tumour genetics converge, the precise molecular site of a tumour suppressor gene is more likely to be revealed. The original observations of Professor Harris (1969) and his co-workers are thus likely to have a strong and lasting impact on the development of this aspect of tumour cell biology.

References

Armstrong, J. F., Pritchard-Jones, K., Bickmore, W. A., Hastie, N. D., and Bard, J. B. L. (1992). The expression of the Wilms' tumor gene, WT1, in the developing mammalian embryo. *Mech. Dev.*, **40**, 85–97.

Call, K. M., Glaser, T. M., Ito, C. Y., Buckler, A. J., Pelletier, J., Haber, D. A., et al. (1990). Isolation and characterization of a zinc finger polypeptide gene at the human chromosome 11 Wilms' tumor locus. *Cell*, **60**, 509–20.

Cowell, F. K., Wadley, R. B., Haber, D. A., Call, K. M., Housman, D. E., and Prichard, F. (1991). Structural rearrangements of the WT1 gene in Wilms' tumor cells. *Oncogene*, **6**, 595–9.

Dracopoli, N. C., Ahadeff, B., Houghton, A. N., and Old, L. J. (1987). Loss of heterozygosity at autosomal and X-linked loci during tumor progression in a patient with melanoma. *Cancer Res.*, **47**, 3995.

Dryja, T. P., Rapaport, J. M., Joyce, J. M., and Peterson, R. A. (1986). Molecular detection of deletions involving band q14 of chromosome 13 in retinoblastomas. *Proc. Natl Acad. Sci., USA*, **83**, 7391.

Fearon, E. R., Cho, K., Nigro, J., Kern, J., Simons, S., Ruppert, J., *et al*. (1990). Identification of a chromosome 18q gene that is altered in colorectal cancers. *Science*, **247**, 49–56.

Fountain, J. W., Karayiorgou, M., Ernstoff, M. S., Kirkwood, J. M., Vlock, D. R., Titus-Ernstoff, L., *et al*. (1992). Homozygous deletions within human chromosome band 9p21 in melanoma. *Proc. Natl Acad. Sci. USA*, **89**, 10557–61.

Gessler, M., Poustka, A., Cavenee, W., Neve, R. L., Orkin, S. H. and Bruns, G. A. P. (1990). Homozygous deletion in Wilm's tumours of a zinc-finger gene identified by chromosome jumping. *Nature*, **343**, 774–8.

Glaser, T. (1988). The line structure and evolution of the eleventh human chromosome. Ph.D. thesis, Masachusetts Institute of Technology, Cambridge, MA.

Glaser, T., Lane, J., and Housman, D. (1990). A mouse model of the aniridia–Wilms' tumor deletion syndrome. *Science*, **250**, 823–7.

Goss, S. J. and Harris, H. (1975). New method for mapping genes in human chromosomes. *Nature*, **255**, 680.

Haber, D. A., Buckler, A. J., Glaser, T., Call, K., Pelletier, J., Douglass, E., and Housman, D. E. (1990). A 25 bp internal deletion identifies the 11p13 Wilms' tumor gene. *Cell*, **61**, 1257–69.

Haber, D. A., Buckler, A. J., Glaser, T., Call, C. K., Pelletier, J., Sohn, R. L., *et al*. (1991). An internal deletion within an 11p13 zinc finger gene contributes to the development of Wilms' tumour. *Cell*, **61**, 1257–69.

Harris, H., Miller, Q. J., Klein, G., *et al*. (1969). Suppression of malignancy by cell fusion. *Nature*, **223**, 363–8.

Huff, V., Miwa, H., Haber, D., Call, K. M., Housman, D. E., Strong, L. C., and Saunders, G. F. (1991). Evidence of WT1 as a Wilms' tumour (WT) gene: intragenic germinal deletion in bilateral WT. *Am. J. of Human Genet*. **48**, 997–1003.

Joseph, L. J., LeBeau, M. M., Jamieson, G. A., *et al*. (1988). Molecular cloning, sequencing and mapping of EGR2, a human early growth response gene encoding a protein with zinc-binding finger structure. *Proc. Natl Acad. Sci. USA*, **85**, 7164–8.

Kao, F. T., Jones, C., and Puck, T. T. (1976). Genetics of somatic mammalian cells: Genetic, immunologic, and biochemical analysis with Chinese hamster cell hybrids containing selected human chromosomes. *Proc. Natl Acad. Sci. USA*, **73**, 193.

Knudson, A. G. (1971). Mutation and cancer: a statistical study. *Proc. Natl Acad. Sci. USA*, **68**, 820–23.

Knudson, A. G. and Strong, L. C. (1972). Mutation and cancer:a model for Wilms' tumour of the kidney. *J. Natl Cancer-Inst.*, **48**, 313–24.

Lewis, W. H., Yeger, H., Bonetta, L., *et al*. (1987). Homozygous deletion of a DNA marker from chromosome 11p13 in sporatic Wilms' tumour. *Genomics*, **3**, 25.

Lewis, W. H., Yeager, H., Bonetta, L., Chan, H. S. L., Kang, J., Junien, C., Cowell, J., Jones, C. and Dafoe, L. A. (1988). Homozygous deletion of a DNA marker from chromosome 11p13 in a sporatic Wilms' tumor. *Genomics*, **3**, 25.

Mitchell, P. J. and Tjian, T. (1989). Transcriptional regulation in mammalian cells by sequence-specific DNA binding proteins. *Science*, **245**, 371–8.

Nardelli, J., Gibson, T. J., Vesque, C., and Charnay, P. (1991). Base sequence discrimination by zinc-finger DNA-binding domains. *Nature*, **349**, 175.

Nehlin, J. O. and Ronne, H. (1990). Yeast repressor is related to the mammalian early growth response and Wilms' tumour finger proteins. *EMBO J.*, **9**, 2891–8.

Pelletier, J., Bruening, W., Kashtan, C. E., Mauer, S. M., Manivel, J. C., Striegel, J. E., *et al.* (1991a). Germline mutations in the Wilms' tumour suppressor gene are associated with abnormal urogenital development in Denys–Drash syndrome. *Cell*, **67**, 437–47.

Pelletier, J., Bruening, W., Li, F. P., Haber, D. A., Glaser, T., and Housman, D. E. (1991*b*). WT1 mutations contribute to abnormal genital system development and hereditary Wilms' tumor. *Nature*, **353**, 431–4.

Pritchard-Jones, K., Fleming, S., Davidson, D., Bickmore, W., Portious, D., Gosden, C., *et al.* (1990). The candidate Wilms' gene is involved in genitourinary development. *Nature*, **346**, 194–7.

Puck, T. T., Wuthier, P., Jones, C., and Kao, F. (1971). Genetics of somatic mammalian cells: lethal antigens as genetic markers for study of human linkage groups. *Proc. Natl Acad. Sci.* **68**, 3102.

Rauscher, F. J., III, Morris, J. F., Tournay, O. E., Cook, D. M., and Curran, T. (1990). Binding of the Wilms' tumour locus zinc finger protein to the EGR1 consensus sequence. *Science*, **250**, 1259–62.

Riccardi, V. M., Sujansky, E., Smith, A. C., and Francke, U. (1978). Chromosomal imbalance in the aniridia – Wilms' tumour association: 11p interstitial deletion. *Pediatrics*, **61**, 604–10.

Rose, E. A., Glaser, T. M., Jones, C. A., Smith, C. L., Lewis, W. H., Call, K. M., *et al.* (1990). Complete physical map of the WAGR region of 11p13 localizes a candidate Wilms' tumor gene. *Cell*, **60**, 495–508.

Sager, R. (1985). Genetic suppression of tumour formation. *Adv. Cancer Res.* **44**, 43–68.

Ton, C. T., Huff, V., Call, K. M., Cohn, S., Strong, L. C., Housman, *et al.* (1991). Smallest region of overlap in Wilms' tumour deletions uniquely implicates an 11p13 zinc finger gene as the disease locus. *Genomics*, **10**, 293–7.

Yunis, J. J. and Ramsey, N. (1978). Retinoblastoma and subband deletion of chromosome 13. *Am. J. Dis. Child.* **132**, 161.

III

Monoclonal antibodies

11. The road to monoclonal antibodies

R. G. H. Cotton

Introduction

This article aims to highlight the contribution of cell fusion to the monoclonal antibody technique, and to describe the scientific background to this technique and the early days of its application. The scientific background will be approached both from the broader context of immunogenetics as well as from the local aspect in Cambridge from 1970 to 1974. This summary illustrates the natural progression of scientific findings and techniques towards the development of the monoclonal antibody technique.

Cell fusion

Henry Harris has told his own story of cell fusion (Harris 1970, 1984, 1987) and others have told of its promise (Anon 1969). However, it is important to record some of the important milestones towards the use of cell fusion as an important, widespread, and routine biological tool, particularly as used in monoclonal antibody production. These milestones are displayed in Table 11.1. Four lines of research converged to produce the cell fusion protocol now used in the monoclonal

Table 11.1. Milestones in cell fusion

Year	Experiment	Authors
1954	Cell fusion by measles virus in culture	Enders and Peebles (1954)
1955	Cloning of mammalian cells	Puck and Marcus (1955)
1958	Cell fusion of Ehrlich tumour cells by HJV virus	Okada (1958)
1958	Selective marker described	Szybalski (1958)
1959	HAT medium described	Hakala and Taylor (1959)
1960	Hybrid overgrowth of two mixed mouse cultures	Barski *et al.* (1960)
1962	HAT system used for cell selection	Szybalski and Szybalski (1962)
1964	Selection of hybrids by HAT	Littlefield (1964)
	Cloning of cells in agar	Sanders and Burford (1964)
1965	Man–mouse hybrids induced by Sendai virus	Harris and Watkins (1965)
1969	Chemical agent for cell fusion (lysolecithin)	Howell and Lucy (1969)
1974	Polyethylene glycol used for cell fusion	Kao and Michayluk (1974)

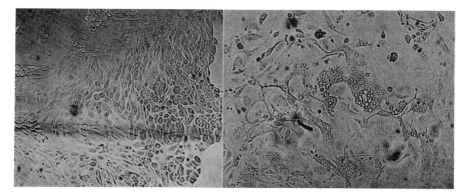

Figure 11.1 Cell fusion of human kidney cells by infection with measles virus. Left panel: uninfected cells. Right panel: infected cells, showing syncytium formation. Reproduced with permission from Enders and Peebles (1954).

antibody technique: cell fusion itself, virally or chemically aided cell fusion, cell marker and selection systems, and somatic cell genetics.

The first clear demonstration of cell fusion in tissue culture was described in 1954 by Enders and Peebles. Infection of human kidney cells by measles virus led to a very characteristic cytopathic effect. This was formation of syncytial giant cells

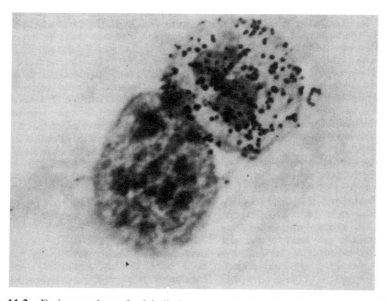

Figure 11.2 Fusion product of a labelled human (HeLa) cell and a mouse cell (Ehrlich) showing the tritium-labelled human nucleus. Reproduced with permission from Harris (1984).

each containing many nuclei which was likely to be a result of cell fusion (Fig. 11.1). In 1958 Okada deliberately fused Ehrlich's tumour cells using HVJ virus (Okada 1958) *in vitro*. This fusion appeared to be related to the haemolytic activity of the virus. Further progress was achieved by Barski and colleagues who (without virus) mixed two different established mouse cell lines and later showed overgrowth of the culture with cells which had a karyotype which was the sum of the two original cell lines (Barski *et al.* 1960). As these cells did not occur in parallel cultures of the two original lines, this provided strong evidence that the lines were indeed hybrid. Sorieul and Ephrussi (1961) confirmed these studies demonstrating that cell fusion was reproducible.

It was five years until the landmark fusion of Harris and Watkins appeared (Harris and Watkins 1965). In this case the cells used were HeLa and mouse Ehrlich ascites tumour cells. UV-inactivated Sendai virus was used to enhance the frequency of fusion. Heterokaryon (hybrid) formation was proven either by noting the morphology of the nuclei present as they were quite distinct in the parental lines or by labelling one of the lines with tritiated thymidine before fusion (Fig. 11.2). Nuclei were shown to fuse in these experiments and there was evidence of cell division after fusion but hybrid lines were not isolated in this study. This paper did more than just show that cells of two species could be hybridized. It provoked an enormous reaction in the press and the scientific literature (Fig. 11.3).

Figure 11.3 Cartoon from the *Daily Mirror*, 15 February 1965, indicating the media perception of the mouse–human hybrid. Permission to reproduce the figure has kindly been granted.

This reaction served to launch cell fusion as a well-known field of scientific endeavour both in its own right and as a tool.

Another important thread for the future field of somatic cell genetics was forming at around the same time as the above-mentioned experiments. The crucial cloning of mammalian cells was described in 1955 by Puck and Marcus. In 1958 a cell line with a selective marker was described (Szybalski 1958). This was a human cell line, D98, that was resistant to 8-azaguanine (AG). This group went on further to demonstrate the HAT system (hypoxanthine aminopterin, thymidine) for isolating revertants of the AG-resistant cell line (Szybalski and Szybalski 1962). The HAT system had been defined in 1959 by Hakala and Taylor.

Littlefield used the HAT system in 1964 to select hybrids between cell lines resistant to AG or 5-bromodeoxyuridine (BUdR) (Littlefield 1964). The cell lines were clonal mouse fibroblast lines and no fusing agent was used; the lines were just grown together for 4 days. Hybrid cell lines had chromosome numbers near the sum of the two parental lines. 1964 also saw the first cloning of cells in agar (Sanders and Burford 1964) where an underlay was poured and the BHK 21 cells poured onto this in semi-solid agar. There was no feeder layer.

The next important step was the development of chemicals that could cause cell membranes to fuse. The first chemical shown to do this was lysolecithin (Howell and Lucy 1969) but the now widely used polyethylene glycol (PEG) was used to fuse protoplasts from plants of different genus in 1974 (Kao and Michayluk 1974). These workers showed fusion of protoplasts of many plants including peas and beans and showed later cell division, a crucial point as the fused cells would be useless if they could not divide.

Myeloma biology to monoclonal antibodies

The history of monoclonal antibodies has been told by Milstein and others on numerous occasions (Milstein 1980, 1985, 1986a,b, 1987; Wade 1982; Randle and Rabbitts 1986). The aim of this section is to trace a direct line from the early days of mouse myeloma studies to the development of the monoclonal antibody technique. Mouse

Table 11.2. Milestones in myeloma biology

Year	Experiment	Authors
1959	Induction of mouse plasma cell tumours	Merwin and Algire (1959)
1962	Induction of myelomas by mineral oil	Potter and Boyce (1962)
1965	Cloning of spleen cells in agar	Pluznik and Sachs (1965)
1967	Continuous myeloma cell culture in suspension	Pettengill and Sorenson (1967)
1968	Myeloma proteins active against haptens	Schubert et al. (1968)
	Myeloma proteins active against polysaccharide	Potter and Leon (1968)
1970	Agar cloning of myeloma cells	Coffino et al. (1970)
		Horibata and Harris (1970)

Table 11.3. Myeloma cell fusion

Year	Experiment	Authors
1970	Mouse myeloma × mouse fibroblastic L cell hybrid; immunoglobulin at reduced level	Periman (1970)
1971	human thymocyte × mouse fibroblast hybrid; some immunoglobulin initially	Parkman et al. (1971)
	Mouse myeloma × mouse 3T3 fibroblast; immunoglobulin negative (< 1 %)	Coffino et al. (1971)
	Human lymphoblast × mouse A9 fibroblast; immunoglobulin negative	Klein and Wiener (1971)
	Mouse myeloma × mouse lymphoma; immunoglobulin positive	Mohit and Fan (1971)
1973	Mouse myeloma × rat myeloma; immunoglobulin of both parents	Cotton and Milstein (1973)
	Mouse myeloma × human peripheral lymphocyte; immunoglobulin of both parents; lymphocyte immortalization	Schwaber and Cohen (1973)
	Human lymphoblast × mouse fibroblast; immunoglobulin of human	Orkin et al. (1973)
	Human lymphoblast × human lymphoblast; immunoglobulin of both parents	Nyormoi et al. (1973, personal communication)
1974	Human lymphoblast × human lymphoblast; no immunoglobulin of either parent	Bloom and Nakamura (1974)
1975	Mouse myeloma × mouse myeloma; immunoglobulin of both parents	Kohler and Milstein (1975)
	Mouse myeloma × mouse spleen cell; immunoglobulin of pre-determined specificity	Kohler and Milstein (1975)
1977	Monoclonal antibodies to proteins in impure preparations	Galfrè et al. (1977)
	Monoclonal antibodies to lymphocyte cell surface proteins	Williams et al. (1977)

myeloma cells were studied with a particular intensity around the 1960s and later because they were a clone of cells producing effectively pure antibody. This could then be readily studied in contrast to the heterogeneous mixture present in animal sera. The crucial developments along this route are listed in Tables 11.2 and 11.3.

The three most important findings which led to the monoclonal antibody technique were

1. the demonstration that the proteins produced by myeloma cells were antibodies binding antigens;

2. the finding that the hybrid between two antibody-producing cells continued to produce antibodies of both parents;

3. a differentiated function of a non-malignant cell could be immortalized by fusion with a neoplastic cell.

Single antibody-producing cells had been found to produce only one antibody (Nossal and Lederberg 1958). Plasma cell tumours also producing a single antibody of mouse and other animals had of course been known for some time, but a key finding was the fact that plasma cell tumours or myelomas could be induced at will (Merwin and Algire 1959). This was done by physical irritation; the introduction of mineral oil for the same purpose was an advance (Potter and Boyce 1962). The continuous culture of myeloma cells was a significant advance (Pettengill *et al.* 1967) and necessary for many of the important future studies to be carried out on immunoglobulin production. The finding that some myeloma proteins were active against some antigens was a great stimulus to the field as this started to convince the sceptics that the myeloma proteins were indeed representative of and indeed real antibodies. Thus, in 1968 Schubert *et al.* and Potter and Leon found amongst their collections some myelomas, antibodies from which reacted with some chemical haptens and polysaccharides respectively.

The ability to clone the myeloma cells was important to many of the experiments to be performed later. Pluznik and Sachs (1965) had cloned spleen cells in agar with the two layer systems but found they needed to grow feeder cells on the base of the plate to obtain colonies. Myeloma cloning was achieved in agar in 1970 (Coffino *et al.* and Horibata and Harris) who used the two layer system but without the need for feeder cells.

The final phase on the road to monoclonal antibodies started in 1970 with the first fusions of myelomas being described (Table 11.3). Considering the activity in deliberate cell hybridization started 5–6 years earlier, such studies seem overdue. These cell fusions were mainly directed at analysis of the control of the differentiated function, immunoglobulin production in the case of myeloma cells. Analysis of such experiments and those involving other differentiated functions was performed by Davis and Adelberg in 1973. Five reports appeared in the series and all involved fusion of immunoglobulin-producing cells with non-immunoglobulin-producing cells. Four papers (Periman 1970; Coffino *et al.* 1971; Klein and Wiener 1971; Parkman *et al.* 1971) fused fibroblasts with immunoglobulin-producing cells. The lack of immunoglobulin production in the first two cases could have been due to extinction due to a repressor, lack of necessary biochemical machinery for immunoglobulin production, or other reasons. Similar explanations could be invoked for the latter cases where initial production on the membrane disappeared after culture (Parkman *et al.* 1971) or the immunoglobulin production was at a greatly reduced level (Periman 1970). This work was the idea of Michael Potter who stimulated Periman to start the fusion of malignant plasma cells in 1967. Even though only a small amount of immunoglobulin was produced, this work could claim to have isolated the first *hybrid* producing an antibody. Periman later attempted to fuse non-producing plasma cells with producing plasma cells but was unsuccessful (P. Periman, personal communication). The final paper in this series (Mohit and Fan 1971) begins to be a little more relevant to the evolution of the monoclonal antibody techniques as at least both cell types are either immunoglobulin-producing or related cells. Immunoglobulin-producing mouse myeloma cells, producing IgG and free κ chains were fused with non-producing mouse lymphoma cells. The hybrid population produced only the κ chain (Mohit and Fan 1971) but subclones with higher chromosome numbers produced similar immunoglobulins to the myeloma parent (Mohit 1971) at around 50 per cent the level of the parent but the level was not strictly quantified.

The next key series of experiments appeared in 1973. These papers looked at allelic exclusion and immortalization. Allelic exclusion refers to the phenomenon of immunoglobulin-producing cells only expressing one of the two possible alleles (Pernis *et al.* 1965; Weiler 1965). This is in contrast to the usual situation where in the case of a disease gene, in the heterozygote both normal and wild-type alleles are expressed. Thus at the time the experiments reported in 1973 started it could not be said whether, after fusion of two immunoglobulin-producing cells, both, one, or neither of the parental immunoglobulins would be produced in the hybrid. In one case a mouse myeloma producing heavy (γ 2a) and light chains (κ) was hybridized with a rat myeloma producing light chain (κ) alone (Fig. 11.4) (Cotton and Milstein 1973). Hybrid clones were found to produce substantial quantities of both parental immunoglobulins, so allelic exclusion was clearly not operating at least for those chains expressed. This was thought to indicate that integration of V and C regions had occurred at an earlier stage of the differentiation pathway. A similar conclusion could be drawn from the experiment of Schwaber and Cohen (1973). They fused an IgA-producing mouse myeloma (not drug-resistant) with human peripheral lymphocytes. After some weeks clones of large cells were seen and isolated. One clone was studied in detail and shown to be a mouse–human hybrid. It was shown to produce the mouse immunoglobulin and human immunoglobulin in the medium as well as having both on the surface of individual cells. The human parent carried a human immunoglobulin on the surface and fusion caused it to be secreted by the hybrid. Thus no allelic exclusion was occurring. Orkin *et al.* (1973) fused an 8-AG-resistant human lymphoblast cell line with a BUdR-resistant mouse fibroblast line. The lymphoblast parent produced λ light chain, as did at least two hybrids. The amount was substantial but not quantified. These workers suggested that the system was suitable for mapping and control studies. Thus the authors suggested that the extinction of differentiated function was not occurring in this system which is in contrast to the work reported above. Bloom and Nakamura (1974) hybridized two human lymphocyte cell lines both producing immunoglobulins (IgG, and IgM with

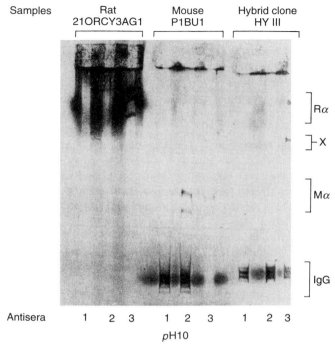

Figure 11.4 Isoelectric focusing of immunoglobulins from parental and hybrid myeloma cells. Hybrids contain all parental immunglobulins as well as a unique band, X, not found in the parents. Permission to reproduce the figure has kindly been granted.

κ chains). Selection was possible after Sendai virus cell fusion because one line was thioguanine-resistant and the other from a citrullinaemic patient. The tetraploid hybrid showed no evidence of extinction or allelic exclusion as all parental immunoglobulins were produced. Nyormoi *et al.* (1973) fused two human lymphoblastoid cell lines for unrelated reasons and did not look at immunoglobulin production; however, they later reported that some of the hybrids were double producers of κ and λ light chains (Rosen *et al.* 1977). These were important findings for the regulatory studies as they showed that within one species there was no allelic exclusion in the case of immunoglobulin production. A further important point established by the work of Schwaber and Cohen (1973) was not highlighted by these workers in this article, namely that they had immortalized a differentiated cell function by cell hybridization. That is, the human peripheral lymphocyte and its specific immunglobulin were immortalized in the hybrid (this could be similarly achieved by Epstein–Barr virus). Thus if we were scanning the literature for the first 'hybridoma', the hybrids produced by these workers would have to be a candidate. However, they did not go on to determine the specificity of the human immunoglobulin or to show its usefulness.

The stage was now set for the introduction of the monoclonal antibody technique. The paper of Kohler and Milstein (1975) reported two experiments, firstly the fusion of two different myelomas of the same species (mouse) and secondly the fusion of a mouse myeloma with mouse spleen cells and immortalization of a spleen cell which had responded to an antigen (red blood cells) when the donor mouse was immunized. The first experiment showed that no allelic exclusion occurred when myelomas of the same species were fused; this extended the findings from human cells (see above) to another species and to the myeloma cell type. This was an important point as allelic exclusion could well have operated with intra-specific hybrids. If it had, future development of the monoclonal antibody technique would have had to have fallen back on inter-specific hybrids. However, this work established an important theoretical point about allelic exclusion and provided the specific experimental prelude to the crucial experiment.

The dream that a cell line like a myeloma line might one day be produced which secretes limitless amounts of antibody of predetermined specificity seems to have surfaced in 1967. Cohn (1967) reported that none of the myelomas he induced produced antibody to the antigen with which the mice were heavily immunized. Around the same time a project was conceived by Mike Fried to work on when he arrived on a fellowship in Rod Porter's Laboratory. Mike Fried came as a postdoctoral fellow for 2 years to Porter's Laboratory in 1968 to work on trying to immortalize the specific immunoglobulin-producing cells from immunized mouse spleens. The first experiment was the fusion of a spleen from a mouse immunized with bacteriophage ϕX174 with a TK$^-$ L cell followed by selection of hybrids in HAT medium. No neutralization of ϕX174 plaques was seen and this was presumed to be because the differentiated function from the spleen was extinguished (see above). Mike Fried then tried myeloma cells but without selective markers and ultimately fused them with spleens from mice immunized with sheep red blood cells. He isolated some haemadsorption positive hybrids which he further worked on but did not publish the data (M. Fried, personal communication).

A completely unrelated line of research in fact yielded the first documented example of a 'hybridoma' by current definitions and practice in 1970 (for review see Sinkovics 1981; Wainwright 1992). It is not clear if this work influenced that of subsequent workers but little reference has been made to it. The work of Dr Sinkovics and colleagues was stimulated by the observation that in some patients with Burkitt's lymphoma, recurrence occurred after 4 years with tetraploid neoplastic cells. This recurrence was hypothesized to result from the fusion of the usual diploid Burkitt's cells and the immune spleen cells, the hybrid then being resistant to the host defence. Having come across a similar situation in mice the authors tested this hypothesis by co-cultivating of the diploid mouse lymphoma line with a splenic explant from a mouse immune to the lymphoma. From this culture grew tetraploid cells which grew invasively in the host (whereas the diploid lymphoma grew not invasively but slowly). Appropriate controls were included (Sinkovics *et al.* 1970*a*). These tetraploid cells were shown to produce immunoglobulins that neutralized leukaemia virus (Sinkovics *et al.* 1970*b*; Trujillo *et al.*

1970). At the time, the workers proposed not that the fused cells might serve as a source of antibodies (see Sinkovics 1981) but that the phenomenon might occur in lymphoproliferative diseases and be the origin of the Reed–Sternberg or the Sezary cell (see Sinkovics 1981). If the group's grant application for further work had been funded and not 'approved without funding' (Sinkovics 1982) we might have had the monoclonal antibody technique several years earlier!

The second experiment of Kohler and Milstein, i.e. fusion of mouse myeloma with mouse spleen cells, has had enormous repercussions throughout biology. The resulting ability to produce limitless amounts of antibody of predefined specificity offers obvious advantages which have been well used since 1975. Considering the number of spleen cells available for fusion there was some surprise that enough of the required specific hybrids were present and could be isolated. This was thought to be due to the fact that stimulated spleen cells must be very susceptible to cell fusion (Kohler and Milstein 1976).

In 1977 two papers showed the way for much of the future work by demonstrating that mixed antigens such as spleen cells with foreign histocompatibility antigens (Galfrè *et al.* 1977) or membranes from rat thymocytes (Williams *et al.* 1977) could be used and had the potential to study individual molecules to which antibodies were directed (Galfre *et al.* 1977) or to study differentiation antigens on the surface of white cells by such techniques as fluorescence-activated cell sorting (Williams *et al.* 1977).

Work in the MRC laboratory 1970–74

The work going on in César Milstein's laboratory in the early 1970s obviously had a great bearing on the later outcomes. The MOPC 21 myeloma had been in continuous spinner culture for some time and various experiments were in progress. In collaboration with George Brownlee and with Tim Harrison, César was characterizing, labelling, isolating, and sequencing mRNA from this cell line (Brownlee *et al.* 1973); Nick Cowan had set up a system for growing this cell line synchronously (Cowan and Milstein 1972) and Nye Svasti was sequencing the myeloma protein (Svasti and Milstein 1972). Kay Adetugbo arrived during my stay to sequence first the MOPC 21 heavy chain and later the variants isolated from the screening work. David Secher had started in September 1970 and was to sequence a human myeloma protein. During purification, heterogeneity had been observed and Secher and Milstein wondered if this heterogeneity might be related to generation of antibody diversity in the combining site. This led to attempts to clone the MOPC 21 line and also isoelectric focusing of the *in vitro* labelled protein products. The cloning was not achieved but the isoelectric focusing showed at least three major bands.

Because I wished to learn somatic cell genetics I chose to attempt to clone the myeloma cells and worked with David Secher on this program. A serendipitous error in calculation, resulting in the use of medium 10 per cent too concentrated, was found to allow the cells to form clones immediately whereas they would not do so in medium of 'correct' ionic strength. After ascertaining that the three bands visible after isoelectric focusing were not due to variants, I devised a scheme to screen for charge-change variants of the cells, looking for evidence that the mutations might cause antibody combining site diversity.

This involved taking myeloma clones with a Pasteur pipette, placing them on a piece of dialysis tubing which was placed on radioactive medium, incubating, and then placing the colonies (still on the dialysis tubing) at the origin of the isoelectric focusing gel. The Pasteur pipette contaminated with myeloma cells from the clone was used to inoculate specific wells of medium so that if a colony was interesting the clone could be studied again. We were able to screen hundreds of clones per week and by the time of our first publication (Cotton *et al.* 1973) had screened over 1000 clones and eventually 7000 were screened, without evidence that the type of mutation we were detecting caused antibody diversity. In the first months I also initiated cell fusion experiments stimulated by papers describing myeloma cell fusion appearing in 1970 and 1971 (Table 11.2). The aim was to examine if allelic exclusion occurred after fusion of two immunoglobulin-producing cells. This idea was given considerable impetus by the fact that I was readily able to obtain inactivated Sendai virus from Abraham Karpus in the same building. I aimed to fuse two mouse cell lines but being unable to isolate an AG-resistant MOPC 21 derivative I used a rat myeloma I made resistant to this drug and found no evidence of allelic exclusion in this hybrid (Cotton and Milstein 1973). After I left in May 1973 Shirley Howe continued under my remote supervision to attempt to achieve the mouse X mouse hybrid; around a year later it looked promising but it was George Kohler, who arrived in October 1974, who was able to isolate an AG-resistant derivative of MOPC 21 and then fuse it (Kohler and Milstein 1975). I also set out to look for variants of a myeloma with a known antigenic target so that I could look for variants in the binding site. I started with MOPC 108 producing an anti-dextran antibody but was unable to clone it for the requisite experiments. George Kohler produced his hybrid to obtain a line with known antigen to pursue this line of enquiry.

The outcome of the variant screening and hybrid experiments which I initiated were reported in the subsequent Cold Spring Harbor Meeting (Milstein *et al.* 1976). On my return to Australia I began to apply somatic cell genetic techniques to the phenylalanine hydroxylase system at the Royal Children's Hospital, Melbourne.

Conclusion

The story of this discovery illustrates three points in particular. The first is that most discoveries are built on at least decades of slow but sure increases in knowledge. However the work of Sinkovics illustrates that the monoclonal antibody technique could have arisen from careful clinical observation and then testing a hypothesis *in vitro* in an animal model. The second is that the monoclonal antibody technique was a technique whose time had come in 1975 but this is perhaps a corollary of the first point. The final and perhaps most important point for fund givers to note is that this important technique finally arose out of an environment where competent scientists were asking *basic* questions and developing tools to do so. This is amply illustrated by the outstanding discoveries from the MRC Laboratory of Molecular Biology and also the Sir William Dunn School of Pathology. This serves to underline the statement of Albert Einstein that 'One can organise to apply a discovery already made but not to make one' but perhaps the MRC *did* arrange *to* make discoveries by establishing a centre of excellence in basic science.

References

Anonymous (1969). Cell fusion: a new gift to biology. *Nature*, **223**, 1039–41.
Barski, G., Sorieul, S., and Cornefert, F. (1960). *Compt. Rend. Acad. Sci.*, **251**, 1825.
Bloom, H. D. and Nakamura, F. T. (1974). Establishment of a tetraploid immunoglobulin producing cell line from the hybridization of two human lymphocyte lines. *Proc. Natl Acad. Sci. USA*, **71**, 2689–92.
Brownlee, G. G., Cartwright, E. M., Cowan, N. J., Jarvis, J. M., and Milstein, C. (1973). Purification and sequence of messenger RNA for immunoglobulin light chains. *Nature New Biol.*, **224**, 236–240.
Coffino, P., Laskov, R., and Scharff, M. D. (1970). Immunoglobulin production: method for detecting and quantitating variant myeloma cells. *Science*, **167**, 186–88.
Coffino, P., Knowles, B., Nathenson, S. G., and Scharff, M. D. (1971). Suppression of immunoglobulin synthesis by cellular hybridization. *Nature New Biol.*, **231**, 87–90.
Cohn, M. (1967) Natural history of the myeloma. *Cold Spring Harbor Symp. Quant. Biol.*, **32**, 211–21.
Cotton R. G. H. and Milstein, C. (1973). Fusion of two imunoglobulin-producing myeloma cells. *Nature*, **224**, 42–3.
Cotton, R. G. H., Secher, D. S., and Milstein, C. (1973). Somatic mutation and the origin of antibody diversity. Clonal variability of the immunoglobulin produced by MOPC 21 cells in culture. *Eur. J. Immunol.*, **3**, 135–40.
Cowan, N. J. and Milstein, C. (1972). Automatic monitoring of biochemical parameters in tissue culture. *Biochem. J.*, **128**, 445–54.
Davis, F. M. and Adelberg, E. A. (1973). Use of somatic cell hybrids for analysis of the differentiated state. *Bacteriol. Rev.*, **37**, 197–214.
Enders, J. F. and Peebles, T. C. (1954). Propagation in tissue cultures of cytopathogenic agents from patients with measles. *Proc. Soc. Exp. Biol. Med.*, **86**, 277–86.
Galfrè, G., Howe, S. C., Milstein, C., Butcher, G. W., and Howard, J. C. (1977). Antibodies to major histocompatibility antigens produced by hybrid cell lines. *Nature*, **266**, 550–52.
Hakala, M. T. and Taylor, E. (1959). The ability of purine and thymine derivatives and of glycine to support the growth of mammalian cells in culture. *J. Biol. Chem.*, **234**, 126–8.
Harris, H. (1970) Cell Fusion, 1–3. *The Dunham Lectures*. Oxford University Press.
Harris, H. (1984). Cell fusion. *BioEssays*, **2**, 176–99.
Harris, H. (1987). *The balance of improbabilities*. Oxford University Press, Oxford.
Harris, H. and Watkins, J. F. (1965). Hybrid cells derived from mouse and man: artificial heterokaryons of mammalian cells from different species. *Nature*, **205**, 640–46.
Horibata, K. and Harris, A. W. (1970). Mouse myelomas and lymphomas in culture. *Exp. Cell Res.*, **60**, 61–77.
Howell, J. I. and Lucy, J. A. (1969). Cell fusion induced by lysolecithin. *FEBS Lett.*, **4**, 147–150.

Kao, K. N. and Michayluk, M. R. (1974). A method for high-frequency intergeneric fusion of plant protoplasts. *Planta (Berl.)*, **115**, 355–67.

Klein, E. and Wiener, F. (1971). Loss of surface-bound immunoglobulin in mouse A9 and human lymphoblast hybrid cells. *Exp. Cell Res.*, **67**, 251.

Kohler, G. and Milstein, C. (1975). Continuous cultures of fused cells secreting antibody of predefined specificity. *Nature*, **256**, 495–7.

Kohler, G. and Milstein, C. (1976). Derivation of specific antibody producing tissue culture and tumour cell lines by cell fusion. *Eur. J. Immunol.* **6**, 511–19.

Littlefield, J. W. (1964). Selection of hybrids from matings of fibroblasts *in vitro* and their presumed recombinants. *Science*, **145**, 709–10.

Merwin, R. M. and Algire, G. H. (1959). Induction of plasma cell neoplasma and fibrosarcomas in Balb/c mice carrying diffusion chambers. *Proc. Soc. Exp. Biol. Med.*, **101**, 437–9.

Milstein, C. (1980) Monoclonal antibodies. *Sci. Am.* **243**, 56–64.

Milstein, C. (1985). From antibody structure to immunological diversification of immune response. *EMBO J*, **4**, 1083–92.

Milstein, C. (1986a). From antibody structure to immunological diversification of immune response. *Science*, **231**, 1261–68.

Milstein, C. (1986b). Sir Frederick Gowland Hopkins Memorial Lecture. *Biochem. Soc. Bull.* **8**, 23–24.

Milstein, C. (1987). Inspiration from diversity in the immune system. *New Scientist*, 21 May, 54–58.

Milstein, C., Adetugbo, K., Cowan, N. J., Kohler, G., Secher, D. S., and Wilde, C. D. (1976). Somatic cell analysis of antibody secreting cells: studies of clonal diversification and analysis by cell fusion. *Cold Spring Harbour Symp. Quant. Biol.*, **41**, 793–803.

Mohit, B. (1971). Immunoglobulin G and free κ-chain synthesis in different clones of a hybrid cell line. *Proc. Natl Acad. Sci. USA*, **68**, 3045–48.

Mohit, B. and Fan, K. (1971). Hybrid cell line from a cloned immunoglobulin-producing mouse myeloma and a nonproducing mouse lymphoma. *Science*, **171**, 75–7.

Nossal, G. J. V. and Lederberg, J. (1958). Antibody production by single cells. *Nature*, **181**, 1419–20.

Nyormoi, O., Klein, G., Adams, A., and Dombos, L. (1973). Sensitivity to EBV virus super infection and IUdR inducibility of hybrid cells formed between a sensitive and a relatively resistant Burkitt lymphoma cell line. *Int. J. Cancer*, **12**, 396–408.

Okada, Y. (1958). The fusion of Ehrlich's tumour cells caused by HVJ virus *in vitro*. *Biken's J.*, **1**, 103–10.

Orkin, S. H., Buchanan, P. D., Yount, W. J., Resner, H., and Littlefield, J. W. (1973). Lambda chain production in human lymphoblast–mouse fibroblast hybrids. *Proc. Natl Acad. Sci. USA*, **70**, 2401–5.

Parkman, R., Hagemeier, A., and Merler, E. (1971). Production of a human gamma-globulin fragment by human thymocyte–mouse fibroblast hybrid cells. *Fed. Proc.*, **30**, 530.

Periman, R. (1970). IgG synthesis in hybrid cells from an antibody-producing mouse myeloma and an L cell substrain. *Nature*, **228**, 1086–7.

Pernis, B., Chippino, G., Kelus, A. S., and Gell, P. G. H. (1965). Cellular localization of immunoglobulins with different allotypic specificities in rabbit lymphoid tissues. *J. Exp. Med.*, **122**, 853–75.

Pettengill, O. S. and Sorenson, G. D. (1967). Murine myeloma cells in suspension culture. *Exp. Cell Res.*, **47**, 608–13.

Pluznik, D. H. and Sachs, L. (1965). The cloning of normal 'mast' cells in tissue culture. *J. Cell. Comp. Physiol.* **66**, 319–24.

Potter, M. and Boyce, C. R. (1962). Induction of plasma-cell neoplasmas in strain BALB/c mice with mineral oil and mineral oil adjuvents. *Nature*, **193**, 1086–7.

Potter, M. and Leon, M. A. (1968). Three IgA myeloma immunoglobulins from the Balb/c mouse: precipitation with pneumococcal C polysaccharide. *Science*, **162**, 369–71.

Puck, T. T. and Marcus, P. I. (1955). Cloning of mammalian cells. *Proc. Natl. Acad. Sci. USA*, **41**, 432–7.

Randle, P. and Rabbitts, T. H. (1986). Sir Frederick Gowland Hopkins Memorial Lecture. *Biochem. Soc. Bull.*, **8**, 23–4.

Rosen, A., Clements, G., and Klein, G. (1977). Double immunoglobulin production in cloned somatic cell hybrids between two human lymphoid cell lines. *Cell*, **11**, 139–47.

Sanders, F. K. and Burford, B. O. (1964). Ascities tumours from BAK.21 cells transformed in vitro by polyoma virus. *Nature*, **201**, 786–9.

Schubert, D., Jobe, J., and Cohn, M. (1968). Mouse myelomas producing precipitory antibody to nucleic acid bases and/or nitrophenyl derivatives. *Nature*, **220**, 882–5.

Schwaber, J. and Cohen, E. P. (1973). Human X mouse somatic cell hybrid clone secreting immunoglobulins of both parental types. *Nature*, **244**, 445–7.

Sinkovics, J. G. (1981). Early history of specific antibody-producing lymphocyte hybridomas. *Cancer Res.*, **41**, 1246–7.

Sinkovics, J. G. (1982). An interesting early observation concerning specific antibody-producing hybridomas. *J. Infectious Dis.*, **145**, 135.

Sinkovics, J. G., Drewinko, B., and Thornell, E. (1970a). Immunoresistant tetrapoloid lymphoma cells. *Lancet*, **i**, 139–40.

Sinkovics, J. G., Trjillo, J. M., Pienta, R. J., and Ahearn, M. J. (1970b). Leukemogenesis stemming from autoimmune disease In *Genetic concepts and neoplasia.* pp. 138–90. Williams and Wilkins, Baltimore, MD.

Sorieul, S. and Ephrussi, B. (1961). Karyological demonstration of hybridization of mammalian cells *in vitro. Nature*, **190**, 653–4.

Svasti, J. and Milstein, C. (1972). The complete amino acid sequence of mouse κ light chain. *Biochem. J.*, **128**, 427–444.

Szybalski, W. (1958). Resistance to 8-azaguanine, a selective genetic marker for a human cell line. *Microb. Genet. Bull.*, **16**, 30.

Szybalski, W. and Szybalski, E. H. (1962). Drug sensitivity as a genetic marker for human cell lines. *University of Michigan Med. Bull.* **28**, 277–93.

Trujillo, J. M., Ahearn, M. J., Pienta, R. J., Gott, C., and Sinkovics, J. G. (1970). Immunocompetence of leukemic murine lymphoblasts: ultrastructure, virus, and globulin production. *Cancer Res.*, **30**, 540–45.

Wade, N. (1982). Hybridomas: the making of a revolution. *Science*, **215**, 1073–5.

Wainwright, M. (1992) The Sinkovics hybridoma—the discovery of the first 'natural hybridoma'. *Perspect Biol. Med.*, **35**, 372–9.

Weiler, E. (1965). Differential activity of allelic gamma-globulin genes in antibody producing cells. *Proc. Natl. Acad. Sci. USA*, **54**, 1765–72.

Williams, A. F., Galfrè, G., and Milstein, C. (1977). Analysis of cell surfaces by xenogenic myeloma-hybrid antibodies: differentiation antigens of rat lymocytes. *Cell*, **12**, 663–73.

IV

Genetic suppression of tumour formation

12. The genetic analysis of human cancer

Eric J. Stanbridge

Another possibility is that in every normal cell there is a specific *arrangement for inhibiting*, which allows the process of division to begin only when the inhibition has been overcome by a special stimulus ... i.e. *definite chromosomes which inhibit division*.

Cells of tumours with unlimited growth would arise if those inhibiting chromosomes were eliminated.

One must also ... assume the existence of chromosomes which promote division.

Theodor Boveri (1914)

The field of cancer research has been replete with theories of causation that span over a century or more of study. Some of the more prevailing theories are presented in Table 12.1. Although at first glance they would seem to be rather disparate, virtually all of these theories have a common unifying theme—the involvement of genetic information (and damage thereof) in the genesis of cancer. This has included both exogenous genetic information—in the form of DNA and RNA tumour virus transforming genes—and endogenous cellular genes, consisting of oncogenes and tumour suppressor genes. It is the analysis of these latter genes that is the focus of this review.

The discovery of oncogenes and tumour suppressor genes

The identification of oncogenes—genes that are thought to promote cellular growth—and tumour suppressor genes, which are thought somehow to negatively regulate cellular growth, was accomplished by rather different experimental approaches. Although, in the chronology of discovery, the identification of the genetic basis of tumour suppression predated that of oncogenes (Harris *et al.* 1969), I will first discuss this latter category of genes.

The implication that cellular genes may play an important role in cancer came from the finding that the retroviral transforming gene, *v-scr*, has a cellular homologue, *c-src* (Stehelin *et al.* 1976). The key series of experiments confirming the existence of cellular oncogenes came from DNA transfection studies, whereby genomic DNA isolated from human cancer cells was transfected into morphologically normal (but aneuploid and immortal) mouse NIH3T3 cells. Discrete foci of morphologically transformed cells were observed which were subsequently found to be neoplastic. Following several rounds of serial transfections the ultimate

Table 12.1. Prevailing theories of cancer

T. Boveri (1914) Evolution of cancers is intimately associated with chromosomal aberrations and chromosomal 'imbalance'.

P. Rous (1911) and **L. Gross (1951)**. Noted that filterable agents—subsequently found to be RNA tumour viruses or retroviruses—cause leukaemias and solid tumours in chickens and rodents.

Eddy (1962), Huebner (1963), Burkitt (1962), zur Hausen (1982), and **Beasley (1981).** A variety of DNA tumour viruses—including SV40, polyoma, adenoviruses, herpes simplex viruses, human papilloma viruses, and hepatitis B virus—have been shown to induce tumours in a variety of mammalian species and/or have been implicated in human cancers.

J. Potts (1775), E. Miller (1949), and **R. Doll (1953)**. These investigators and many others recognized that a variety of environmental conditions—including exposure to chemical carcinogens and ionizing radiation, diet, and smoking—were directly related to an increased cancer incidence to exposed populations.

A. Haddow (1938). Speculated that DNA damage and somatic mutation may play critical roles in the induction of cancer, thereby implicating cellular genes.

R. Weinberg (1981) and **G. Cooper (1980).** Provided the first direct evidence that a class of cellular genes—termed oncogenes—were involved in promotion of cellular growth and neoplastic transformation.

H. Harris (1969) and **E. Stanbridge (1976).** Using the technique of somatic cell fusion these investigators provided the initial evidence for tumour suppression and its genetic basis in both rodent (Harris) and human (Stanbridge) cell systems—leading to the eventual identification and cloning of tumour suppressor genes.

---

transformed populations of 3T3 cells were found to contain discrete fragments of human genomic DNA (Krontiris and Cooper 1981; Shih *et al.* 1981). These fragments, when cloned, were found to contain the transforming activity and to share strong sequence similarity to *v-ras* genes present in the family of sarcoma retroviruses, e.g. Moloney sarcoma virus (MSV) (Der *et al.* 1982). Sequence analysis of the exons of the *c-ras* oncogenes showed that they differed in only a single nucleotide from the homologue that was cloned from DNA obtained from normal tissue (Santos *et al.* 1982). Thus, it was established that activated oncogenes with transforming potential were derived from normal proto-oncogene progenitors and that activation could occur as a consequence of a single point mutation.

These exciting findings were followed by an avalanche of reports confirming the original observation and also extending the number of candidate oncogenes to include other retroviral transforming gene members, e.g. *myc*, *myb*, etc. (Varmus 1984). Other oncogenes were identified using sequence identity strategies to clone and characterize various superfamilies of oncogenes. Many of these candidate oncogenes appear to encode proteins that are involved in signal cascades, ultimately resulting in stimulation of cell proliferation. These gene products include growth factors (e.g. *c-sis* or platelet-derived growth factor (PDGF)), growth factor receptors (e.g. epidermal growth factor receptor and PDGFR), G-proteins (e.g. *ras*), kinases (e.g. *c-src* and *raf*1), phosphatases, and transcription factors (e.g. *myc* and *myb*). Detailed biochemical characterization has provided evidence for an intricate networking among these factors, resulting in signal transduction cascades that culminate in mitogenesis. The implication of this is that any one member of a cascade, when acting aberrantly—as a consequence of mutation or inappropriate expression—may interfere with normal growth constraints (Cantley *et al.* 1991; Hunter 1991; Lewin 1991). It was, in fact, suggested that a single 'dominantly acting' oncogene could cause cancer. This is a naive postulate from an epidemiological point of view since it is clear that the vast majority of cancers arise as a consequence of multiple 'hits' or genetic events (Doll 1978). The erroneous claim that a single event of oncogene activation will induce cancer was based upon transformation of aneuploid NIH3T3 cells. Subsequent studies with primary rodent embryo fibroblasts indicated the need for two 'co-operating' oncogenes, e.g. *myc* and *ras*, to induce transformation (Parada *et al.* 1984). As we shall see later in this discussion, even this notion will require re-evaluation.

The claim of 'dominantly-acting' oncogenes would seem to leave little intellectual room for the involvement of other genetic elements in the genesis of cancer. However, studies predating the discovery of oncogenes by a decade or more clearly demonstrated the phenomenon of tumour suppression and its genetic basis (Harris *et al.* 1969). Tumour suppression was identified using the techniques of somatic cell fusion and involved the fusion of malignant cells with normal cells. These studies had a rather chequered history and led to considerable scepticism regarding the veracity of tumour suppression. Initial experiments in this area involved fusions between rodent cells. When Barski and Cornefert (1962) fused mouse cells of high and low malignant potential they noted that the resulting hybrids were highly malignant, a finding in keeping with the notion of 'dominantly acting' oncogenes. However, Harris and colleagues, in a more extensive and critical series of experiments, found that tumorigenicity was suppressed (Harris *et al.* 1969; Harris 1971). The discrepancy between the two sets of experimental findings was due to the extreme chromosomal instability of intraspecific rodent hybrid cells and required careful monitoring of the chromosomal status of the resultant hybrids. This problem plagued the field for some years and acceptance of the phenomenon of tumour suppression and its genetic basis was slow in coming.

Nevertheless, Harris and colleagues made a number of seminal observations and conclusions. For example: 'If our results are to be expressed in Mendelian terms, then one would say that malignancy behaves as if it were a recessive character' (Wiener *et al.* 1971); 'We propose that the malignant phenotype results from some stable genetic ... loss' (Wiener *et al.* 1974). As we shall see these predictions have been vindicated.

The problem of chromosome instability was eventually overcome by using intraspecies human cell hybrids, in which there is stable retention of chromosomes from both parental origins (Stanbridge 1976; Stanbridge *et al.* 1982). The stability of chromosome retention also allowed for the identification of specific chromosomes associated with tumour-suppressing activity (Stanbridge *et al.* 1981), again providing evidence for a genetic basis to this phenomenon.

In most instances, in both rodent and human cell systems, there was a clear separation between control of expression of the transformed versus tumorigenic phenotypes (Straus *et al.* 1976; Stanbridge and Wilkinson 1978). Generally speaking, the non-tumorigenic hybrids retained expression of most phenotypic traits of transformation encountered *in vitro*, including growth in low concentrations of serum and colony formation in soft agar. Although these studies would argue for dominance of the transformed state, later experiments indicated that the phenotype of cellular senescence is also dominant in many situations (Pereira-Smith and Smith 1988). Thus, transformed, non-tumorigenic hybrid cells have presumably lost the genetic information required to orchestrate the programme of cellular senescence and further genetic (chromosomal) loss is required to abrogate tumour-suppressing activity and allow for the emergence of tumorigenic segregants.

Because of these constraints, i.e. tumour suppressor genes and growth suppressor genes introduced into cancer cells lead to either no change in the *transformed* phenotype or to complete cessation of proliferation, functional assays for tumour suppressor genes (analogous to the 3T3 transformation assay for oncogenes) have not been forthcoming. Thus, molecular analyses of these genetic elements have lagged far behind those of oncogenes and again led to a rather tardy acceptance of their existence. This attitude has changed remarkably in the last few years.

The experimental observations of somatic cell hybrids clearly indicated that cancer cells have suffered genetic defect(s), and that presumably correction of those defect(s) via introduction of normal genetic information (via somatic cell fusion) leads to restoration of normal cellular growth control. This notion of loss of genetic information associated with malignant behaviour had been implicit for many years on the basis of cytogenetic analysis of non-random chromosome alterations—including deletions—associated with many human cancers (Heim and Mitelman 1989).

The first molecular evidence for genetic deletions associated with human cancers came from allelotyping, using the powerful technology of restriction fragment length polymorphism (RFLP) analysis. Taking advantage of the regional localization of chromosome deletions identified by cytogenetic analysis, Cavenee and colleagues (1983) provided clear evidence for loss of heterozygosity (LOH) in the 13q14 region in retinoblastoma. This seminal finding has led to a subfield of scientific enquiry that has identified dozens of candidate loci exhibiting LOH in many human cancers. In a significant proportion of cancers multiple discrete chromosome regions show LOH, indicating the loss of function of multiple tumour suppressor genes in a single given tumour. The implications of these findings will be addressed later.

A critical point to be made here is that RFLP analyses that identify LOH at discrete chromosomal regions reveal the existence of *candidate* tumour suppressor gene loci only. Confirmation requires functional evidence, which ultimately must include cloning and characterization of candidate tumour suppressor genes.

As I have mentioned above, functional assays are not available to assist in the cloning of tumour suppressor genes. The technology that has been used successfully is

Table 12.2. Candidate human tumour suppressor genes

Tumour suppressor gene	Cancer(s) involved	Chromosome location	Candidate function
RB	Retinoblastoma, osteosarcoma, bladder and prostate carcinoma (and others)	13q14	Complexes with cellular proteins; involved in cell cycle control
p53	Many	17p13	Transcriptional factor; involved in induction of apoptosis
WT1	Wilms' tumour	11p13	Transcriptional factor
NF-1	Neurofibroma	17q11	GTPase activating protein (GAP)
DCC	Colon carcinoma	18q21	Cell adhesion; communication
MCC	Colon carcinoma	5q21	G protein activation
APC	Colon carcinoma	5q21	Cell adhesion; communication
PTP	Lung carcinoma	3p21	Protein tyrosine phosphatase
Krev-1	None identified	1p12–p13	Unknown

positional cloning, utilizing chromosome walking techniques. The region investigated is generally preordained by RFLP/LOH analysis. The first tumour suppressor gene to be cloned in this fashion was the retinoblastoma (RB) gene (Friend *et al.* 1986). This is a particularly arduous procedure, usually involving several years of invested time. Thus, relatively few candidate tumour suppressor genes have been cloned to date. However, as noted in Table 12.2, the functions of these candidate tumour suppressor gene products—deduced from sequence analysis—are remarkably similar to those of oncogene products. They include transcription factors in particular (p53, pRb, and WT-1), cell attachment molecules (DCC), G-protein like molecules (K*rev*-1), and GAP molecules (NF-1/neurofibromin) (Marshall 1991).

Functional evidence for oncogenes and tumour suppressor genes

As information concerning the functional characteristics of oncogenes and tumour suppressor genes accumulates, several important questions arise. Perhaps the most compelling issue is that of 'dominantly acting' oncogenes. If such genes do, indeed, function in a dominant fashion then one must question how tumour suppressor genes would also be dominantly acting. Let us first deal with the oncogene issue.

The notion of 'dominantly acting' oncogenes has arisen from studies of their transforming activity in rodent cells. Single activated oncogenes, e.g. *myc* and *ras*, readily transform aneuploid NIH3T3 cells, whereas co-operation between *myc* and *ras* is required for neoplastic transformation of primary cultures of rodent fibroblasts. In both situations the oncogene products would seem to override normal cellular controls. However, the picture is actually not so straightforward. When one examines the karyotypes of the neoplastically transformed cells it is seen that significant chromosomal rearrangements have occurred—an indication of further genetic alterations. Studies with transgenic mice also reflect a similar complexity. Transgenic mice bearing oncogenes whose expression is driven from tissue-specific promoters have a high incidence of tumours in that tissue. However, the tumours are monoclonal although all cells within the tissue express the transgene (Hanahan 1988). Presumably other genetic alterations are necessary in order for the committed cell to progress to its final neoplastic state.

Studies with human cells raise the same issues. Whereas the combination of *myc* and *ras* is sufficient to obtain reproducible transformation of primary rodent cells, transfection of the same combination of oncogenes into normal human diploid cells has no transforming effect and the cells eventually senesce (Sager 1984; Stanbridge 1989). Thus, the expression of these activated cellular oncogenes is not sufficient to induce immortality or neoplastic transformation of human cells. In an attempt to produce a phenotypic effect following the introduction of an activated oncogene, Boukamp and colleagues (1990) transfected an activated c-Ha-*ras* oncogene into immortalized, non-tumorigenic human keratinocytes (HaCaT cell line). HaCaT cell clones expressing the c-Ha-*ras* gene formed benign papillomas or, in some instances, locally invasive squamous cell carcinomas when they were implanted in immune deficient nude mice. It was concluded that expression of the c-Ha-*ras* oncogene allowed the cells a *limited* proliferative advantage *in vivo*. This limited degree of cell cycling—which may be aberrant—sets up conditions for further genetic alterations, e.g. chromosome loss via non-disjunction, that may then predispose the cell to neoplastic progression. Such changes could be envisaged to include activation of other oncogenes and/or inactivation of tumour suppressor genes. What is clear, however, is that expression of an activated oncogene is not sufficient to endow an immortalized cell with neoplastic properties.

If expression of an activated oncogene in some way precipitates other genetic changes which are critical for neoplastic progression, then one must ask whether continued expression of the activated oncogene is required for *maintenance* of the malignant phenotype. We have recently obtained evidence that continued expression is not required. The human fibrosarcoma cell line, HT1080, is pseudodiploid and heterozygous for an activated N-*ras* oncogene (Paterson *et al.* 1987). We have managed to delete the activated allele of N-*ras* leaving the normal allele intact. Although the N-*ras*^{del} cell grow more slowly in culture they continue to form tumours in nude mice (M. J. Anderson and E. J. Stanbridge, unpublished observation). These results are at variance with those of Paterson and colleagues (1987) who reported that expression of activated N-*ras* is directly responsible for the neoplastic behaviour of HT1080.

Loss of function of tumour suppressor genes in human cancer and therefore their importance in negatively controlling malignant behaviour, is a little more clear cut. Following the cloning of the RB gene it was quickly shown that retinoblastomas have suffered inactivation or loss of both alleles of the RB locus via a variety of mechanisms (Weinberg 1992). These findings provided confirmation of Knudson's now classic two-hit theory of retinoblastoma, where he proposed that all types of retinoblastoma involve two separate mutations (Knudson 1971). In sporadic retinoblastoma both mutations occur somatically in the same retinal precursor cell, whereas in heritable retinoblastoma one of the mutations is germinal and the second somatic. It was further shown that when wild-type RB cDNA is transfected into retinoblastoma cells and into other cancer cells that lack a functional RB gene, it exerts a strong growth suppressing effect *in vitro* and tumour-suppressing effect *in vivo* (Huang *et al*. 1988).

The biochemical characterization and functional activities of RB are addressed elsewhere in this symposium. Suffice it to say that RB is intimately involved in control of the cell cycle (Dyson and Harlow 1992). It probably also functions as a transcription factor in a complex with other cellular proteins (Livingston 1992) and may be involved in commitment to differentiation. This latter point is of considerable interest since it has been shown that the biological mechanism of tumour suppression in somatic cell hybrids is the induction of terminal differentiation (Peehl and Stanbridge 1982; Harris 1990).

The only other tumour suppressor gene that has unequivocally been shown to possess tumour-suppressing properties is the p53 gene. A detailed description of this gene and its biochemistry is beyond the scope of this review and is dealt with elsewhere in this symposium. However, several properties are germane to this discussion. p53 has had a chequered history. It was first identified as an oncogene because the initial cloned p53 cDNA was shown to co-operate with *ras* in a cell transformation assay (Parada *et al*. 1984). It was subsequently found that this was a mutant p53, having suffered a point mutation. When wild-type p53 cDNA was eventually cloned and tested it was found to function as a potent growth suppressor gene, both in rodent and human cells (Finlay *et al*. 1989; Baker *et al*. 1990). An obvious conclusion to be drawn is that the point mutation destroyed the tumour suppressor activity of p53 which also then acquired a new oncogene function (Levine 1992). An alternative explanation is that it represents a 'dominant-negative' mutation (Herskowitz 1987). It has been deduced that p53 is a transcriptional factor which functions as a dimer or oligomer. It has been postulated that when a mutant p53 polypeptide complexes with a wild-type p53 molecule it, in some way, neutralizes the normal function of the p53. There is some evidence to support this notion in rodent cells (Levine 1992) but not in human cells (Chen *et al*. 1990; Goyette *et al*. 1992).

p53 is coming under increasing scrutiny because deletions and mutations in this gene are encountered at high frequencies in a large variety of human cancers

(Hollstein et al. 1991). It would thus seem to serve as a common checkpoint in cell cycle control and its loss of function contributes significantly to neoplastic progression. To a lesser extent RB dysfunction is also seen in several human cancers, including virtually all retinoblastomas.

In the vast majority of cancers where p53 alterations are found, both alleles are involved. Loss of function again occurs via a variety of mechanisms—most commonly a combination of allele loss and point mutations. The mutations can be found throughout the gene, but are also clustered in several hot-spots spanning exons 5–8 (Levine 1992). Restoration of wild-type p53 function, via cDNA transfection, in cancer cells that are null or mutant for p53 results in abrupt cessation of growth and apoptosis (Baker et al. 1990; Mercer et al. 1990; Casey et al. 1991).

Rather remarkably, although a number of candidate tumour suppressor genes have been cloned (see Table 12.2), functional evidence is forthcoming only for p53 and RB. Other experimental procedures have been used to provide functional evidence for tumour suppressor genes in the absence of the cloned gene. In particular, monochromosome transfer, via microcell fusion, has been used successfully to identify tumour suppressor activity associated with specific chromosomes. These studies are summarized in Table 12.3. In the majority of cases tumour suppressor activity, predicted from RFLP allelotyping and assigned to a particular chromosome, was indeed associated with the chromosome in question (reviewed in Stanbridge 1992). In addition, novel candidate tumour suppressor loci were uncovered.

Table 12.3. Growth suppression *in vitro* or tumour suppression *in vivo* associated with the transfer of single human chromosomes via Microcell fusion

Tumour cell line	Tumour suppressing chromosome		Candidate tumour suppressor gene
	Expected	Observed	
HeLa (cervical carcinoma)	11	11, 4*[a]	Not identified
SiHa (cervical carcinoma)	–	11	Not identified
Retinoblastoma, osteosarcoma, prostate carcinoma, bladder carcinoma	13	13**[b]	RB
Renal cell carcinoma	3	3	Not identified
Wilms' tumour	11p13	11p15	WT2
Colon carcinoma	5, 17, 18	5, 17*[a], 18	FAP/APC, p53, DCC
Endometrial carcinoma	–	1, 6, 9	Not identified
Melanoma	6	6	Not identified
Neuroblastoma (NGP)	1	17	Not identified (not p53)
HT1080 (fibrosarcoma)	1	1, 11, 17	p53 (17)

[a]Growth-suppressing effect *in vitro*.
[b]Growth-suppressing effect *in vitro* and tumour-suppressing effect *in vivo*.
Details of these experiments are discussed in Stanbridge (1992).

I wish to discuss only one example here because it illustrates several important points that have arisen during this discourse. As indicated earlier, certain human cancers seem to accumulate LOH—and by inference loss of tumour suppressor gene activity—at multiple sites, implicating loss of function of multiple tumour suppressor genes during neoplastic progression. The most intensely studied cancer in this regard is colorectal carcinoma (see Fig. 12.1). Discrete stages of progression, exemplified by hyperplastic epithelium → benign adenomas → carcinoma *in situ* → metastasis, have been documented extensively by pathologists. Molecular analyses of these discrete stages have identified both activation of oncogenes (usually Ki-*ras*) and inactivation of tumour suppressor genes (Fearon and Vogelstein 1990). Three candidate tumour suppressor loci have been identified on chromosomes 5q21, 17p13, and 18q21, respectively. Candidate tumour suppressor genes have been cloned from these regions; namely APC/FAP (5q21), DCC (18q21), and p53 (17p13) (Fearon *et al.* 1990; Groden *et al.* 1991; Kinzler *et al.* 1991). Only p53 cDNA has unequivocally been shown to possess tumour-suppressing activity.

Thus, a given colorectal carcinoma may have accumulated a series of genetic alterations, including activation of the Ki-*ras* oncogene and inactivation of multiple tumour suppressor genes. An important question, therefore, is whether one is able to reverse the malignant behaviour of the carcinoma cells by correcting any single defect, e.g. by introduction of wild-type p53, or is it necessary to correct some combination, or even the full spectrum of genetic defects? In the case of the tumour suppressor genes the rather surprising and exciting answer is that correction of any single defect, using a combination of chromosome transfer and cDNA transfection, restores cell growth control *in vivo* (Tanaka *et al.* 1991; Goyette *et al.* 1992). A hierarchy of efficacy is seen, where p53 again acts as a potent inducer of apoptosis *in vitro*. Both chromosome 5 (APC/FAP) and chromosome 18 (DCC) transfer have little effect on *in vitro* cell proliferation but exhibit dramatic tumour-suppressing effects *in vivo* when the cells are implanted into nude mice.

Although inactivation of the Ki-*ras* oncogene has not been attained, our experience with HT1080 fibrosarcoma and N-*ras* deletion would suggest that it would have little effect. In that regard it has previously been shown that suppression of the tumorigenic phenotype via somatic cell fusion occurs despite continued expression of activated oncogenes (Geiser *et al.* 1986).

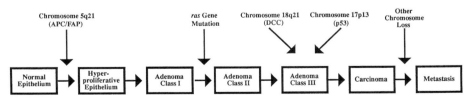

Fig. 12.1 A model for neoplastic progression in colon carcinoma. Tumorigenesis proceeds via a series of genetic alterations, including activation of a *ras* oncogene and loss of function of several tumour suppressor genes on chromosomes 5 (FAP/APC), 17 (p53), and 18 (DCC).

Functional significance of oncogenes and tumour suppressor genes in human cancer

In Table 12.4 I have summarized those properties of oncogenes and tumour suppressor genes that I consider germane to their functional relevance in cancer progression.

Activated oncogenes are found in approximately 20 per cent of human cancers—a figure that is probably conservative. Although more than 100 'oncogenes' have been identified, usually by sequence similarity, very few seem to play a functional role in human cancer; *ras*, *myc*, and *myb* are most frequently encountered. Thus, it is safe to assume that oncogenes play a role in cancer induction. The real question is, what is that role? I believe that the extrapolation of findings in experimental rodent systems to the human condition has been misguided and misleading, particularly where those extrapolations have been undertaken without obtaining supporting experimental data using human materials. What we do know from studies of oncogene effects on human cells is that activated oncogene expression is not sufficient to induce immortalization or neoplastic transformation and may not be necessary to maintain the neoplastic properties of cancer cells. At the moment we are left with the speculation that oncogene activation may play a rather auxiliary role by providing conditions (e.g. aberrant cell cycling) that allow for increased frequencies of genetic alterations which incrementally are critical for neoplastic progression.

Table 12.4. Functional significance of oncogenes and tumour suppressor genes in human cancer

Oncogenes	Tumour suppressor (TS) genes
1. Activated oncogenes (predominantly *ras* and *myc*) have been found in approximately 20 per cent of cancers.	1. Alterations (loss of function) observed in the majority of cancers.
2. The expression of activated oncogenes is *not sufficient* to endow a cell with malignant properties	2. Restoration of defective TS gene functions in cancer cells results in tumour suppression.
3. The expression of an activated oncogene is *not necessary* for maintenance of malignant behaviour	3. Germ-line mutations in TS genes have been identified in a number of familial cancers: retinoblastoma (RB), Li–Fraumeni (p53), Denys–Drash (WT1), familial adenomatous polyposis (FAP/APC), and neurofibromatosis (NF-1).
4. No germ-line mutations in oncogenes have been found in familial cancers.	

It is likely that for tumour suppressor genes constitute critical checkpoints in neoplastic progression. They seem to be recessive genes, i.e. inactivation of both alleles is required for loss of function (this may not always be the case: cf. NF-1 in neurofibromatosis and neurofibrosarcomas). Loss of function of tumour suppressor genes, most notably p53, is found in the majority of cancers and at very high frequencies in particular cancers, e.g. colorectal carcinomas. Restoration of normal function of tumour suppressor genes in cancer cells with defects in those genes leads to restoration of normal growth control. In this regard the field of tumour suppressor gene enquiry has probably benefited from the fact that the majority of studies have utilized human cells (Stanbridge 1989).

From these points it can be seen that I consider that inactivation of tumour suppressor genes are critical events in tumour progression. Perhaps the most compelling argument is that in familial cancer pedigrees, exhibiting high frequencies of early onset cancers, none have been found to contain germ-line mutations of oncogenes whereas several have been shown to carry germ-line mutations or deletions of tumour suppressor genes, e.g. Li–Fraumeni syndrome and p53 (Malkin *et al.* 1990).

Having made the case for tumour suppressor genes let me hasten to say that we are still at the tip of the iceberg with respect to our knowledge of the genetic basis of cancer. One illustration will suffice to bring this point home. Recently transgenic 'knock-out' mice have been developed that have both alleles of the p53 gene inactivated. The mice develop normally, are disease-free for several months and then exhibit a high incidence of tumours that are monoclonal in origin (Donehower 1992). RB 'knock-out' mice have also been developed. In this case the homozygous deletion is lethal (W. H. Lee, personal communication). We still have much to learn.

References

Baker S. J., Markowitz, S., Fearon, E. R., Wilson, J. K. V., and Vogelstein, B. (1990). Suppression of human colorectal carcinoma cell growth by wild-type p53. *Science*, **249**, 912–15.

Barski, G. and Cornefert, F. (1962). Characteristics of 'hybrid'-type clonal cell lines obtained from mixed cultures *in vitro*. *J. Natl Cancer Inst.*, **28**, 801–21.

Beasley R. P., Whang, L.-Y., Lin, C.-C., and Chien, C.-S. (1981). Hepatitis B and liver cancer. *Lancet* **ii**, 1129–33.

Boukamp, P., Stanbridge, E. J., Foo, D. Y., Cerutti, P., and Fusenig, N. (1990). Lack of correlation between activated c-Ha-*ras* oncogene expression and malignant transformation or altered differentiation potential of the human HaCaT keratinocyte cell line. *Cancer Res.*, **50**, 2840–47.

Boveri, T. (1914) *Zur Frage der Erstehung Maligner Tumoren.* Fischer, Jena. English translation published in 1929 by Williams and Wilkins (Baltimore, MD).

Burkitt, D. (1962). A children's cancer dependent on climatic factors. *Nature*, **194**, 232–4.

Cantley, L. C., Auger, K. R., Carpenter, C., Duckworth, B., Graziani, A., Kapeller, R., and Soltoff, S. (1991). Oncogenes and signal transduction. *Cell*, **64**, 281–302.

Casey, G., Lo-Hsueh, M., Vogelstein, B., and Stanbridge, E. J. (1991). Growth suppression of human breasts cancer cells by the introduction of a wild-type p53 gene. *Oncogene*, **6**, 1791–7.

Cavenee, W. K., Dryja, T. P., Phillips, R. A., Benedict, W. F., Godbout, R., Gallie, B. L., et al. (1983). Expression of recessive alleles by chromosomal mechanisms in retinoblastoma. *Nature*, **305**, 779–84.

Chen, P.-L., Chen, Y., Bookstein, R., and Lee, W.-H. (1990). Genetic mechanisms of tumour suppression by the human p53 gene. *Science*, **250**, 1576–9.

Cooper, G. M., Okenquist, S., and Silverman, L. (1980). Transforming activity of DNA of chemically transformed and normal cells. *Nature*, **284**, 481–21.

Der, C. J., Krontiris, T. G., and Cooper, G. M. (1982). Transforming genes of human bladder and lung carcinoma cell lines are homologous to the ras genes of Harvey and Kirsten sarcoma viruses. *Proc. Natl Acad. Sci. USA*, **79**, 3637–40.

Doll, R. (1953). Bronchial carcinoma: incidence and aetiology. *Br. Med. J.*, **2**, 521–7, 585–90.

Doll, R. (1978). An epidemiological perspective of the biology of cancer. *Cancer Res.*, **38**, 3573–83.

Donehower, L. A., Harvey, M., Slagle, B. L., McArthur, M. J., Montgomery C. A., Jr, Butel, J. S. and Bradley, A. (1992). Mice deficient for p53 are developmentally normal but susceptible to spontaneous tumours. *Nature*, **356**, 215–21.

Dyson, N. and Harlow, E. (1992). Adenovirus E1A targets key regulators of cell proliferation. *Cancer Survs*, **12**, 161–96.

Eddy B. E. (1962). Tumors produced in hamsters by SV_{40}. *Fed. Proc.* **21**, 930–5.

Fearon, E. R. and Vogelstein, B. (1990). A genetic model for colorectal tumorigenesis. *Cell*, **61**, 759–67.

Fearon, E. R., Cho, K. R., and Nigro, J. M., Kern, S. E., Simons, J. W., Rupperl, J. M., et al. (1990). Identification of a chromosome 18q gene that is altered in colorectal cancers. *Science*, **247**, 49–56.

Finlay, C. A., Hinds, P. W., and Levine, A. J. (1989). The p53 proto-oncogene can act as a suppressor of transformation. *Cell*, **57**, 1083–93.

Friend, S. H., Bernards, R., Rogelj, S., Weinberg, R. A., Rapaport, J. M., Albert, D. M., and Dryja, T. P. (1986). A human DNA segment with properties of the gene that predisposes to retinoblastoma and osteosarcoma. *Nature*, **323**, 643–6.

Geiser, A. G., Der, C. J., Marshall, C. J., and Stanbridge E. J. (1986). Suppression of tumorigenicity with continued expression of the c-Ha-*ras* oncogene in EJ bladder carcinoma–human fibroblast hybrid cells. *Proc. Natl Acad. Sci. USA*, **83**, 5209–13.

Goyette, M. C., Cho, K., Fasching, C. L., Levy, D. B., Kinzler, K. W., Paraskeva, C., et al. (1992). Colorectal cancer: progression is associated with multiple tumour suppressor gene defects but reversion to the nontumorigenic state is accomplished by correction of any single defect via chromosome transfer. *Mol. Cell. Biol.*, **12**, 1387–95.

Groden, J., Thliveris, A., Samowitz, W., Carlson, M., Gelbert, L., Albertson, H., et al. (1991). Identification and characterization of the familial adenomatous polyposis coli gene. *Cell*, **66**, 589–600.

Gross, L. (1951). 'Spontaneous' leukemia developing in C3H mice following inoculation in infancy with AK-leukemic extracts or AK-embryos. *Proc. Soc. Exp. Biol. Med.* **7**, 27–32.

Haddow A. (1938). The influence of carcinogenic substances on sarcomata induced by same and other compounds. *J. Pathol. Bacteriol.*, **37**, 581–91.

Hanahan, D. (1988). Dissecting multistep tumorigenesis in transgenic mice. *Annu. Rev. Genet.*, **22**, 479–519.

Harris, H. (1971). Cell fusion and the analysis of malignancy: the Croonian lecture. *Proc. Roy. Soc. Lond., B*, **179**, 1–20.

Harris, H. (1990). The role of differentiation in the suppression of malignancy. *J. Cell Sci.*, **97**, 5–10.

Harris, H., Miller, O. J., Klein, G., Worst, P., and Tachibana, T. (1969). Suppression of malignancy by cell fusion. *Nature*, **223**, 363–8.

Heim, S., and Mitelman, F. (1989). Primary chromosome abnormalities in human neoplasia. *Adv. Cancer Res.*, **52**, 2–44.

Herskowitz, I. (1987). Functional inactivation of genes by dominant negative mutations. *Nature*, **329**, 219–22.

Hollstein, M., Sidransky, D., Vogelstein, B., and Harris, C. C. (1991) p53 mutations in human cancers. *Science*, **253**, 49–53.

Huang, H.-J. S., Yee, J. K., Shew, J.-Y., Chen, P.-L., Bookstein, R., Friedman, T., et al. (1988). Suppression of the neoplastic phenotype by replacement of the RB gene in human cancer cells. *Science*, **242**, 1563–66.

Huebner, R. J., Rowe, W. P., Turner, H. C., and Lane, W.T. (1963). Specific adenovirus complement-fixing antigens in virus-free hamster and rat tumors. *Proc. Natl Acad. Sci. USA*, **50**, 379–84.

Hunter, T. (1991). Cooperation between oncogenes. *Cell*, **64**, 249–70.

Kinzler, K. W., Nilbert, M. C., Su, L.-K, Vogelstein, B., Bryan, T. M., Levy, D. B., et al. (1991). Identification of FAP locus genes from chromosome 5q21. *Science*, **253**, 661–5.

Knudson, A. G., Jr (1971). Mutation and cancer: statistical study of retinoblastoma. *Proc. Natl Acad. Sci, USA*, **68**, 820–23.

Krontiris, T. G. and Cooper, G. M. (1981). Transforming activity of human tumour DNAs. *Proc. Natl Acad. Sci. USA*, **78**, 1181–4.

Levine, A. J. (1992). The p53 tumour suppressor gene and product. *Cancer Survs.*, **12**, 59–80.

Lewin, B. (1991). Oncogenic conversion by regulatory changes in transcription factors. *Cell*, **64**, 303–12.

Livingston, D. M. (1992). Functional analysis of the retinoblastoma gene product and of RB–SV40 T antigen complexes. *Cancer Survs*, **12**, 153–60.

Malkin, D., Li, F., Strong, L. C., Fraumeni, J. F., Jr, Nelson, C. E., Kim, D. H., et al. (1990). Germ line p53 mutations in a familial syndrome of breast cancer, sarcomas, and other neoplasms. *Science*, **250**, 1233–8.

Marshall, C. J. (1991). Tumor suppressor genes. *Cell*, **64**, 313–26.

Mercer, W. E., Shields, M. T., Amin, M., Sauve, G. J., Appella, E., Romano, J. W., et al. (1990). Negative growth regulation in a glioblastoma tumor cell line that conditionally expresses human wild-type p53. *Proc. Natl Acad. Sci. USA*, **87**, 6166–70.

Miller, J. A., Sapp, R. W., and Miller, E. C. (1949). Carcinogenic activities of certain halogen derivatives of 4-dimethylaminoazo-benzene in the rat. *Cancer Res.*, **9**, 652–60.

Parada, L. F., Land, H., Weinberg, R. A., Wolf, D., and Rotter, W. (1984). Cooperation between gene encoding p53 tumour antigen and *ras* in cellular transformation. *Nature*, **312**, 649–51.

Paterson, H., Reeves, B., Brown, R., Hall, A., Furth, M., Bos, J., et al. (1987). Activated N-*ras* controls the transformed phenotype of HT1080 human fibrosarcoma cells. *Cell*, *51*, 803–12.

Peehl, D. M., and Stanbridge, E. J. (1982). The role of differentiation in the control of tumorigenic expression in human cell hybrids. *Int. J. Cancer*, **30**, 113–20.

Pereira-Smith, O. M., and Smith, J. R. (1988). Genetic analysis of indefinite division in human cells: identification of four complementation groups. *Proc. Natl Acad. Sci, USA*, **85**, 6042–6.

Potts, P. (1775). *Chirugical observations relative to the cataract, polypus of the nose, cancer of the scrotum, the different kinds of ruptures, and the mortification of toes and feet.* London (1963). A facsimile of the relevant section of this work may be found in the *Natl Cancer Inst. Monogr.*, **10**, 7–13.

Rous, P. (1911). Transmission of a malignant new growth by means of cell-free filtrate. *JAMA*, **56**, 198.

Sager, R. (1984). Resistance of human cells to oncogenic transformation. *Cancer Cells*, **2**, 487–93.

Santos, E. G., Tronick, S. R., Aaronson, S. A., Pulciani, S., and Barbacid, M. (1982). T24 human bladder carcinoma oncogene is an activated form of the normal human homologue of Balb- and Harvey-MSV transforming genes. *Nature*, **198**, 343.

Shih, C., Padhy, L. C., Murray, M., and Weinberg, R. A. (1981). Transforming genes of carcinomas and neuroblastomas introduced into mouse fibroblasts. *Nature*, **290**, 261–4.

Stanbridge, E. J. (1976). Suppression of malignancy in human cells. *Nature*, **260**, 17–20.

Stanbridge, E. J. (1989). An argument for using human cells in the study of the molecular genetic basis of human cancer. In *Cell transformation and radiation-induced cancer*, (ed. K. Chadwick, C. Seymour, and B. Barnhart), pp. 1–10. Adam Hilger, New York.

Stanbridge, E. J. (1992). Functional evidence for human tumour suppressor genes: chromosome and molecular genetic studies. *Cancer Survs.*, **12**, 5–24.

Stanbridge, E. J., and Wilkinson, J. (1978). Analysis of malignancy in human cells: malignant and transformed phenotypes are under separate genetic control. *Proc. Natl Acad. Sci. USA*, **75**, 1466–9.

Stanbridge, E. J., Flandermeyer, R. R., Daniels, D. W., and Nelson-Rees, W. A. (1981). Specific chromosome loss associated with the expression of tumorigenicity in human cell hybrids. *Somat. Cell Mol. Genetic.*, **7**, 699–712.

Stanbridge, E. J., Der., Doersen, C. J., Nishimi, R. Y. Peehl D. M. Weissman, B. E., *et al.* (1982) Human cell hybrids: analysis of transformation and tumorigenicity. *Science*, **215**, 252–9

Stehelin, D., Varmus, H. E., Bishop, J. M., and Vogt, P. K. (1976). DNA related to the transforming gene(s) of avian sarcoma viruses is present in normal avian DNA. *Nature*, **260**, 170–73.

Straus, D. S., Jonasson, J., and Harris, H. (1976). Growth *in vitro* of tumour and cell x fibroblast hybrids in which malignancy is suppressed. *J. Cell Sci.*, **25**, 73–86.

Tanaka, K., Oshimura, M., Kikuchi, R., Seki, M., Hayashi, T., and Miyaki, M. (1991). Suppression of tumorigenicity in human colon carcinoma cells by introduction of normal chromosome 5 or 18. *Nature*, **349**, 340–2.

Varmus, H. E. (1984). The molecular genetics of cellular oncogenes. *Annu. Rev. Genet.*, **18**, 553–612.

Weinberg, R. A. (1981). Use of transfection to analyze genetic information and malignant transformation. *Biochem. Biophys. Acta*, **651**, 25–35.

Weinberg, R. A. (1992). The retinoblastoma gene and gene product. *Cancer Surv.*, **12**, 43–58.

Wiener, F., Klein, G. M., and Harris, H. (1971). The analysis of malignancy by cell fusion III. Hybrids between diploid fibroblasts and other tumour cells. *J. Cell Sci.*, **8**, 682–92.

Wiener, F., Klein, G. M., and Harris, H. (1974). The analysis of malignancy by cell fusion V. Further evidence of the ability of normal diploid cells to suppress malignancy. *J. Cell Sci.*, **15**, 177–83.

zur Hausen, H. (1982). Human genital cancer: synergism between two virus infections or synergism between a virus infection and initiating events. *Lancet*, **2**, 1370–2.

13. Genes in control of cell proliferation and tumorigenesis in *Drosophila*

Bernard M. Mechler

Introduction

Cancer is generally considered as a failure in the normal progression of differentiation. As a result, the cancer cells escape the mechanisms controlling normal growth and proliferate. When the overgrowth of neoplastic cells reaches a critical threshold, complex syndromes arise and, ultimately, lead to the death of the organism. A genetic basis for cancer is now well established and includes the discovery of proto-oncogenes, whose activation or altered expression is associated with the cancerous state and tumour suppressor genes which appear to be lost or functionally inactivated.

About 25 years ago, pioneering studies in the field of *Drosophila*, mouse somatic cell, and human genetics revealed that neoplasia may result from recessive mutations in regulatory genes controlling cell growth and differentiation (Gateff and Schneiderman 1967, 1969; Harris *et al.* 1969; Knudson 1971).

Over the past two decades the fruit-fly *Drosophila* has become the organism of choice for molecular and genetic investigations of eukaryotic biology. Its emergence as an animal model system is closely related to the rapid advances in recombinant DNA technology and ideas established by decades of classical genetics and embryology. One of the principal reasons for the choice of *Drosophila* resides in the genetic legacy of earlier studies, bequeathed by Morgan and his colleagues. Over 80 years of investigations have provided powerful and extremely accurate analytical techniques designed for this organism which are unsurpassed by any other animal system. A further advantage of *Drosophila* is the size and complexity of its genome which is intermediate between those of prokaryotes and mammals. Such a combination makes *Drosophila* particularly good for studying molecular biology of development, provides tools for dissecting the components of the regulatory pathways which underlie cell proliferation and differentiation, and ultimately may enable us to elucidate the biochemical functions of the macromolecules that participate in these pathways. Knowledge on these mechanisms is accumulating rapidly: a recent survey of the genome of *Drosophila* (Lindsley and Zimm 1992) reports genetic and molecular data on more than 4000 genes. As more genes are molecularly cloned, more knowledge about the signals governing developmentally expressed genes is likely to be acquired.

Although the existence of tumour suppressor genes was first postulated more than two decades ago, it has only been in recent years that insights into their identity and function have emerged. Eight human tumour suppressor genes have been so far identified, but there may be many more. There are at least 50 distinguishable human hereditary cancers, each possibly attributable to a different gene (Knudson 1985). In *Drosophila*, more than 40 genes have so far been identified genetically by mutations causing tissue overgrowth (Gateff and Mechler 1989; Mechler and Strand 1990; Mechler 1991; Watson *et al.* 1991) and recent genetic analyses have revealed that mutations in many more genes may induce specific tissue overgrowth during *Drosophila* development.

Cell proliferation during *Drosophila* development

Before being committed to neoplastic growth, a cell must be able to duplicate all its essential parts. This ability is normally limited to undifferentiated cells that are either in a proliferative phase or quiescent but can readily be recruited into mitosis.

Drosophila displays a very precise programme of cell proliferation during its development. (Campos-Ortega and Hartenstein 1985). The embryo is initially a syncytium in which 13 rapid cycles of synchronous nuclear division occur at approximately 10 min intervals. Then the nuclei migrate to the cortex and become cellularized at the interphase of cycle 14. From this developmental phase the cell cycle lengthens and the following divisions occur in complex mitotic domains according to a specific temporal programme which is co-ordinated by a complex pattern of gene expression and leads to the morphogenesis of specific tissues within the embryo. After 24 h, the embryo hatches and contains about 50 000 cells which can be subdivided into two classes—those that form the larval tissues and grow by expansion of the cell volume with endoreplication of DNA in the absence of mitosis and those that will constitute the imaginal cells which are destined to form the adult organism and are not themselves necessary for the survival of the larva. These imaginal cells resume their proliferation in the middle of the first larval instar and continue to divide throughout the three larval instars up to the larva–pupa transition, arresting only as terminal differentiation occurs. In the adult, only the germ-line cells and their associated gonadial tissues are actively dividing, as well as some haematopoietic cells. Thus, manifestation of cell overproliferation can become noticeable at the end of both major periods of cell proliferation: late embryogenesis and the larva–pupa transition phase. Mutations giving rise to tumorous growth have been identified at the end of either one or the other period.

Identifying overgrowth mutations in *Drosophila*

In *Drosophila*, overgrowth mutations are usually recessive because dominant mutations will be eliminated almost immediately after they have arisen if they interfere with the life of the individual. Only dominant conditional mutants, such as temperature-sensitive (ts) mutants, which behave normally at one temperature, e.g.

below about 20 °C, and display abnormal development at another temperature, for example, above 25 °C, can be perpetuated. However, the search for such dominant ts mutations requires intensive genetic investigations and for this reason, few ts mutants have been isolated in *Drosophila*. Presently, only one dominant ts mutation (*Tum1*) is known to cause tissue overgrowth (Hanratty and Ryerse 1981); all other mutations are recessive (Gateff and Mechler 1989; Mechler and Strand 1990; Mechler 1991).

Another advantage of *Drosophila* is that recessive mutations can easily be maintained in permanent cultures without constant control and selection due to the availability of a system of balancer chromosomes. Balancer chromosomes contain multiple inversions preventing crossing over and themselves carry other mutations which are homozygous lethals during embryogenesis. Thus, when a chromosome carrying an overgrowth mutation is placed in conjunction with a balancer chromosome, such genetical configurations produce only heterozygous viable and fertile animals and, in addition, provide a constant supply of homozygous animals with overgrown tissues.

Three different experimental approaches can be used for identifying mutations causing tissue overgrowth: direct observation of the overgrown organs by dissection of the mutant animals, production of tumours following transplantation of mutant tissues into the abdomen of adult flies, and induction of growth abnormalities in mosaic tissues following somatic recombination.

The first approach has been most productive for identifying mutations causing tissue overgrowth (Gateff and Mechler 1989; Mechler and Strand 1990; Mechler 1991). The large majority of overgrowth mutations so far identified affects the development of the imaginal tissues and, thus, the tissues which grow during the larval life. The mutant animals are easily recognized because the growth of the tumorous tissues is usually accompanied by developmental arrest at the end of the larva–pupa transition. As a consequence the larval life of the mutant animals is extended over several days and the tumorous tissues can reach a considerable mass which is readily observed upon dissection.

Mutant larvae with brain or imaginal disc tumours become either bloated or 'giant' and transparent and those with blood neoplasia become opaque white or completely melanized. In blood neoplasia numerous haemocytes are released from the haematopoietic organs, invade the entire larval body, and destroy most other tissues.

By contrast, mutations affecting control of cell proliferation in embryos do not give rise to visible tumours and, in this case, neoplastic growth can only be assayed by transplantation (Gateff and Schneiderman 1974). For this purpose, fragments of embryonic tissues are implanted into the abdomen of adult flies. Under these conditions, the neoplastic cells proliferate rapidly and kill the host in 3–14 days. By contrast, the normal tissues grow only moderately and exert no deleterious effect on the host. This procedure provides an efficient system for recognizing autonomously growing tissues and, particularly, neoplastic tissues, but may be insufficient for detecting potentially tumorous tissues with limited capacity for autonomous cell proliferation.

The third approach allows the identification of mutations which are usually zygotic lethals but have the capacity to produce abnormal cell proliferation when homozygosity is induced by mitotic recombination. In this assay mitotic recombination is induced in imaginal tissues either by ionizing radiation (for review see Becker (1979)), or by the activation of a heat-inducible site-specific yeast recombinase gene inserted in the *Drosophila* genome (Golic 1991). As a result, a single cell may become homozygously mutated in a background of heterozygous cells and, depending upon the developmental phase, this founder cell will generate a clone of cells which colonize one or several organs. If recombination occurs during the early larval stage, the ensuing clone will be confined only to one organ such as a wing, a leg, or an eye where growth abnormalities can easily be scored. Analysis by somatic recombination has allowed the identification of a series of mutations which can cause a global increase in the number of imaginal cells and can alter their pattern of differentiation (Schubiger and Palka 1987; Diaz-Benjumea and Garcia-Bellido 1990; Garcia-Bellido and de Celis 1992). However, none of these mutations can produce a massive growth of tissues displaying neoplastic properties.

The tumour phenotype

Drosophila tumours can be classified into two broad categories: neoplasia and hyperplasia (Gateff 1978; Bryant 1987; Gateff and Mechler 1989; Mechler 1991).

The differences between these two types of overgrowth (discussed below) are particularly visible in imaginal discs, which are distinct groups of undifferentiated cells present in the larvae and give rise to various parts of the adult body during metamorphosis. These groups of cells can be identified in the newly hatched larvae as discs each containing approximately 40 cells. They grow throughout larval life and form folded sacs of epithelia which are essentially constituted of a monolayer of columnar cells. Other internal organs with proliferating cells, such as the brain hemispheres and the haematopoietic organs, may also be affected by neoplasia or hyperplasia. However, due to the structure of these organs, it is difficult, by morphological criteria, to assign tumours in these organs into one or the other category.

Neoplastic growth

Neoplasia of imaginal discs are characterized by a massive proliferation of cells which disrupts the monolayered epithelial structure of the imaginal discs converting them into amorphous masses of tissues resembling tumours. The imaginal discs present in the cephalic region of the larvae frequently fuse with one another and the mass of tumorous discs may reach several times their normal size. In addition, the neoplastic cells have lost their apical–basal polarity and are cuboidal rather than columnar in shape. Other tissues, such as the brain hemispheres and the haematopoietic organs, can also give rise to neoplasia.

Like vertebrate tumours *Drosophila* neoplasms exhibit a range of characteristics, including cell overproliferation, altered cell morphology, loss of differentiation capacity, invasiveness, and transplantability. These features can help to define neoplasia in *Drosophila*, but only some tumours display all of them (for review, see Gateff and Mechler (1989)).

The main features of *Drosophila* neoplasms are the overgrowth of the neoplastic tissues to form amorphic structures that have lost their capacity to differentiate. Further grades in the tumour phenotype can be established. For example, malignant neoplasia grow aggressively after transplantation and kill the host, whereas benign neoplasia are unable to proliferate or grow only moderately. Furthermore, benign neoplasia form usually compact tumours which remain limited within the organ of origin and do not invade the surrounding tissues. However, some tumours which should be classified as benign because they are unable to grow autonomously after transplantation exhibit a very aggressive pattern of growth *in situ*, invading the surrounding tissues.

Brain neoplasia are characterized by gross enlargement of the two optic lobes resulting from the uncontrolled proliferation of optic neuroblasts and ganglion mother cells of the anlagen of the adult optic centres of the larval brain. In tumorous brains these cells do not differentiate optic neurones and display a disorganized pattern of growth.

Neoplastic growth of the haematopoietic tissues is characterized by a massive outgrowth of the five to seven pairs of haematopoietic organs located along the dorsal heart vessel, behind the brain hemispheres. In several mutants, large numbers of undifferentiated haemocytes are released into the haemolymph where they behave aggressively, invading all larval tissues and causing their destruction, whereas in other mutants, the haemocytes remain confined to the haematopoietic organs which expand massively.

Hyperplastic growth

Hyperplasia of the imaginal discs are characterized by a massive proliferation of the imaginal cells which maintain a columnar shape with a normal apical–basal polarity. During the prolonged larval life the hyperplastic discs can grow to several times their normal size and retain a folding pattern, even though this is abnormal. In addition, the discs often appear to be duplicated. Furthermore, the hyperplastic disc cells retain some capacity to differentiate, displaying upon differentiation a restricted pattern of adult cuticular structures. In some of these mutants, other proliferating tissues, such as the brain hemispheres, and the imaginal rings of the foregut, hindgut, and salivary glands, exhibit variable overgrowth. However, when transplanted into adult hosts, the hyperplastic tissues are unable to grow. Imaginal discs are the only tissues that have been shown to give rise to hyperplastic growth although it is not excluded that some other tumourous tissues may fall into this category. Like human tumours, the *Drosophila* tumours are difficult to classify because of their lack of invariant properties. Presently, we can only define *Drosophila* tumours on the basis of *in vivo* structural and behavioural characteristics of the overgrown cells. We still lack good markers, molecular or biochemical, for characterizing the various forms of tumorous growth. An ultimate goal of our investigations is to elucidate the molecular function of the genes that control cell proliferation and to determine whether a common mechanism may underlie each category of tissue overgrowth.

Tumour suppressor genes in *Drosophila*

In *Drosophila*, mutagenesis screens and the analysis of spontaneously occurring mutations have allowed the identification and genetic mapping of a series of tumour suppressor and genes controlling tissue overgrowth. Known genes described in the literature are listed in Table 13.1. Tissue overgrowth appears at different developmental stages and in distinct tissues: two mutations give rise to potential neoplasia during embryogenesis, 19 produce visible malignancies during the larval develop-

Table 13.1. Neoplastic and hyperplastic mutants in *Drosophila melanogaster*

Mutant	Symbol	Locus	Cytological location
A. Neoplastic mutants			
1. Mutants producing embryonic neoplasms[a]			
Df (1) Notch	N	1-3, 0	3B4
shibire[ts1]	*shi*[ts1]	1-52, 2	13F-14A
2(a) brain and imaginal disc neoplasms[b]			
lethal (2) giant larvae	*l(2)gl*	2-0.0	21A
2(b) Brain neoplasia			
lethal (2) giant discs	*lgd*	2-42, 7	32A-E
lethal (2) 37Cf	*brat*	2-53, 9	37C5-7
lethal (3) malignant brain tumour	*l(3)mbt*	3-93	97F11
2(c) Imaginal disc neoplasia			
lethal (1) discs large-1	*dlg*	1-34, 8	10B8-9
lethal (2) tumorous discs	*tid*	2-104	59F5-8
2(d) Blood cell neoplasia			
lethal (1) malignant blood neoplasm	*l(1)mbn*	1-27, 6	8D8-9
Tumorous lethal blood neoplasmc	*Tum*[l]	1-34, 5	
lethal (3) malignant blood neoplasm-1	*l(3)mbn-1*	3-13, 3	64F4-5
lethal (3) malignant blood neoplasm-2	*l(3)mbn-2*	3-	
lethal (1) air[1]	*air*[1]		1C3-4; 1D3-E1 2B15; 2B17-18
lethal (1) air [2]	*air*[2]	1E3; 1E4	
lethal (1) air[6]	*air*[6]	1	5A1; 2; 5E8
lethal (1) air[7]	*air*[7]	7A6; 7A8	
lethal (1) air[8]	*air*[8]	7C4; 7C9	
lethal (1) air[9]	*air*[9]	7C9; 7D1 7D5; 7D10	
lethal (1) air[11]	*air*[11]	1-29.0	(8A5; 9A2) (11A7; 13F10)
lethal (1) air[13]	*air*[13]	1-35.4	(8A5; 9A2) (11A7; 13F10)
lethal (1) air[15]	*air*[15]	1-55.6	(11A7; 13F10)

Table 13.1. *continued*

Mutant	Symbol	Locus	Cytological location
B. **Hyperplastic mutants** (imaginal disc overgrowth mutants)			
lethal (2) fat	fat	2–12	24D8
lethal (2) giant discs	lgd	2–42.7	32AE
lethal (3) c43	hyd	3–49.0	85E
lethal (3) disc overgrown	l (3) dco	3–	100A1.2–B1
C. **Germline tumours**			
female sterile (1) 231	fs (1) 231	1–	
female sterile (1) 1621	fs (1) 1621	1–11.7	4F1–5A1
ovarian tumors	otu	1–23.2	7F1
fused	fu	1–59.5	17C3–D2
female sterile (2) of Bridges	fs (2) B	2–5	
narrow	nw	2–79.68	54A–55A
benign (2) gonial cell neoplasm	b (2) cgn	2–106.7	60.A3–7
bag-of-marbles	bam	3–85	96C
D. **Mutants with tissue overgrowth in mitotic recombination clones**			
shifted	shf	1–17, 9	6A3–F11
fused	fu	1–59.5	17C3–D2
net	net	2–0, 0	21B
knot	kn	2–72, 3	51C–E
disrupted	dsr	2–90	
plexus	px	2–100, 5	58F
extramacrochaetae	emc	3–0, 0	61D1–2
warts	wts	3–	100A1.2–B1

[a] The mutant insects develop neoplasia during the embryonic development and die during embryogenesis.
[b] The mutant insects develop neoplasia during the larval development and die as third-instar larvae
[c] Dominant temperature-sensitive mutant.

ment in either the brain hemispheres, the imaginal discs, or the haematopoietic organs, and eight affect the germ-line. Four mutations cause well-defined hyperplasia of the imaginal discs. Capacity for tissue overgrowth in mitotic clones has been found in the case of eight mutations.

All these genes, with the exception of *Tum1* which displays a ts dominance (Hanratty and Ryerse 1981), act as recessive determinants of tissue overgrowth and are classified as tumour suppressor genes or anti-oncogenes because their normal function is to control cell proliferation and/or differentiation. In *Drosophila* convincing demonstration of tumour suppression can be obtained by introducing an intact allele of a cloned tumour suppressor gene into the genome of homozygous mutated animals and by showing restoration of normal growth and reversion to a normal phenotype (Opper *et al.* 1987). This type of experiment has proved successful in the case of the *lethal(2)giant larvae (l(2)gl)* gene. Introduction of a normal copy of this tumour suppressor gene into the genome of animals deficient for this gene can prevent the occurrence of tumorigenesis and restore a complete development of these animals. This result shows that the *l(2)gl* gene behaves as a tumour suppressor (Opper *et al.* 1987).

Thus, 40 genes causing tissue overgrowth are presently known in *Drosophila*. Their identification has occurred erratically over the last 25 years and, until recently, no systematic search for mutations causing either neoplastic transformation of embryonic tissues or overgrowth of imaginal tissues has been undertaken. However, the recent development of derivatives of the transposable P-element allowing the tagging of specific genes and their direct molecular cloning can provide a valuable approach for identifying new tumour suppressor genes and estimating their numbers. It would be interesting to know whether future mutagenesis experiments will constantly result in the isolation of already known genes—indicating that the *Drosophila* genome is saturated for mutations affecting tumour suppressor genes—or whether they will reveal new genes and, in this case, how many? The answer to these questions may help us to estimate the number of genes controlling cell growth and tumorigenesis in *Drosophila*.

Recent P-element-mediated mutagenesis experiments on the X-chromosome of *Drosophila melanogaster* have led to the isolation of nine complementation groups producing overgrowth of the haematopoietic organs (Watson *et al.* 1991). In this experiment, 1100 X-linked lethal mutations were examined and each new tumour suppressor gene is only represented by a single mutant allele. This indicates clearly that the X-chromosome has not yet been saturated for mutations affecting tumour suppressor genes. Furthermore, in an ongoing experiment involving P-mediated mutagenesis of the second chromosome, Kiss and his group at Szeged have already found 22 new mutations causing overgrowth of imaginal tissues. These mutations are distributed into 19 complementation groups, 17 of which are represented by single mutant alleles (I. Kiss, personal communication). Cytogenetic and genetic analyses have shown that these mutations affect genes which are distinct from the five previously known tumour suppressor genes present on the second chromosome. Since only half of the collection of 2700 second chromosome lethals has been screened, one can expect that additional mutants affecting tumour suppressor genes will be discovered in the other half of the collection. It is also reasonable to envisage that some of these as yet undiscovered mutants will correspond to unidentified genes. The results of this analysis show again that the second chromosome of *Drosophila* has not yet been fully saturated for P-element mutations causing overgrowth of imaginal tissues.

Altogether, 11 tumour suppressor genes controlling hyperplasia and neoplasia of imaginal tissues are presently known on the X-chromosome and 22 on the second chromosome. Since the X-chromosome represents approximately 20 per cent of the size of the *Drosophila* genome, and the second chromosome 40 per cent, it is reasonable to envisage that at least 55 tumour suppressor genes may be present in *Drosophila*, but this number could certainly be larger—perhaps as high as 100–200.

Analysis of the mutants isolated by Kiss and co-workers at Szeged revealed a series of new phenotypes which have not yet been described, such as simultaneous

Table 13.2. Cloned tumour suppressor genes in *Drosophila*

Gene	Overgrowth	Tissue	Encoded protein	Reference
l(2)gl	Neoplasia	Brain imaginal discse	Cytosolic and inner cell-membrane associated protein at regions of cell–cell contact	Mechler et al. (1985) Jacob et al. (1987) Strand et al. (1991) Strand and B. M. Mechler (personal communication)
dgl	Neoplasia	Imaginal discs	Guanylate kinase homologue associated with septate junctions	Woods and Bryant (1991)
brat	Neoplasia	Brain	Novel	Hankins (1990)
air8	Neoplasia	Haemotopoietic organs	S6 ribosomal protein	Watson et al. (1992)
shi	Neoplasia	Embryonic tissues	Dynamin, GTP-binding and microtubule-binding protein	Poodry (1990) van der Bliek and Meyerowitz (1991) Chen et al. (1991)
fat	Hyperplasia	Imaginal discs	Giant cadherin-like cell adhesion molecule	Mahoney et al. (1991)
hyd	Hyperplasia	Imaginal discs	Novel	A. Shearn (personnel communication)
bam	Ovarian tumour		Novel	McKearin and Spradling (1990)
otu	Ovarian tumour		Novel	Steinbauer et al. (1989)
fu	Ovarian tumour and cuticular overgrowth		Ser-Thr kinase	Preat et al. (1990)
emc	Cuticular overgrowth		Non-basic helix–loop-helix protein	Garrel and Campuzano (1991)

overgrowth of three distinct tissues: brain hemispheres, imaginal discs, and haemotopoietic organs, or overgrowth limited to only some imaginal discs, such as the labial, clypeo-labral, and antennal discs but surprisingly not the eye disc which is intimately associated with the antennal disc.

Mutations in tumour suppressor genes cause complex biological effects that not only affect the tissues with potential tumorous growth but may also have deleterious effects on other tissues. In these respects, the work of Hadorn's group has shown that the *l(2)gl* mutation affects the development of numerous tissues (i.e. atrophy of the male germ-line and underdevelopment of the imaginal cells of the salivary glands and the gut) before the appearance of the tumorous growth (for review see Hadorn (1961)). These mutations thus have pleiotropic effects, neoplasia or hyperplasia being the most striking feature. However, the other types of damage should not be forgotten because they may help to understand the function of these genes. In view of the complex biological effects caused by these mutations, the molecular cloning of tumour suppressor genes is an essential step for understanding the primary genetic changes that lead to these abnormalities and particularly to tumorigenesis. At the molecular level the first known tumour suppressor gene of any species, the *l(2)gl* gene of *Drosophila melanogaster*, was isolated in 1985 and 10 more genes controlling cell overgrowth have since been characterized (see Table 13.2).

The *lethal(2)giant larvae* gene

Genetics and development biology of *l(2)gl*

The best known tumour suppressor gene of *Drosophila* is the *l(2)gl* gene. Homozygous mutations in the *l(2)gl* gene lead to neoplastic transformation of the neuroblasts and ganglion mother cells of the adult optic centres in the larval brain and neoplasia of the imaginal discs (Gateff and Schneiderman 1967, 1969, 1974). In the mutant animals these neoplasia first become visible in the third larval instar. The brain hemispheres and imaginal discs grow to several times their normal size during the extended life of the *l(2)gl* larvae.

The *l(2)gl* gene was discovered by Bridges in 1933 (Bridges and Brehme 1944) and was intensively studied by Hadorn and his collaborators (for review, see Hadorn (1961)). The *l(2)gl* mutation was first recognized as a lethal factor producing pleiotropic effects with degeneration of the imaginal discs of the head, thorax, and genitalia, aplasia of germ cells in the testes but not of the ovaries, and underdevelopment of the salivary glands, fat bodies, and prothoracic cells of the ring gland, which is the source of ecdysone. Lack of ecdysone leads to considerable delay of pupation, and the mutant animals die as giant larvae or pseudopupae. However, atrophy of the ring gland is apparently not the primary cause of the mutation but appears to result from the invasive growth of the malignant neuroblasts within the brain hemispheres, which disrupts essential functions required for the maturation of the ring gland. This is supported by observations that transplantation of a normal ring gland into *l(2)gl*-deficient larvae (Hadorn 1937) or injection of ecdysone (Karlson and Hauser 1952) can induce pupation but cannot fully rescue the development of the mutant larvae. Furthermore, ecdysone deficiency is not sufficient to cause neoplastic growth (Garen *et al.* 1977). Therefore, the *l(2)gl* phenotype does not seem to result from a deficiency in ecdysone or its receptor (Richards 1976).

l(2)gl was first mapped distally to position 21C at the left end of the second chromosome (Lewis 1945). Molecular cloning has shown that *l(2)gl* is located at position 21A and is the first known gene to be at this end of the chromosome (Mechler *et al.* 1985). On its distal side *l(2)gl* is flanked by telomeric repetitive sequences. Numerous spontaneous mutant alleles have been isolated from wild *Drosophila* populations in the Soviet Union (Golubovsky 1978, 1980) and in California (Green and Shepherd 1979).

Molecular analysis of the chromosomal structure of these mutant alleles has revealed that almost all consist of terminal deletions that have removed the *l(2)gl* gene or part of it (Mechler *et al.* 1985). Examination of naturally occurring variants has shown that approximately 1 per cent of flies in these widely separated populations had mutations at the *l(2)gl* locus (Golubovsky 1980). The high frequency of *l(2)gl* mutations and the preponderance of deletions suggest that the *l(2)gl* gene is associated with an unstable chromosomal region, possibly linked to telomeric variation (Mechler *et al.* 1985).

Isolation of the l(2)gl gene

The *l(2)gl* gene was isolated by molecular cloning in 1984–85 (Mechler 1984; Mechler *et al.* 1985) and shown to encompass 13 kb of DNA at the extremity of the left arm of the second chromosome. The initial identification depended on detection of deletions involving large segments at the left end of chromosome 2 and could be visualized by *in situ* hybridization on polytene chromosomes of salivary glands or by means of Southern blotting analysis. The limits of the gene were determined by mapping 20 distinct chromosomal rearrangements and by relating the structural disruptions to one of the transcription units contained within the cloned region. This procedure identified a 13 kb transcription unit encoding two classes of overlapping transcripts of about 4.5 and 6 kb, respectively. In all examined *l(2)gl* mutants, this transcription unit was structurally altered (Mechler *et al.* 1985; Lützelschwab *et al.* 1986).

Positive identification of the biologically active *l(2)gl* was obtained by reintegrating a 13.1 kb DNA fragment containing the putative *l(2)gl* gene into the genome of *l(2)gl*-deficient animals by P-mediated germ-line transformation and showing that this sequence fully rescues the development of the insects that would otherwise have succumbed with brain and imaginal disc neoplasia (Jacob *et al.* 1987; Opper *et al.* 1987).

The l(2)gl protein

The entire 13 kb genomic DNA segment and several cDNAs have been sequenced (Jacob *et al.* 1987). This analysis showed that the *l(2)gl* gene encodes a polypeptide of 1161 amino acids with a molecular weight of about 127 kDa. Polyclonal and monoclonal antibodies prepared against *l(2)gl* fusion proteins or synthetic peptides recognize a polypeptide of 130 kDa in size (Lützelschwab *et al.* 1987; Merz *et al.* 1990) which can be metabolically labelled with ^{32}P (Strand *et al.* 1991). No other post-translational modifications have been detected so far (Strand *et al.* 1991).

Although *l(2)gl* does not appear to encode a membrane protein, having no likely signal sequence or membrane-spanning domain, others have found similarities between the *l(2)gl* sequence and the extracellular domains of the L-CAM family (Lützelschwab *et al.* 1987) or the cadherin family of cell adhesion molecules

(Klämbt et al. 1989). Recent biochemical investigations and cell fractionation studies have revealed, however, that *l(2)gl* encodes an intracellular protein which is predominantly free in the cytoplasm and becomes attached to the inner face of the plasma membrane at contact sites between cells (Merz et al. 1990; Strand et al. 1991). Amino acid sequence analysis and comparison with protein sequences show the presence of several amino acid motifs, such as reiterated heptad units of hydrophobic amino acid residues and a repeated motif showing distant homology with motifs present in β-subunits of G-proteins (Dalrymple et al. 1989; Mechler et al. 1991). Such motifs indicate that the *l(2)gl* polypeptide may interact with other proteins and/or form multimeric complexes with itself.

The presence of potential motifs for protein–protein interaction has instigated studies to determine whether any domain of the *l(2)gl* polypeptide can promote its multimerization or its interaction with other proteins. Recent analysis has shown that the segments of the *l(2)gl* protein which contain such motifs can indeed induce multimerization of a fused protein A (R. Jakobs, D. Strand, and B. Mechler, in preparation). In addition, the regular spacing of the hydrophobic amino acid residues as well as the other motifs playing a potential role in protein–protein interaction are strongly conserved during evolution and can be found in *l(2)gl* homologues of other dipteran species (Török et al. 1993).

l(2)gl gene expression

The spatio-temporal patterns of *l(2)gl* transcription and protein expression have been studied by *in situ* hybridization and immunostaining. Both analyses have shown ubiquitous gene expression during early embryogenesis.

The *l(2)gl* transcripts are first found in all embryonic cells from the blastoderm stage (2.5 h of development) up to completion of the germ-band extension (about 8.5 h). During this period *l(2)gl* expression is uniform and relatively intense over all embryonic cells. At the time of the dorsal closure (about 10 h) *l(2)gl* expression becomes gradually restricted to the epithelial cells of the midgut, where it persists until the end of embryogenesis (Mechler et al. 1989; Merz et al. 1990).

Immunostaining with anti-*l(2)gl* antibodies has shown that the pattern of protein expression generally follows the pattern of transcription. During early embryogenesis, the *l(2)gl* protein is uniformly distributed in all embryonic cells. Inside the cell the *l(2)gl* protein is found essentially in the cytoplasm and can also be detected in association with the cell periphery, presumably bound to the plasma membrane (Merz et al. 1990; D. Strand, A. Kalmes, and B. M. Mechler, in preparation). Cell fractionation studies have confirmed this intracellular distribution. A large proportion of the *l(2)gl* protein is recovered in a soluble form, whereas about 20 per cent is tightly associated with membranes (D. Strand, A. Kalmes, and B. M. Mechler, in preparation).

In late embryonic stages, the *l(2)gl* protein is restricted to midgut and salivary gland epithelia. Cell fractionation studies indicated that about 50 per cent of the *l(2)gl* protein is bound to membranes.

On the plasma membranes of polyhedral cells of the midgut, microscopical examination revealed that the *l(2)gl* protein is present only on domains of the plasma membrane that face contiguous cells and is absent from basal or apical membranes. This particular intracellular distribution suggests that the *l(2)gl* protein may play a role in signal transduction by interacting with plasma membrane components that establish contacts or junctions with other cells. However, the function of *l(2)gl* in the midgut epithelium is not yet understood, because the tissue exhibits no phenotypic defect in the absence of *l(2)gl* gene activity. In the midgut, *l(2)gl* seems to exert a dispensable or redundant function whose absence leads to a weak phenotype that remains unnoticed by comparison with the serious defects seen in mutant larvae.

Biology of l(2)gl function

The results of molecular analyses indicate that the critical period of gene expression for preventing tumorigenesis is associated with the early embryonic period of *l(2)gl* transcription when the gene is ubiquitously expressed in all cells, and especially in the progenitors of cells that are neoplastically transformed in mutant animals. Furthermore, explants of *l(2)gl* embryonic tissues grow malignantly in adult hosts (Gateff and Schneiderman 1974), suggesting further that the critical period for commitment to tumorous growth takes place during embryogenesis. However, as mentioned above, *l(2)gl* gene expression occurs during two phases of *Drosophila* development.

The respective contributions of each period of *l(2)gl* expression were analysed by inducing the appearance of *l(2)gl$^-$* clones in otherwise heterozygous animals and by studying the fate of the clones during development.

Analysis of the *l(2)gl* mosaic animals (Mechler 1991) showed that neoplastic growth occurs in clones of cells that have lost the *l(2)gl* gene in the preblastoderm syncytial embryos before any expression of the *l(2)gl* gene. Clones produced at later embryonic stages, but still during the embryonic phase of *l(2)gl* expression, do not display the neoplastic phenotype but are unable to complete differentiation. Finally, when *l(2)gl$^-$* clones arise during larval stages, the *l(2)gl$^-$*-deficient tissues show normal or nearly normal development. With the indication that the *l(2)gl* gene produces an intracellular protein and that the *l(2)gl* activity is cell autonomous, these data show that the critical period for the establishment of *l(2)gl* neoplasia is during early embryogenesis, when intense *l(2)gl* activity occurs in all embryonic cells.

The cytoplasmic localization of the *l(2)gl* protein and its association with the inner surface of the plasma membrane suggest that p127 may act as a transducer in a signal pathway linking plasma membrane bound receptors to intracellular effector proteins. Consistent with this idea is the recent finding that immunocomplexes of *l(2)gl* protein display a strong serine/threonine kinase activity (D. Strand, A. Kalmes and B. M. Mechler, personnel communication). Studies are currently under way to identify the nature and specificity of the kinase associated with *l(2)gl*.

Present investigations are concerned with the mechanism by which the *l(2)gl* protein becomes attached to the plasma membrane and the relevance of this binding to *l(2)gl* function. Recent analysis has shown that the *l(2)gl* protein expressed in *Spodoptera frugiperda* cells infected with a recombinant *l(2)gl* baculovirus mimics p127 behaviour in *Drosophila* embryos. When the *Spodoptera* cells make contact, the *l(2)gl* protein becomes associated with plasma membrane domains facing contiguous cells, whereas in single cells, the protein appears to be homogeneously distributed within the cytoplasm (D. Strand, A. Kalmes, and B. M. Mechler, in preparation).

Acknowledgements

I thank Mrs Karin Helm for excellent secretarial help and Dr Dennis Strand for editorial assistance. The studies reported here were supported by grants of the Deutsche Forschungsgemeinschaft, the Swiss Cancer League, the European Commission, and general funds of the Deutsches Krebsforschungszentrum.

References

Becker, H. J. (1976). Mitotic recombination. In *The genetics and biology of Drosophila*, Vol. IC, (ed. M. Ashburner and E.Novitski), pp. 1019–87. Academic Press, London.
Bridges, C. B. and Brehme, K. F. (1944). *The mutation of Drosophila melanogaster*. Carnegie Institute of Washington. Publication no. 552.
Bryant, P. J. (1987). Experimental and genetic analysis of growth and cell proliferation in *Drosophila* imaginal discs. In *Genetic regulation of development*, (ed. W.F. Loomis), pp. 339–372. Alan R. Liss, New York.
Campos-Ortega, J. A. and Hartenstein, V. (1985). *The Embryonic development of Drosophila melanogaster*. Springer Verlag, Berlin.
Chen, M. S., Obar, R. A., Schroeder, C. C., Austin, T. W., Poodry, C. A., Wadsworth, S. C., and Vallee, R. B. (1991). Multiple forms of dynamin are encoded by *shibire*, a *Drosophila* gene involved in endocytosis. *Nature*, **351**, 583–6.
Dalrymple, M. A., Petersen-Bjorn, S., Friesen, J. D., and Beggs, J. D. (1989). The product of the *PRP4* gene of *S. cerevisiae* shows homology to β subunits of G proteins. *Cell*, **58**, 811–12.
Diaz-Benjumea, F. J. and Garcia-Bellido, A. (1990). Genetic analysis of the wing vein pattern of *Drosophila*. *Roux's Arch. Dev. Biol.*, **198**, 336–54.
Garcia-Bellido, A. and de Celis, J. F. (1992). Developmental genetics of the venation pattern of *Drosophila*. *Annu. Rev. Genet.*, **26**, 275–302.
Garen, A., Kauvar, L., and Lepesant, J. A. (1977). Roles of ecdysone in *Drosophila* development. *Proc. Natl Acad. Sci. USA*, **74**, 5099–102.
Garrel, J. and Campuzano, S. (1991). The helix–loop–helix domain, a common motif for bristles, muscles and sex. *BioEssays*, **10**, 493–8.
Gateff, E. (1978). Malignant neoplasms of genetic origin in the fruit fly *Drosophila melanogaster*. *Science*, **200**, 1446–59.
Gateff, E. and Mechler, B. M. (1989). Tumor-suppressor genes of *Drosophila melanogaster*. *CRC Crit. Rev. Oncogen.*, **1**, 221–45.
Gateff, E. and Schneiderman, H. A. (1967). Developmental studies of a new mutation of *Drosophila melanogaster*: lethal malignant brain tumor *l(2)gl4*. *Am. Zool.*, **7**, 760.
Gateff, E. and Schneiderman, H. A. (1969). Neoplasms in mutant and cultured wild-type tissues of *Drosophila*. *Natl Cancer Inst. Monogr.*, **31**, 365–97.
Gateff, E. and Schneiderman, H. A. (1974). Developmental capacities of benign and malignant neoplasms of *Drosophila*. *Wilhelm Roux' Archiv für Entwicklung und Organogenese*, **176**, 23–65.
Golic, K. G. (1991). Site-specific recombination between homologous chromosomes in *Drosophila*. *Science*, **252**, 958–61.

Golubovsky, M. D. (1978). The 'lethal giant larvae'—the most frequent second chromosome lethal in natural populations of *D.melanogaster*. *Drosophila Information Service*, **53**, 179.

Golubovsky, M. D. (1980). Mutational process and microevolution. *Genetica*, **52/53**, 139–49.

Green, M. M. and Sheperd, S. H. Y. (1979). Genetic instability in *Drosophila melanogaster*: the induction of specific chromosome 2 deletions by MR element. *Genetics* **92**, 823–32.

Hadorn, E. (1937). An accelerating effect of normal 'ring glands' on puparium formation in lethal larvae of *Drosophila melanogaster*. *Proc. Natl Acad. Sci. USA*, **23**, 478–84.

Hadorn, E. (1961). *Developmental genetics and lethal factors*. Methuen, London.

Hankins, G. R. (1990). Analysis of a *Drosophila* neuroblastoma gene. PhD Dissertation, University of Virginia, Charlottesville.

Hanratty, W. P. and Ryerse, J. S. (1981) A genetic melanotic neoplasm of *Drosophila melanogaster*. *Dev. Biol.*, **83**, 238–49.

Harris, H., Miller, O. J., Klein, G., Worst, P., and Tachibana, T. (1969). Suppression of malignancy by cell fusion. *Nature*, **223**, 363–8.

Jacob, L., Opper, M., Metzroth, B., Phannavong, B., and Mechler, B. M. (1987). Structure of the *l(2)gl* gene of *Drosophila* and delimitation of its tumor suppressor domain. *Cell*, **50**, 215–25.

Karlson, P. and Hauser, G. (1952). Über die Wirkung des Puparisierungshormons bei der Wildform und der Mutante *lgl* von *Drosophila*. *Z. Naturforsch.*, **7b**, 80–83.

Klämbt, C., Müller, S., Lützelschwab, R., Rossa, R., Totzke, F., and Schmidt, O. (1989). The *Drosophila melanogaster l(2)gl* gene encodes a protein homologous to the cadherin cell-adhesion molecular family. *Dev. Biol.*, **133**, 425–36.

Knudson, A. G., Jr (1985). Hereditary cancer, oncogenes, and antioncogenes. *Cancer Res.*, **45**, 1437–43.

Knudson, G. A. (1971). Mutation and cancer: statistical study of retinoblastoma. *Proc. Natl Acad. Sci. USA*, **68**, 820–23.

Lewis, E. B. (1945) The relation of repeats to position effect in *Drosophila melanogaster*. *Genetics*, **36**, 137–66.

Lindsley, D. L. and Zimm, G. G. (1992). *The genome of Drosophila melanogaster*. Academic Press, San Diego.

Lützelschwab, R., Müller, G., Wälder, B., Schmidt, O., Fürbass, R., and Mechler, B. (1986). Insertion mutation inactivates the expression of the recessive oncogene *lethal(2)giant larvae* of *Drosophila melanogaster*. *Mol. Gen. Genet.*, **204**, 58–63.

Lützelschwab, R., Klämbt, C., Rossa, R., and Schmidt, O. (1987). A protein product of the *Drosophila* recessive tumour gene *l(2)giant gl*, potentially has cell adhesion properties. *EMBO J.*, **6**, 1791–7.

McKearin, D. M. and Spradling, A. C. (1990). *Bag-of-marbles*: a *Drosophila* gene required to initiate both male and female gametogenesis. *Genes Dev.*, **4**, 2242–51.

Mahoney, P. A., Weber, U., Onofrechuck, P., Biessmann, H., Bryant, P. J., and Goodman, C. S. (1991). The *fat* tumor suppressor gene in *Drosophila* encodes a novel member of the cadherin gene family. *Cell*, **67**, 853–68.

Martin, P., Martin, A., and Shearn, A. (1977). Studies of $1(3)C^{43hs1}$ a polyphasic, temperature sensitive mutant of *Drosophila melanogaster* with a variety of imaginal disc defects. *Dev. Biol.*, **55**, 213–32.

Mechler, B. (1984). Molecular cloning of the recessive oncogene *lethal(2)giant larvae* of *Drosophila melanogaster*. *Eur. J. Cell Biol.*, **33** (5), 23.

Mechler, B. M. (1991). The fruitfly *Drosophila* and the fish *Xiphosphorus* as model systems for cancer studies. *Cancer Surv.*, **9**, 505–27.

Mechler, B. M. and Strand, D. (1990) Tumor suppression in *Drosophila*. In *Tumor suppressor genes*, (ed. G. Klein), pp. 123–44. Marcel Dekker, New York.

Mechler, B. M., McGinnis, W., and Gehring, W. J. (1985). Molecular cloning of *lethal(2)giant larvae*, a recessive oncogene of *Drosophila melanogaster*. *EMBO J.*, **4**, 1551–7.
Mechler, B. M., Török, I., Schmidt, M., Opper, M., Kuhn, A., Merz, R., and Protin, U. (1989). Molecular basis of the regulation of cell fate by the *lethal(2)giant larvae* tumour suppressor gene of *Drosophila melanogaster*. In *Genetic analysis of tumour suppression*, (ed. G. Bock and J. Marsh), Ciba Foundation Symposium, Vol. 142, pp 166–80. Wiley, Chichester.
Mechler, B. M., Strand, O., Kalmes, A., Merz, R., Schmidt, M. and Török, I. (1991). *Drosophila* as a model system for molecular analysis of tumorigenesis. *Env. Health Persp.*, **93**, 63–71.
Merz, R., Schmidt, M., Török, I., Protin, U., Schuler, G., Walther, H. P., et al. (1990). Molecular action of the *l(2)gl* tumor suppressor gene of *Drosophila melanogaster*. *Env. Health Persp.*, **88**, 163–7.
Opper, M., Schuler, G., and Mechler, B. M. (1987). Hereditary suppression of *lethal(2)giant larvae* malignant tumor development in *Drosophila* by gene transfer. *Oncogene*, **1**, 91–6.
Poodry, C. A. (1990). *shibire*, neurogenic mutant of *Drosophila*. *Dev. Biol.*, **138**, 464–72.
Preat, T., Therond, P., Lamour-Isnard, C., Limbourg-Bouchon, B., Tricoire, H., Erk, I., et al. (1990). A putative serine/threonine protein kinase encoded by the segment-polarity gene fused of *Drosophila*. *Nature*, **347**, 87–9.
Richards, G. P. (1976). The *in vitro* induction of puffing in salivary glands of the mutant *l(2)gl* of *Drosophila melanogaster* by ecdysone. *Roux's Arch. Dev. Biol.*, **179**, 339–48.
Schubiger, M. and Falka, J. (1987). Changing spatial patterns of DNA replication in the developing wing of *Drosophila*. *Dev. Biol.*, **123**, 145–53.
Steinbauer, N. R., Walsh, R. C., and Kalfayan, L. J. (1989). Sequence and structure of the *Drosophila melanogaster* ovarian tumor gene and generation of an antibody specific for the ovarian tumor protein. *Mol. Cell. Biol.*, **9**, 5726–32.
Strand, D., Török, I., Kalmes, A., Schmidt, M., Merz, R., and Mechler, B. M. (1991). Transcriptional and translational regulation of the expression of the *l(2)gl* tumor suppressor gene of *Drosophila melanogaster*. *Adv. Enzyme Regul.*, **31**, 339–50.
Török, I., Hartenstein, K., Kalmes, A., Schmitt, R., Strand, D., and Mechler, B. M. (1993). The *l(2)gl* homologue of *Drosophila pseudoobscura* suppresses tumorigenicity in transgenic *Drosophila melanogaster*. *Oncogene*, **8**, 1537–49.
van der Bliek, A. M. and Meyerowitz, E. M. (1991). Dynamin-like protein encoded by the *Drosophila shibire* gene associated with vesicular traffic. *Nature*, **351**, 411–14.
Watson K. L., Johnson, T. K., and Denell, R. B. (1991). *Lethal(1)aberrant immune response* mutations leading to melanotic tumor formation in *Drosophila melanogaster*. *Dev. Genet.*, **12**, 173–87.
Watson, K. L., Konrad, D. K., Woods, D. F., and Bryant, P. J. (1992). The *Drosophila* homolog of the human S6 ribosomal protein is required for tumor suppression in the haematopoietic system. *Proc. Natl Acad. Sci. USA.*, **89**, 11302–6.
Woods, D. F. and Bryant, P. J. (1989). Molecular cloning of the *lethal(2)discs large-1* oncogene of *Drosophila*. *Dev. Biol.*, **134**, 222–35.
Woods, D. F. and Bryant, P. J. (1991). The *discs-large* tumor suppressor gene of *Drosophila* encodes a guanylate kinase homolog localized at septate junctions. *Cell*, **66**, 451–64.

14. Genetics and molecular biology of tumour formation in Xiphophorus

Joachim Altschmied and Manfred Schartl

Genes and cancer

The fact that genes cause cancer has never been appreciated accordingly. Although as early as 1914 Theodor Boveri had postulated that 'cancer cells have lost certain features of their normal phenotype' and that 'this loss is the consequence of an abnormal chromosomal constitution' (Boveri 1914), his chromosome theory of cancer could not be verified for quite some time because neither the technical tools nor the experimental systems were available. In 1928, K. H. Bauer took up some of Boveri's ideas and postulated that cancer is due to mutations (Bauer 1928). It was the pioneering work of Henry Harris which led to the development of an experimental system that has made it possible to define and characterize the targets of carcinogenic mutations. Using somatic cell hybrids of cancer cells fused with normal cells, genes that are functionally impaired in the tumour cell and that suppress the neoplastic phenotype in the normal cell were detected (Harris *et al.* 1969). Such genes were then also recognized in certain hereditary forms of human cancer like retinoblastoma (Knudson 1971) and Wilm's tumour (Koufos *et al.* 1985). Some of these paradigmatic tumour suppressor genes have been characterized at the molecular level (e.g. Friend 1986; Gessler *et al.* 1990; Malkin *et al.* 1990). Besides the tumour suppressor genes, which are negative regulators of growth and behave as recessive genes by definition, another group of cancer genes has been detected from a totally different line of evidence. These genes are said to act dominantly because they acquire the potential to transform a cell neoplastically following their mutational activation. Such dominant oncogenes were first noticed in acutely transforming retroviruses and many such activated oncogenes have since been found in different animal and human tumours (Bishop 1987). Careful evaluation of the existing data has led to the generally accepted view that most cancers that become apparent as a disease arise not from a single genetic change but from the accumulation of multiple genetic changes in a cell (see Weinberg 1989) including dominant activation of oncogenes as well as inactivation of tumour suppressor genes in the same sequence of events (Fearon and Vogelstein 1990). The main difficulty in analyzing such a complex situation is that in most cases only the *endpoint* of the multistep process can be investigated. The biochemical and molecular biological changes in a tumour cell compared with whatever is regarded as its normal counterpart are numerous. In most cases it appears impossible to decide which of these differences are due to the first genetic change instrumental in the causation of neoplastic transformation and which are required for tumour progression and metastasis. The sequence of the secondary events is difficult to determine. Finally, some of the changes observed in the cancer cell may be totally irrelevant for establishing and maintaining the neoplastic phenotype: they may simply reflect features of the chaotic molecular biology of the cancer cell. Genetic model systems have the advantage that at least the gene(s) that cause the primary event of neoplastic transformation are clearly defined by classical genetics. Such systems include tumour formation in the fruit-fly Drosophila (see Chapter 13), hereditary tumours in hybrids of Xiphophorus fish, and—at a lower level of genetic resolution—hereditary kidney tumours of the rat (Eker and Mossige 1961).

Genetic control of spontaneous melanoma formation in Xiphophorus

Fish of the genus Xiphophorus inhabit fresh water biotopes of the Atlantic drainage of Mexico, Honduras, and Guatemala. In several species some individuals (from 1 to 40 per cent of a given population) exhibit spot patterns composed of large, intensely black pigment cells. These cells have been termed macromelanophores to distinguish them from micromelanophores, the normal sized black pigment cells that make up the uniform greyish body colouration (Gordon 1927). More than 60 years ago it was discovered independently by Gordon, Kosswig, and Häussler that certain hybrids of platyfish carrying macromelanophore patterns (*Xiphophorus maculatus*) and of unspotted swordtail (*X. helleri*) spontaneously develop malignant melanoma (Gordon 1927, Häussler 1928; Kosswig 1928) (Fig. 14.1).

Shortly thereafter it was recognized that the occurrence of tumours in these hybrids is due to a single locus (the macromelanophore locus) of *X. maculatus* that 'interacted' with the *X. helleri* genome (Kosswig 1929; Gordon 1931). This interaction was later defined as the effect of modifying genes. It has been debated over the decades whether these modifying genes are 'intensifiers' contributed by *X. helleri* to the hybrid offspring genome or if they are 'suppressors', originally present in the *X. maculatus* genome, but are eliminated by the selective breeding process through substitution of the corresponding platyfish chromosomes by those from the swordtail (Gordon 1958; Atz 1962; Kosswig 1965; Zander 1969; Kallman 1970). Supposing multiple modifier genes and extrapolating Boveri's ideas, Breider explained melanoma formation in the hybrids by the loss of 'inhibitory' genes that suppress species-specifically the macromelanophore genes (Breider 1952). The currently generally accepted explanation for the observed phenomenon of hereditary melanoma by Anders and co-workers formalizes such considerations on the basis of numerous genetic experiments (Anders 1990).

In a typical crossing experiment a female *X. maculatus* which carries the X-chromosomal macromelanophore locus *Sd* (spotted dorsal, small spots in the dorsal fin) is mated to *X. helleri*, which does not carry the corresponding locus. The F_1

Fig. 14.1. Melanoma-bearing hybrid of Xiphophorus (genotype: $Tu\text{-}Sd/\text{--}; \text{--}/\text{--}$).

hybrid shows enhancement of the Sd phenotype. Backcrossing of the F_1 hybrid to *X. helleri* results in offspring that segregate, giving 50 per cent which have not inherited the Sd locus and are phenotypically like the *X. helleri* parental strain and 50 per cent which carry the macromelanophore locus and develop melanoma. The severity of melanoma ranges from very benign in some individuals (phenotype like the F_1 hybrids) to highly malignant in others. Highly malignant melanomas of such fish become invasive and exophytic and are fatal to the individual. They even grow progressively following transplantation to thymusaplastic (nude) mice (Schartl and Peter 1988).

Based on a variety of such classical crossing experiments a genetic model has been developed to explain tumour formation in Xiphophorus (Ahuja and Anders 1976) (see Fig. 14.2). The macromelanophore locus was formally equated to a sex chromosomal melanoma oncogene locus, whose critical constituent was designated 'tumour gene' (Tu). Melanoma formation was then attributed to the uncontrolled activity of Tu. In non-tumourous fish Tu activity was proposed to be negatively controlled by cellular regulatory genes or tumour suppressor genes (R genes, corresponding to the repressing modifying genes mentioned above). For the crossing experiment outlined above this means that *X. maculatus* contains the $Tu\text{-}Sd$ locus on the X-chromosome and the corresponding major R on an autosome, while *X. helleri* is proposed not to contain this particular Tu locus and its corresponding R. According to the model, backcrossing of the Tu-containing hybrids to *X. helleri* results, in effect, in the progressive replacement of R-bearing chromosomes from *X. maculatus* by R-free chromosomes of *X. helleri*. The stepwise elimination of regulatory genes is thought to allow expression of the Tu phenotype, leading to benign melanoma if one functional allele of R is still present or malignant melanoma if R is absent.

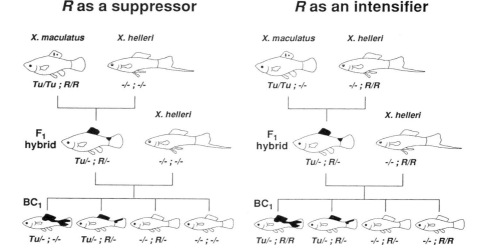

Fig. 14.2. Models to explain the classical crossing experiment leading to hybrids that develop melanoma.

It should be noted, however, that there is another scenario that is also compatible with the crossing data, namely that *Tu* activity is due to the presence of intensifying genes contributed by *X. helleri* chromosomes to the hybrid genome (see Fig. 14.2). To our knowledge, there is no crossing experiment that would help to decide between these two possibilities. Reintroduction of suppressor genes or diluting out activating genes, respectively, by crossing hybrids bearing malignant melanomas to parental *X. maculatus* was shown to lead to a reversion of the malignant phenotype resulting in totally tumour-free fish in the succeeding backcross generations using *X. maculatus* as the recurrent parent (Anders *et al.* 1984). This demonstrates that the melanoma oncogene *Tu* itself remains structurally unaltered during the process of activation via hybridization.

Depending on the macromelanophore pattern of the parental platyfish, the melanomas of the hybrids are localized in different body compartments. For example, in crosses where the *Sd* locus was introduced into the hybrid genome, melanomas spread from the dorsal fin, while in fish with an *Sp* (spotted) locus, which causes macromelanophore spots on the flanks of the parental platyfish, melanomas of the corresponding hybrids originate exactly from this region. This pattern information and the neoplastic transformation locus *Tu* are extremely closely linked and only a few mutants exist which affect the original macromelanophore pattern and, consequently, the melanoma compartment (Anders *et al.* 1973; Kallman 1975). According to Kallman's work the different patterns are due to a series of alleles of macromelanophore genes each harbouring the capacity for melanoma formation (Kallman 1975) while Anders *et al.* proposed that a single *Tu* locus is closely linked to a series of non-allelic, compartment-specific genes that suppress the appearance of macromelanophores in the various body regions. The different patterns would then be due to mutational impairment of one of the *cis*-acting compartment genes (Anders 1991). The implication from both explanations is that the gene that determines the macromelanophore phenotype is identical to the dominant oncogene.

Siciliano and Perlmutter (1972) noted a considerable maternal effect on development of melanoma in *Tu-Sd*-bearing hybrids. The nature of this effect seen early in the life of the hybrid fish is to augment the expression of enhanced pigmentation, melanosis, or malignant melanoma. The maternal effect was observed only in those hybrids whose female parent had a closer relationship to the swordtail than hybrids of the reciprocal cross. Whether this phenomenon represents genomic imprinting of 'modifier genes' or a hormonal effect (see below) due to the different combinations of platyfish and swordtail sex chromosomes and sex chromosome homologues (see Kallman 1984) has to be elucidated.

Epigenetic modulation of the melanoma phenotype

Besides a genetic control through the action of regulatory genes like *R*, the malignancy of melanoma can also be influenced by a variety of epigenetic factors some of which are androgens. In certain genotypes melanoma formation starts earlier in males and leads to a higher malignancy (Siciliano *et al.* 1971) and in other genotypes only mature males develop melanoma (Anders *et al.* 1984). Treatment with testosterone has dramatic effects on melanoma formation depending on the genotype of the fish and the developmental stage of the tumour. Melanoma induction, enhancement of malignancy, and melanoma regression were observed using the same treatment protocol but different hybrid genotypes and *Tu* alleles. All the various effects could be explained by promotion of pigment cell differentiation by androgens (Schartl *et al.* 1982).

Another substance that has been found to modulate the malignancy of melanoma in Xiphophorus is dinitrochlorbenzene (DNCB). Application of this compound led to tumour regression (Scholz 1977). DNCB is known to stimulate the immune system and thus may enhance an immune response against the melanoma cells. The same mechanism may explain the observation that raising tumour-bearing back-cross hybrids under hyperthermic conditions also results in suppression of melanoma formation (Perlmutter and Potter 1988). The high temperature in the environment most likely induces some kind of 'artificial fever' in the poikilothermic animals.

Besides steroids and the immune system many more factors may exist that modulate the neoplastic phenotype. Due to the fact that the genetic factors determining malignancy are defined and lead to tumours of high pathophysiological uniformity, melanoma formation in Xiphophorus provides a unique system for studies on epigenetic modulatory factors and their mechanism of action. This hopefully will also offer new perspectives for therapeutic approaches.

Hereditary ocular and thyroidal tumours in Xiphophorus

Gordon reported on a strain of *X. montezumae* that spontaneously developed thyroidal tumours at high incidence (Gorbman and Gordon 1951). A genetic susceptibility was obvious; however, no further genetic analyses were performed. Ocular tumours appeared in two broods of hybrids of platyfish and swordtails (Gordon 1947) similar to the case for melanoma. An autosomal recessive gene (*oc*) was defined that contained the genetic information for tumour formation but 'tumour-modifier' genes seemed also to be involved. Unfortunately, all these strains were lost and could not be analysed further.

Genetic factors in the formation of carcinogen-induced tumours

Besides the above-mentioned tumours of hereditary aetiology, a large variety of tumours of neurogenic, epithelial, and mesenchymal origin arose after treatment of fish with chemical carcinogens or X-rays. The surprising finding was that purebred descendants of feral populations of five Xiphophorus species developed tumours only at very low rates. When the interspecies hybrids of F_1 and further filial generations as well as the backcross generations were similarly treated, the tumour rates were increased up to 10-fold (Anders *et al.* 1991). A similar observation was made for N-methyl-N'-nitro-N-nitrosoguanidine (MNNG)-induced melanoma but not hepatoma in the medaka fish (*Oryzias latipes*) (Hyodo-Taguchi and Matsudaira 1987). The genetic loci possibly involved are so far ill-defined and there are no molecular data that would help towards an explanation of these phenomena.

Reverse genetic approaches towards isolation of *Tu* and *R*

In order to understand the molecular basis of hereditary melanoma, isolation and characterization of the genes involved was attempted. Until then no candidate gene product of *Tu* and *R* had been characterized precluding cloning by conventional recombinant DNA technology. We therefore applied a strategy that has been termed 'reverse genetics' (Orkin 1986) or more recently 'positional cloning' (Collins 1992) to isolate the melanoma-inducing gene of *X. maculatus* and gene(s) encoded by *R*. Our strategy included the following steps:

1. determination of the chromosomal location of the genes in question;

2. identification and cloning of a molecular marker sequence for the corresponding loci;

3. cloning of the *R*- and *Tu*-containing region by chromosome walking or jumping;

4. identification and isolation of a candidate gene;

5. verification that the candidate gene is indeed responsible for the *Tu* or *R* phenotype.

The *R* locus has so far resisted molecular cloning because of the lack of phenotypic markers, characterized mutants or alleles resulting in a paucity of information on its map position and genetic behaviour. A distantly linked (approximately 30 cM) isozyme marker (EST-1) (Siciliano *et al.* 1976; Ahuja *et al.* 1980; Förnzler *et al.*

1991) locates R to linkage group V of Xiphophorus (Morizot and Siciliano 1983). So far a large number of probes with polymorphism information content have been found that can now be tested for linkage to R in informative platyfish/swordtail backcross hybrids (Förnzler et al. 1991).

The situation was much more promising for the cloning of the Tu-encoded gene(s) thanks to the availability of many mutants, cross-overs in the vicinity, phenotypic markers, and many different Tu alleles. The chromosomal localization of Tu had been clearly defined by recombination and mutation analyses as residing within the distal portion of the sex chromosomes (Anders et al. 1973; Kallman 1975). The most critical step then was to identify a molecular marker sequence. One of several approaches (Schartl et al. 1993) was to use heterologous oncogene/proto-oncogene probes for Southern hybridizations. The rationale for this strategy was that most oncogenes/proto-oncogenes of higher vertebrates fall into one of several classes of multigene families. The members of such gene families share highly conserved regions, e.g. kinase domains and DNA-binding domains. Under conditions of reduced stringency in Southern hybridization, a molecular probe containing such a conserved region detects not only all members of the gene family of the same organism but also those from distantly related species (Mäueler et al. 1988a,b; Hannig et al. 1991). As many sequences are identified in such experiments with a single probe, these are very informative with respect to the detection of RFLPs. In addition, it appeared not totally fantastic to expect that the sought dominant melanoma oncogene of Xiphophorus might be a member of one of the known oncogene/proto-oncogene multigene families.

From all probes tested, the viral erbB (v-erbB) probe was most informative. It is derived from the B oncogene of avian erythroblastosis virus and represents an oncogenically activated version of the avian epidermal growth factor receptor (EGFR) gene (Ullrich et al. 1984). A probe that encompasses most of the highly conserved kinase domain detects in EcoRI digests, besides other strongly hybridizing bands, two weaker bands that are only present in the DNA of fish carrying a sex chromosomal Tu locus, one of 6.5 kb cosegregating with Y-chromosomal Tu loci and one of 5 kb, cosegregating with X-chromosomal Tu loci (Schartl 1988). In linkage analysis employing more than 500 individual fish no recombinant between this RFLP and the Tu locus was found (Schartl 1988, 1990; Zechel et al. 1988; Wittbrodt et al. 1989) indicating that this sequence is either intimately linked to Tu or even an integral part of the locus. The 5 kb band was cloned and found to detect besides the Y-chromosomal 6.5 kb band a third hybridizing sequence of 7 kb which was invariably present in DNA of all fish irrespective of the presence or absence of a Tu locus (Adam et al. 1988; Schartl 1988).

Cloning and characterization of the Tu locus encoded gene, Xmrk

The 5.0 kb X-chromosomal genomic fragment corresponding to the Tu-linked RFLP was used to isolate a cDNA clone from the Xiphophorus melanoma cell line PSM (Wittbrodt et al. 1989). The predicted protein has all the features of a typical growth factor receptor with an extracellular ligand binding domain, a transmembrane region, and an intracellular domain containing all 11 structural motifs diagnostic for a protein tyrosine kinase moiety. In addition there are 12 putative glycosylation sites, all located in the potential extracellular domain and an N-terminal signal peptide of 25 amino acids.

Comparison of the primary structure of the protein with other structures in sequence databases revealed marked similarity with receptor tyrosine kinases (RTKs), most prominently with the human EGFR (Ullrich *et al.* 1984). Isolation of other EGFR-related sequences from Xiphophorus showed clearly the presence of an EGFR homologue distinct from the *Tu* gene in the fish genome demonstrating that this gene encodes a novel member of the subclass I RTK family. Based on these findings the *Tu* gene was designated X*mrk* for **X**iphophorus **m**elanoma **r**eceptor **k**inase.

X*mrk* proto-oncogene and oncogenic copies in the Xiphophorus genome

The initial RFLP analysis had revealed the presence of three different copies of X*mrk* in the Xiphophorus genome, which exist as independent loci (Schartl 1990). All of them were isolated and partially characterized. As far as investigated the three genes share a common exon/intron arrangement. The identical arrangement and exon sizes are also found in the EGFR, HER2, and *erb*B3 genes, again pointing towards the close evolutionary kinship of these genes (Adam *et al.* 1991).

All genotypes of Xiphophorus contain one copy of X*mrk* on each sex chromosome, termed INV. INV is also present in other poeciliid fish species and is not associated with the tumour phenotype. Hence, it represents the corresponding proto-oncogene of *Tu*. The remaining two copies (named X or Y according to their chromosomal location), which are only found in animals carrying the *Tu* locus are tied to the macromelanophore spot pattern loci that give rise to melanomas in appropriate crossings. Thus, they are regarded as oncogenic versions of the INV copy (Schartl and Adam 1992).

The nucleotide sequences of all three X*mrk* genes are highly conserved (Adam *et al.* 1991). Only a few sequence differences (< 1 per cent) are found between the proto-oncogene and the two oncogenes in a total of 18 kb of genomic sequences including introns. At present it is not clear whether the observed sequence differences or possible mutations in the so far not analysed extracellular, transmembrane and juxtamembrane domains in the X and Y oncogenes contribute to the process of neoplastic transformation.

Evidence that the additional X- or Y-chromosomal copies of X*mrk* are the crucial, i.e. melanoma-inducing, constituent of the *Tu* locus is provided by 'loss-of-function mutations' of *Tu*. Such mutations arise spontaneously with a very low frequency ($< 10^{-5}$) in the progeny of melanoma-bearing hybrids and are characterized by their inability to develop spontaneous melanoma. One such mutant was found to carry an insertion within an exon of the X-chromosomal X*mrk* copy, rendering the carriers incapable of developing hereditary melanomas (Wittbrodt *et al.* 1989). This demonstrates clearly that the X*mrk* oncogene is absolutely required for tumorigenesis and thus is the critical constituent of the *Tu* locus.

Evaluation of the sequence differences of the three X*mrk* genes with respect to their phylogeny and linkage data (Schartl 1990) strongly supports a model involving a gene duplication which created a new copy of the INV gene. This copy was translocated 2 cM away from the original on the Y-chromosome during the duplication event and at a later stage transferred to the X-chromosome by a sex-chromosomal crossing over.

Transcriptional activation of the X*mrk* oncogene

The X*mrk* genes give rise to transcripts of different size, one of 5.8 kb and one of 4.7 kb; the longer one is derived from the proto-oncogenic INV copy (Wittbrodt *et al.* 1989; Adam *et al.* 1991). This transcript is highly abundant in the form of maternal mRNA in unfertilized eggs and is differentially expressed during organogenesis. In adult non-tumourous animals, expression of the INV gene is restricted to low levels in the skin, fins, and gills. Comparable amounts of RNA are observed in melanomas. Expression in the tumours is not influenced by the presence or the absence of the *R* locus, as the transcript is found at similar levels in melanomas of differing malignancy.

In contrast, expression of the X- and Y-copies of X*mrk* is restricted to one cell type and dependent on the presence or absence of the regulatory locus *R*. The 4.7 kb mRNA is exclusively detected in melanoma. It is important to note that the level of the oncogene transcript here corresponds to the malignancy of the tumours: the amount is low in benign and very high in malignant melanomas. This correlation shows clearly that low level expression of the oncogenic X*mrk* copies eventually carrying 'activating mutations' in the protein is not sufficient for the full-blown melanoma phenotype. It seems that high level expression is the primary cause of the neoplastic transformation.

Autophosphorylation of the X*mrk* protein

Ligand-induced autophosphorylation is thought to be a crucial step in the process of receptor-mediated signalling in the RTK family. Defects in this process lead to either a loss of biological activity, or constitutive activation, or loss of control.

The fact that as yet no ligand for the X*mrk* gene product is known has hampered the analysis of the biochemical properties and biological functions of the receptor. In an approach towards elucidating the role of the protein in normal and neoplastic tissue the Xmrk extracellular domain was replaced by the ligand-binding domain of the closely related human EGFR and the ligand-induced activity of the chimeric HER–Xmrk protein was examined in cells of different origin (Wittbrodt *et al.* 1992).

Ligand-dependent tyrosine kinase activity was observed *in vitro* and in living cells after transient or stable transfection of the chimera into mammalian and fish cells. In contrast, the wild-type Xmrk oncoprotein always displayed significant autophosphorylation in the transfected mammalian cells. The most likely explanation for the observed low level autophosphorylation of Xmrk is the presence of (an) activating mutation(s) in the extracellular or transmembrane domain of the Xmrk oncoprotein both of which are substituted by HER sequences in the chimera. Similar phenomena have been observed in carcinogen-induced tumours of rats, where the product of the *neu* oncogene was found to bear an activating mutation in the transmembrane domain (Bargmann and Weinberg 1988; Stern *et al.* 1988) and in v-*fms*, where point mutations in the extracytoplasmic portion lead to an intracellular activation of the kinase (Roussel *et al.* 1988; Woolford *et al.* 1988). However, such potential 'activating mutations' in the oncogenic copies of X*mrk* would also be present in the wild-type fish, which never develop tumours. There these mutations cannot be effective. They could only elicit their effects in the hybrid fish where the X*mrk* oncogenes are activated through loss of a suppressor locus or gain of an activator locus. Thus, such mutations cannot be the primary cause of the appearance of the melanomas in the hybrids.

When analysed in Xiphophorus cell lines or biopsies from various melanoma and tissues the Xmrk protein has so far been detected only in melanoma cells, where it is the most abundant tyrosine-phosphorylated protein (Wittbrodt *et al.* 1992). This points towards a highly active receptor, whose targets in the intracellular signal transduction are so far unknown.

Cellular substrates for the X*mrk* protein

The X*mrk* gene codes for a tyrosine kinase type cell surface receptor putting it into a family of proteins which upon ligand-dependent activation induce similar primary responses within cells (for review see Ullrich and Schlessinger (1990)). They activate their substrates by phosphorylating them on tyrosine residues. To identify intracellular targets of the Xmrk protein and compare them with substrates of the closely related EGFR, 293 cells were transiently transfected with plasmids containing the HER–X*mrk* or HER gene and assayed for tyrosine-phosphorylated proteins by western blotting using anti-phosphotyrosine antibodies (Wittbrodt *et al.* 1992).

The Xmrk protein and HER seem to share some common substrates as a protein of 72 kDa can be detected in cells containing either of the two receptors. However, human (PLCγ) phospholipase C γ, a prominent target of HER, could not be classified as a major substrate for the Xmrk kinase as judged by these experiments. Several potential substrates are specific for the Xmrk tyrosine kinase; the two most prominent ones, which become phosphorylated after EGF treatment of HER–X*mrk*-expressing cells, have an apparent molecular weight of 105 and 140 kDa. Further analyses of these specific targets will provide important clues to the processes leading to tumorigenesis following activation of X*mrk*.

Transforming ability of X*mrk*

To investigate the transforming ability of the Xmrk protein, NIH3T3 cells were infected with replication-defective retroviruses carrying the cDNAs for HER–Xmrk and Xmrk (Wittbrodt *et al.* 1992). The chimera produced foci of transformed cells at a rate of one focus per virus after stimulation with EGF. This focus formation assay clearly demonstrates that the Xiphophorus RTK can interact with the other components of the mammalian signal transduction apparatus and that the Xmrk protein has an extremely strong transforming potential even in a heterologous cell system.

In contrast to HER–Xmrk the native Xmrk protein did not show any transforming ability in this type of assay. This may indicate that the observed low level of constitutive tyrosine kinase phosphorylation at least in the heterologous cells is not sufficient to elicit a tumorigenic effect. It provides further evidence that 'activating mutations' in the protein alone are not sufficient for the oncogenic function of X*mrk* in the neoplastic transformation of pigment cells.

Other oncogenes possibly involved in the neoplastic phenotype

A generally accepted view is that cancer is the result of a multistep process. At first glance the situation in the Xiphophorus melanoma system appears to be simpler. It seems as if the activation of a single gene is sufficient to induce the neoplastic phenotype of the pigment cells. However, overexpression of X*mrk* leads to complex alterations in the cell, not all of which may be simply explained as physiological consequences of enhanced RTK activity. The so far analysed members of the family of *src*-related cytoplasmic protein tyrosine kinases in Xiphophorus, namely X*src*, X*yes*, and X*fyn*, are highly expressed in the melanomas. X*yes* and X*fyn* mRNAs are more abundant in tumours than in adult brain, which is the organ of preferential expression in non-tumorous fish (Hannig *et al.* 1991). The situation is even more intriguing with X*src*, where more details are known. Its expression is specific for the transformed state: non-transformed pigment cells do not contain detectable levels of the transcript (Raulf *et al.* 1989). The enzymatic activity of the X*src*-encoded protein, $pp60^{Xsrc}$, correlates with the degree of malignancy of the melanomas. As the X*src* gene is not associated structurally with the *Tu* locus, it might represent a 'secondarily' activated oncogene.

A new upstream region acquired by the X*mrk* oncogene

Where do the dramatic differences in the levels of the proto-oncogene and oncogene mRNA originate?

Differential northern analysis demonstrated that the observed size difference of 1.1 kb of the INV transcript on the one hand, and the X- and Y-transcripts on the other hand, could not be attributed to different 3' end formation. Primer extension studies and comparison of the full-lengths cDNAs revealed intriguing differences in the 5' regions of the X*mrk* genes.

In addition, colinearity of INV and the oncogenes is only found 3' to codon 10 in exon 1, while 5' of this point there is no similarity at all. Southern analysis with the divergent 5' regions as probes confirmed the diversity of the proto-oncogene and oncogene in these parts of the genes. It turned out that the known upstream region of the oncogene is identical to that of another gene present in all Xiphophorus

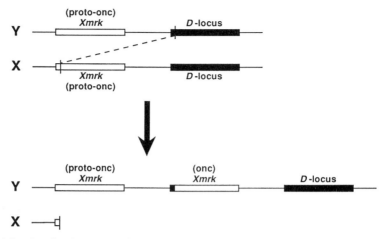

Fig. 14.3. Duplication/recombination model for the generation of the X*mrk* oncogene.

genotypes, irrespective of whether they contain a *Tu* locus or not. These findings indicate that the oncogene has acquired a new promoter as compared with the INV locus. This promoter contains TATA- and CAAT-like sequences (Adam *et al.* 1993) and thus represents the 'non-housekeeping gene' promoter type (Dynan 1986) in contrast to the GC-rich sequences driving transcription of the closely related RTKs like the human EGFR (Ishii *et al.* 1985) or the rat *HER2* gene (Suen and Hung 1990).

The composite structure of the X*mrk* oncogene can be explained simply by extending the previously introduced gene duplication model. A non-homologous recombination event must have taken place between the X*mrk* proto-oncogene and a locus providing the new promoter, therefore this locus was designated *D* (for donor) (see Fig. 14.3).

The breakpoint for the X*mrk* gene lies within exon 1 in the signal peptide which is cleaved off in the mature Xmrk receptor tyrosine kinase. Hence, the mature oncogene and proto-oncogene proteins could have been primarily identical, without excluding the possibility that the oncogene has acquired activating mutations in the course of evolution. Such mutations, however, would appear functional only in the hybrids where the X*mrk* oncogene is deregulated, as parental fish with an identical oncogene are melanoma-free.

This suggests that the genetically defined regulatory locus *R* is either directly or indirectly involved in the control of transcription from the new oncogene promoter and was pre-existing. It is the properties of this accidentally acquired promoter that lead to overexpression of X*mrk* in pigment cells of hybrid genotypes.

Perspective

The identity of the X*mrk* gene product as a RTK provokes further questions, the answers to which should help to understand the role of X*mrk* overexpression in the process of tumour formation. It will be important to identify the ligand and the cellular substrates of Xmrk that are involved in the signal transduction as well as the genes that control X*mrk* activity and those that are activated following X*mrk* stimulation.

The cumulative evidence points towards a major role of the dramatically increased levels of the X*mrk* oncogene transcript in melanoma formation. It suggests that the elevated RNA level in melanomas is due to an increase in transcription caused by the newly acquired oncogene promoter. Structural and functional comparison of the X*mrk* oncogene promoter with that of the *D* locus from which it is obviously derived might point towards important regulatory elements involved in the control of X*mrk* transcription. The fact that the *D* locus is seemingly repetitive (D. Förnzler and M. Schartl, unpublished) poses a problem, since the possibility exists that different copies of this locus might contain different promoter sequences. In this case knowledge of the actual donor sequence involved in the recombination event leading to the generation of the X*mrk* oncogene is required.

So far it has not been possible to decide if the *R* locus is a tumour suppressor gene or an activator of *Tu*; the latter possibility cannot be excluded based on formal genetics. The finding that transcriptional control might be the mechanism through which *R* regulates X*mrk* emphasizes the importance of identifying the factors that control transcription of the oncogenic X*mrk* copies. These analyses will help to decide which model for the mode of action of *R* is correct and might lead to the characterization of the *R* gene. If the product of *R* turns out to be involved in the transcriptional control of X*mrk* it could be a transcription factor itself or interact with such a protein either directly or by altering transcription factor activity through post-translational modifications.

Although all evidence points towards transcriptional regulation of the X*mrk* oncogene probably directly through the *R* locus, there are other possibilities for the function of the *R* gene product. It could stabilize or destabilize the oncogene transcript and thereby lead to the increased mRNA level observed in melanomas, or *R* could code for a ligand, or affect the kinase activity of the receptor, or represent another molecule even further downstream in the signal transduction cascade.

The identification and characterization of the structural and functional components of the Xiphophorus melanoma system will provide essential information as to how a gene can lead to tumour formation in individuals where this gene becomes deregulated.

References

Adam, D., Wittbrodt, J., Telling, A., and Schartl, M. (1988). RFLP for an EGF-receptor related gene associated with the melanoma oncogene locus of *Xiphophorus maculatus*. *Nucleic Acids Res.*, **16**, 7212.

Adam, D., Mäueler, W., and Schartl, M. (1991). Transcriptional activation of the melanoma inducing X*mrk* oncogene in Xiphophorus. *Oncogene*, **6**, 73–80.

Adam, D., Dimitrijevic, N., and Schartl, M. (1993). Tumor suppression in Xiphophorus by an accidentally acquired promoter. *Science*, **259**, 816–19.

Ahuja, M. R. and Anders, F. (1976). A genetic concept of the origin of cancer, based in part upon studies of neoplasms in fishes. *Prog. Exp. Tumor Res.*, **20**, 380–97.

Ahuja, M. R., Schwab, M., and Anders, F. (1980). Linkage between a regulatory locus for melanoma cell differentiation and an esterase locus in Xiphophorus. *J. Heret.*, **71**, 403–7.

Anders, A., Anders, F., and Klinke, K. (1973). Regulation of gene expression in the Gordon—Kosswig melanoma system I, II. In *Genetics and mutagenesis of fish*, (ed. J. H. Schröder), pp. 33–63. Springer-Verlag, New York.

Anders, F., Schartl, M., Barnekow, A., and Anders, A. (1984). Xiphophorus as an *in vivo* model for studies on normal and defective control of oncogenes. *Adv. Cancer Res.*, **42**, 191–275.

Anders, F. (1990). A biologist's view of human cancer. *Modern trends in human leukemia*, 8, (ed. R. Neth *et al.*) XXII–XLV. Springer Verlag, Berlin.

Anders, F. (1991). Contributions of the Gordon—Kosswig melanoma system to the present concept of neoplasia. *Pigment. Cell. Res.*, **3**, 7–29.

Anders, A., Gröger, H., Anders, F., Zechel, C., Smith, A., and Schlatterer, B. (1991). Discrimination of initiating and promoting carcinogen in fish. *Ann. Rev. Vet.*, **22**, 273–94.

Atz, J. W. (1962). Effects of hybridization on pigmentation in fishes of the genus Xiphophorus. *Zoologica*, **47**, 153–81.

Bargmann, C. I. and Weinberg, R. A. (1988). Increased tyrosine kinase activity associated with the protein encoded by the activated *neu* oncogene. *Proc. Natl Acad. Sci. USA*, **85**, 5394–8.

Bauer, K. H. (1928). *Mutationstheorie der Geschwulstentstehung*. Springer-Verlag, Berlin and New York.

Bishop, J. M. (1987). The molecular genetics of cancer. *Science*, **235**, 305–11.

Boveri, T. (1914). *Zur Frage der Entstehung maligner Tumoren*. Gustav Fischer Verlag, Jena.

Breider, H. (1952). Über Melanosarkome, Melaninbildung und homologe Zellmechanismen. *Strahlentherapie*, **88**, 619–39.

Collins, F. S. (1992). Positional cloning: let's not call it reverse anymore. *Nature Genet.*, **1**, 3–6.

Dynan, W. S. (1986). Promoters for housekeeping genes. *Trends Genet.*, **2**, 196–7.

Eker, R. and Mossige, J. (1961). A dominant gene for renal adenomas in the rat. *Nature*, **189**, 858–9.

Fearon, E. R. and Vogelstein, B. (1990). A genetic model for colorectal tumorigenesis. *Cell*, **61**, 759–67.

Förnzler, D., Wittbrodt, J., and Schartl, M. (1991). Analysis of an esterase linked to a locus involved in the regulation of the melanoma oncogene and isolation of polymorphic marker sequences in Xiphophorus. *Biochem. Genet.*, **29**, 509–24.

Friend, S. J., Bernards, R., Rogelj, S., Weinberg, R. A., Rapaport, J. M., Alberts, D. M., *et al.* (1986). A human DNA segment with properties of the gene that predisposes to retinoblastoma and osteosarcoma. *Nature*, **323**, 643–6.

Gessler, M., Poustka, A., Cavenee, W., Neve, R. L., Orkin, S. H., and Bruns, G. A. P. (1990). Homozygous deletion in Wilms' tumours of a zinc-finger gene identified by chromosome jumping. *Nature*, **343**, 774–8.

Gorbman, A. and Gordon, M. (1951). Spontaneous thyroidal tumors in the swordtail *Xiphophorus montezumae*. *Cancer Res.*, **11**, 184–7.

Gordon, M. (1927). The genetics of viviparous top-minnow Platypoecilus: the inheritance of two kinds of melanophores. *Genetics*, **12**, 253–83.

Gordon, M. (1931). Hereditary basis of melanosis in hybrid fishes. *Am. J. Cancer*, **15**, 1495–519.

Gordon, M. (1947). Genetics of ocular-tumor development in fishes (preliminary report). *J. Natl Cancer Inst.*, **7**, 87–92.

Gordon, M. (1958). A genetic concept for the origin of melanomas. *Ann. NY Acad. Sci.*, **71**, 1213–22.

Hannig, G., Ottilie, S., and Schartl, M. (1991). Conservation of structure and expression of the c-*yes* and *fyn* genes in lower vertebrates. *Oncogene*, **6**, 361–9.

Häussler, G. (1928). Über Melanombildungen bei Bastarden von *Xiphophorus maculatus* var. *rubra*. *Klin. Wochenschr.*, **7**, 1561–2.
Harris, H., Miller, O. J., Klein, G., Worst, P., and Taschibana, T. (1969). Suppression of malignancy by cell fusion. *Nature*, **223**, 363–8.
Hyodo-Taguchi, Y. and Matsudaira, H. (1987). Higher susceptibility to N-methyl-N'-nitro-N-nitrosoguanidine-induced tumorigenesis in an interstrain hybrid of the fish, *Oryzias latipes* (Medaka). *Jpn J. Cancer Res.* (Gann), **78**, 487–93.
Ishii, S., Imamoto, F., Yamanashi, Y., Toyoshima, K., and Yamamoto, T. (1985). Characterization of the promoter region of the human c-*erb*B-2 proto-oncogene. *Proc. Natl Acad. Sci. USA*, **84**, 4374–8.
Kallman, K. D. (1970). Different genetic basis of identical pigment patterns in two populations of platyfish, *Xiphophorus maculatus*. *Copeia*, **3**, 472–87.
Kallman, K. D. (1975). The platyfish, *Xiphophorus maculatus*. In *Handbook of genetics*, 4, (ed. R. C. King), pp. 8–132. Plenum Press, New York.
Kallmann, K. D. (1984). A new look at sex determination in Poeciliid fishes. In *Evolutionary genetics of fishes*, (ed. B. J. Turner), pp. 95–171. Plenum Press, New York.
Knudson, A. G. (1971). Mutation and cancer: statistical study of retinoblastoma. *Proc. Natl Acad. Sci. USA*, **68**, 820–3.
Kosswig, C. (1928). Über Kreuzungen zwischen den Teleostiern *Xiphophorus helleri* und *Platypoecilus maculatus*. *Z. Indukt. Abstammungs-Vererbungsl.* **47**, 150–8.
Kosswig, C. (1929). Das Gen in fremder Erbmasse. *Züchter*, **1**, 152–7.
Kosswig, C. (1965). Genetische Grundlagen des Polymorphismus. *Zool. Anz.*, **175**, 21–50.
Koufos, A., Hansen, M. F., Copeland, N. G., Jenkins, N. A., Lampkin, B. C., and Cavenee, W. K. (1985). Loss of heterozygosity in three embryonal tumors suggests a common pathogenetic mechanism. *Nature*, **316**, 330–4.
Mäueler, W., Raulf, F., and Schartl, M. (1988a). Expression of proto-oncogenes in embryonic, adult, and transformed tissue of Xiphophorus (Teleostei: Poeciliidae). *Oncogene*, **2**, 421–30.
Mäueler, W., Barnekow, A., Eigenbrodt, E., Raulf, F., Falk, H. F., Telling, A., et al. (1988b). Different regulation of oncogene expression in tumor and embryonal cells of Xiphophorus. *Oncogene*, **3**, 113–22.
Malkin, D., Li, F. P., Strong, L. C., Fraumeni, J. F., Nelson, C. E., Kim, D. H., et al. (1990). Germ line p53 mutations in a familial syndrome of breast cancer, sarcomas, and other neoplasms. *Science*, **250**, 1233–8.
Morizot, D. C. and Siciliano, M. J. (1983). Linkage group V of platyfishes and swordtails of the genus Xiphophorus (Poeciliidae): linkage of loci for malate dehydrogenase-2 and esterase-1 and esterase-4 with a gene controlling the severity of hybrid melanomas. *J. Natl Cancer Inst.*, **71**, 809–13.
Orkin, S. J. (1986). Reverse genetics and human disease. *Cell*, **47**, 845–50.
Perlmutter, A. and Potter, H. (1988). Hyperthermic suppression of a genetically programmed melanoma in hybrids of fishes: genus Xiphophorus. *J. Cancer Res. Clin. Oncol.*, **114**, 339–62.
Raulf, F., Mäueler,W., Robertson, S. M., and Schartl, M. (1989). Localization of cellular *src* mRNA during development and in the differentiated bipolar neurons of the adult neural retina in Xiphophorus. *Oncogene Res.*, **5**, 39–47.
Roussel, M. F., Downing, J. R., Rettenmier, C. W., and Sherr, C. J. (1988). A point mutation in the extracellular domain of the human CSF-1 receptor (c-*fms* proto-oncogene product) activates its transforming potential. *Cell*, **55**, 979–88.
Schartl, A., Schartl, M., and Anders, F. (1982). Promotion and regression of neoplasia by testosterone promoted cell differentiation in Xiphophorus and Girardinus. *Carcinogenesis*, **7**, 427–34.

Schartl, M. (1988). A sex chromosomal restriction-fragment-length marker linked to melanoma-determining *Tu* loci in Xiphophorus. *Genetics*, **119**, 679–85.

Schartl, M. and Peter, R. U. (1988). Progressive growth of fish tumors after transplantation into thymus-aplastic (*nu/nu*) mice. *Cancer Res.*, **48**, 741–4.

Schartl, M. (1990). Homology of melanoma-inducing loci in the genus Xiphophorus. *Genetics*, **126**, 1083–91.

Schartl, M. and Adam, D. (1992). Molecular cloning, structural characterization and analysis of transcription of the melanoma oncogene of Xiphophorus. *Pigment. Cell Res. Suppl.*, **2**, 173–80.

Schartl, M., Wittbrodt, J., Mäueler, W., Raulf, F., Adam, D., Hannig, G., et al. (1993). Oncogenes and melanoma formation in *Xiphophorus* (Teleostei: Poeciliidae). In *Trends in ichthyology*, (ed. J. H. Schröder and M. Schartl) pp. 79–92. Blackwell, Oxford, England.

Scholz, A. (1977). Untersuchungen zur Dinitrochlorbenzol (DNCB)-induzierten Regression kreuzungsbedingter Melanome der lebendgebärenden Zahnkarpfen (Poeciliidae). Staatsexamensarbeit, Universität Giessen.

Siciliano, M. J., Perlmutter, A., and Clark, E. (1971). Effects of sex on the development of melanoma in hybrid fish of the genus Xiphophorus. *Cancer Res.*, **31**, 725–9.

Siciliano, M. J. and Perlmutter, A. (1972). Maternal effect on development of melanoma in hybrid fish of the genus Xiphophorus. *J. Natl Cancer Inst.*, **49**, 415–20.

Siciliano, M. J., Morizot, D. C., and Wright, D. A. (1976). Factors responsible for platyfish–swordtail hybrid melanoma—many or few? In *Melanomas: basic properties and clinical behaviour*, (ed. V. Riley), pp. 47–54. Karger, Basel.

Stern, D. F., Kamps, M. P., and Lao, H. (1988). Oncogenic activation of p185neu stimulates tyrosine phosphorylation *in vivo*. *Mol. Cell. Biol.*, **8**, 3969–73.

Suen, T. and Hung, M. (1990). Multiple *cis*- and *trans*-acting elements involved in regulation of the *neu* gene. *Mol. Cell. Biol.*, **10**, 6306–15.

Ullrich, A., Coussens, L., Hayflick, J. S., Dull, T. J., Gray, A., Tam, A. W., et al. (1984). Human epidermal growth factor receptor cDNA sequence and aberrant expression of the amplified gene in A431 epidermoid carcinoma cells. *Nature*, **309**, 418–25.

Ullrich, A. and Schlessinger, J. (1990). Signal transduction by receptors with tyrosine kinase activity. *Cell*, **61**, 203–12.

Weinberg, R. A. (1989). Oncogenes, antioncogenes, and the molecular bases of multistep carcinogenesis. *Cancer Res.*, **49**, 3713–21.

Wittbrodt, J., Adam, D., Malitschek, B., Mäueler, W., Raulf, F., Telling, A., et al. (1989). Novel putative receptor tyrosine kinase encoded by the melanoma-inducing *Tu* locus in Xiphophorus. *Nature*, **341**, 415–21.

Wittbrodt, J., Lammers, R., Malitschek, B., Ullrich, A., and Schartl, M. (1992). The X*mrk* receptor tyrosine kinase is activated in Xiphophorus malignant melanoma. *EMBO J.*, **11**, 4239–46.

Woolford, J. W., McAuliffe, A., and Rohrschneider, L. R. (1988). Activation of the feline c-*fms* proto-oncogene: multiple alterations are required to generate a fully transformed phenotype. *Cell*, **55**, 965–77.

Zander, C. D. (1969). Über die Entstehung und Veränderung von Farbmustern in der Gattung Xiphophorus (Pisces). *Mitt. Hamburg Zool. Mus. Inst.*, **66**, 241–71.

Zechel, C., Schleenbecker, U., Anders, A., and Anders, F. (1988). v-*erb*B related sequences in Xiphophorus that map to melanoma determining Mendelian loci and overexpress in a melanoma cell line. *Oncogene*, **3**, 605–17.

15. Loss-of-function mutations in human cancer

Webster K. Cavenee

The breadth of biological questions that have been addressed through the technique of somatic cell fusion is indeed impressive. Perhaps the greatest challenges for the approach are situations in which the question is directed at problems of tissue organization or where intertypic cellular activities are required.

One such situation is the growth of tumours *in vivo* where escape from positional and growth regulatory controls is required. There is a growing realization that the group of diseases collectively called cancer arise largely through genetic alteration. This interpretation receives a good deal of circumstantial support through the study of human populations. Specific types of tumours can aggregate in families and sometimes occur with the formal transmission characteristics of Mendelian autosomal dominant predisposition (Mulvihill 1977). It is reasonably easy to construct models for this when the tumours occur singly and are of homogeneous types (Knudson 1971), but less so when seemingly unrelated tumour types appear to be the phenotypic manifestation of the trait. Even in individuals with homogeneous initial disease, second tumours of entirely different types can occur at distant body sites. In fact, in some instances predisposition to the development of specific, often rare, tumours can be manifest as generalized organismal malformation. In any of these cases, if the tumours are rare, a statistical argument can be put forward that such associations are so unlikely as to suggest an aetiological relationship.

Cytogenetic analysis (Mittelman 1985) shows that various chromosomal aberrations of the germ-line appear to result increased propensities for the development of tumours. Tumours often have specific chromosomal rearrangements and, sometimes, such aberrations resemble those which, when inherited, predispose to similar disease. The major challenge which arises from these observations is to separate those chromosomal derangements that are causal from those that are caused by the neoplastic process. The somatic nature of most cancers requires that these genetic abnormalities be acquired during the replication of cells from the specifically affected organ. This would seem to make it likely that they occur during the process of mitotic duplication and segregation of chromosomes from progenitor to daughter cells. Of course, many aberrations in this process would be lethal while others would confer no particular selective advantage. The rare events comprising viable, advantageous, and transforming mitotic abnormalities could conceivably represent the molecular underpinnings of the neoplastic process.

How, then, do we define the specificity and function of the genetic lesions suggested by the foregoing? One pioneering approach used cell fusion to determine the genetic behaviour of the phenotypes associated with cell transformation. These incisive experiments basically involved examination of the malignant potential of the rodent cell hybrids formed by fusion of normal (or low malignancy) cells with those of greater malignancy (Harris *et al.* 1969). The results indicated that the cell hybrids behaved similarly to the parent of lesser malignancy, leading to the inference that malignancy could be suppressed. The implication of this is that the genetic lesions giving rise to cell transformation were genetically recessive and likely to be due to the elimination of some normal function. A great deal of elegant work has been done with this system and its human analogues and is reviewed elsewhere in this volume.

Although the cell fusion approach has proven to be experimentally powerful, it has several limitations. First, the act of cell fusion creates a gene dosage problem since the hybrids contain essentially tetraploid numbers of each chromosome. Secondly, cell lines do not exist for most human tumour types and many which do are not easily cloned thus several limiting the generality which can be gained from the experiments. Thirdly, the extension of interpretations gained through the analysis of transformed cell traits (or tumorigenic behaviour when implanted into animals) to the natural initiation and growth of tumours *in vivo* may be inappropriate. Finally, the assays do not allow a discrimination between those lesions involved in predisposition and those which are active in later stages of malignant progression. In this chapter, I will discuss some of the work of my colleagues and I have done using human cancers to address these deficiencies.

Genetics of cancer predisposition

In experimental chemical carcinogenesis, the earliest event in tumorigenesis is termed the initiating event. In the human population such initiations may be transmitted as inherited predisposition. At least 50 different forms of human cancer have been observed to aggregate in families as well as to have corresponding sporadic forms. Obviously, these individuals represent a valuable resource in attempts to define the targets of initial genotoxic damage. In many of these cases the aggregation occurs with a pattern consistent with the transmission of an autosomal dominant Mendelian trait. This interpretation is, however, at odds with some lines of evidence. First, if a single mutation was sufficient in and of itself to elicit a tumour, then families segregating for autosomal dominant forms of cancer would be expected to have no normal tissue in the diseased organ. This expectations is in direct contrast to the clinical observation of discrete tumour foci amidst normal, functional tissue in such individuals. Secondly, epidemiological analyses by Knudson (1971) of sporadic and familial forms of several cancer types have indicated that the conversion of a normal cell to a tumour cell requires multiple events.

This latter point leads to a consideration of the development of cancer as a clonal evolution of cells that have undergone a series of genetic alterations that confer growth advantages at specific stages in the process which is outlined in Fig. 15.1(A)

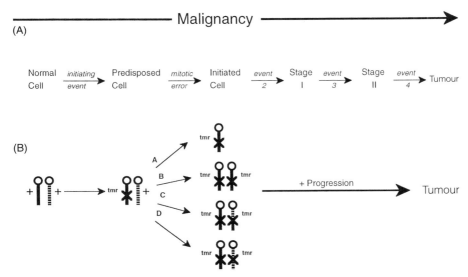

Fig. 15.1. (A) Accumulation of genetic damage culminates in malignancy. The pathway can initiate in a single somatic cell in sporadic cases or in any cell in heritable cases. (B) A model for chromosomal mechanisms which can accomplish the conversion of a normal cell to a cell that is homozygous for inactivation of a tumour (*TMR*) locus. Predisposition occurs either by inheritance or by somatic occurrence of a mutation which converts a wild-type (+) allele to an inactive allele (*TMR*). A tumour could then occur by elimination of the remaining wild-type allele by nondisjunction (A), nondisjunction/duplication (B), mitotic recombination (C), or regional aberration (D). Additional genetic damage is required for the progression of homozygously defective cells to frank neoplasia.

(Nowell 1976). Our work over the past few years has been directed at defining these genetic events. In the case of predisposing lesions, we have taken advantage of the prescient observations of Knudson (1971) which suggested the requirement for as few as two mutations for entry of a precursor cell into the neoplastic pathway. In this model, hereditary cases have inherited a germinal mutation which does not, in itself, cause the tumour but rather predisposes each precursor cell to a further transforming event. In this model, the non-hereditary cases would also result from two mutations except that these events would have to occur in the same somatic cell. Thus, the two forms of the disease could be viewed as resulting from the same two-step process, at the level of the aberrant cell, the difference being the inheritance or somatic occurrence of the first mutation.

We proposed (Cavenee *et al.* 1983) that the second step in tumorigenesis in both heritable and sporadic tumours involves somatic alteration of the normal allele at the 'tumour locus' in a way that unmasks the mutant allele. Thus, the first mutation in this process, although it may be inherited as an autosomal dominant trait at the organismal level, would be expected to have the properties of a recessive defect at the level of the individual precursor cell. In this model, (Fig. 15.1(B)) the heritable form of the disease arises as a germinal mutation of the *TMR* locus and is inherited by an individual who, therefore, is an obligate heterozygote (*tmr*/+) at the *TMR* locus in each of his somatic and germ cells. A subsequent event in any of his target cells which results in homozygosity for the mutant allele (that is, mutant at the *TMR* locus on both chromosome homologues) will result in a tumour clone. The chromosomal mechanisms which could accomplish this loss of constitutional heterozygosity include: mitotic nondisjunction with loss of the wild-type chromosome, which would results in homozygosity at all loci on the chromosome; mitotic nondisjunction with duplication of the mutant chromosome which results in homozygosity at all loci on the chromosome; or mitotic recombination between the chromosomal homologues with a breakpoint between the *TMR* locus and the centromere, which would result in heterozygosity at loci in the proximal region and homozygosity throughout the rest of the chromosome including the *TMR* locus. Regional events such as gene conversion, deletion, or mutation must also be considered. Heritable and sporadic retinoblastoma could each arise through the appearance of homozygosity at the *TMR* locus, the difference being two somatic events in the sporadic case as compared with one germinal and one somatic event in the heritable case.

The test of this hypothesis was first made with the childhood eye tumour, retinoblastoma. This disease had many advantages for the analysis. One was the finding of constitutional deletions involving the chromosome region 13q14 in normal tissues of 3–5 per cent of children with bilateral disease (Francke 1976). Another was that evidence had been obtained suggesting that the same locus was involved in retinoblastoma cases lacking an apparent chromosomal deletion through the demonstration of tight genetic linkage between the disease and isoforms of esterase D, a moderately polymorphic locus mapped to 13q14 (Sparkes *et al.* 1983).

In order to be able to distinguish chromosomes we used chromosome-specific, single copy segments of the human genome, isolated in recombinant DNA form to recognize polymorphisms at the corresponding chromosomal locus. Sequence variation in restriction endonuclease recognition sites, giving rise to restriction fragment length polymorphism (RFLP) in the locus defined by the probe, are revealed as distinct bands on an autoradiogram; they represent alleles of the locus (one from the paternally derived and one from the maternally derived chromosomal homologue) and behave as Mendelian codominant alleles in family studies. These RFLP markers can be used as linkage markers in inherited disorders, including retinoblastoma. If a disease locus is located close to a polymorphic RFLP marker locus it is likely to cosegregate with that marker in a family. Therefore, the genotype of DNA markers can be used to infer the genotype at the retinoblastoma locus, and thus to predict if the offspring has inherited the predisposition. Chromosome segregation during tumorigenesis can also be determined in each patient by comparing the child's constitutional and tumour genotypes at each of these marker loci.

Recombinant DNA segments were isolated from human chromosome 13 and used to determine somatic changes in the germ-line genotypes of many retinoblastoma cases. The data have shown that the predicted chromosomal events are common as they were detected in about three-quarters of all retinoblastoma tumours. These

Table 15.1. Loss of heterozygosity for loci on chromosome 13q in retinoblastomas

		Alleles present at locus					
Patient	Tissue[a]	D13S1	D13S7	D13S4	D13S5	D13S3	Mechanism
(A) Sporadic							
Retin LA69	N	**1,2**[b]	1,1	**1,2**	**1,2**	1,1	Chromosome
	T	**2**	1	**1**	**2**	1	loss
Retin 409	N	2,2	**1,2**	2,2	1,1	**1,2**	Isotrisomy
	T	2,2,2	**1,1,1**	2,2,2	1,1,1	**2,2,2**	
Retin 412	N	**1,2**	1,1	**1,2**	1,1	2,2	Mitotic
	T	**1,2**	1,1	**2,2**	1,1	2,2	recombination
(B) Heritable							
Retin KS2H	N	2,2	**1,2**	2,2	1,1	1,1	Isodisomy
	T	2,2	**2,2**	2,2	1,1	1,1	
Retin 462F	N	**1,2**[c]	1,1	**1,2**	**1,2**	1,1	Mitotic
	T	**1,2**	1,1	**2,2**	**1,1**	1,1	recombination

[a]N, normal; T, tumour.
[b]Alleles designated in bold type are combinations that were heterozygous in constitutional tissue.
[c]D13S6 examined, not D13S1.

rearrangements fell into four different classes as illustrated in Fig. 15.1. For example in one series, 20 of 33 tumours had one constitutional allele missing at all informative loci along the whole chromosome; 19 of these tumours contained two intact chromosomes 13 as determined either by cytogenetic analysis of the tumour cells or densitometric quantification of the autoradiographic signal of the remaining alleles. Therefore, the loss of alleles along the chromosome must have involved two separate events; a nondisjunction resulting in loss of one chromosomal homologue followed by a duplication of the remaining homologue or an abnormal mitotic segregation of the chromosomes resulting in isodisomy as shown for Retin 409 in Table 15.1(A). In one case, data consistent with the sole loss of chromosome 13 were obtained. Evidence for mitotic recombination between the chromosome homologues was provided in four of the 33 tumours (one example is Retin 412, Table 15.1(A)). The constitutional genotype was maintained at all informative loci in nine of the 33 tumours, and therefore, in these cases, the mechanism of attainment of homozygosity could not be determined. These studies strongly suggest that the second event in tumour initiation is comprised of a specific chromosomal rearrangement involving physical loss of the balancing wild-type allele at the *RB1* locus. This inference was corroborated (Cavenee *et al.* 1985) by examining cases of heritable retinoblastoma and showing that the chromosome 13 homologue retained in the tumours was derived from the affected parent, as would be predicted. Two examples are shown in Table 15.1(B).

It is noteworthy that, although the unmasking of predisposing mutations at the *RB1* locus occurs in mechanistically similar ways in sporadic and heritable retino-

Table 15.2. Loss of heterozygosity for loci on chromosome 13q in osteosarcomas

Patient	Tissue[a]	Alleles present at locus					Mechanism
		D13S1	D13S7	D13S4	D13S5	D13S3	
(A) Sporadic							
Osteo 03	N	**1,2**[b]	1,1	2,2	–[c]	**1,2**	Isodisomy
	T	**2,2**	1,1	2,2	–	**1,1**	
Osteo 06	N	1,2	1,1	1,1	**1,2**	1,1	Isodisomy
	T	1,1	1,1	1,1	**1,1**	1,1	
Osteo 09	N	2,2	1,1	2,2	2,2	**1,2**	Isodisomy
	T	2,2	1,1	2,2	2,2	**1,1**	
(B) Second primary to retinoblastoma							
Rb1-1	N	**1,2**	1,1	**1,2**	**1,2**	2,2	Translocation, isodisomy
	T	**1,1**	1,1	**1,1,2**	**1,2,2**	2,2,2	
Rb108	N	**1,2**	1,1	2,2	1,1	**1,2**	Translocation isodisomy
	T	**1,1,2**	1,1	2,2	1,1	**1,1**	

[a]*N, normal; T, tumour.
[b]Alleles designated in bold type are combinations that were heterozygous in constitutional tissue.
[c]–, not determined.

blastoma cases, only the latter carry the initial mutation in each of their cells. Heritable cases also seem to be at greatly increased risk for the development of further primary tumours, particularly osteogenic sarcomas. This high propensity may not be merely fortuitous but may be genetically determined by the predisposing *RBl* mutation. This notion of a pathogenetic causality in the clinical association between these two rare tumour types was tested by determining the constitutional and osteosarcoma genotypes at RFLP loci on chromosome 13; representative data are shown in Table 15.2. Osteosarcomas arising in retinoblastoma patients had become specifically homozygous around the chromosomal region carrying the *RBl* locus (Table 15.2(B)). Furthermore, these same chromosomal mechanisms were observed in sporadic osteosarcomas (Table 15.2(A)), suggesting a genetic similarity in pathogenetic causality. These findings are of obvious relevance to the interpretation of human mixed cancer families as they suggest differential expression of a single pleiotropic mutation in the aetiology of clinically associated cancers of different histological types, perhaps augmented by differing progressional events (Fig. 15.2).

A likely explanation for the association between retinoblastoma and osteosarcoma is that both tumours arise subsequent to chromosomal mechanisms which unmask recessive mutations either in one common locus that is involved in normal

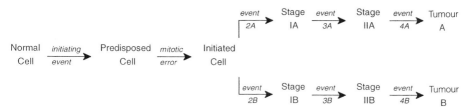

Fig. 15.2 Tumours of different types could arise subsequent to a common predisposing mutation if progressional damage of different types is either tissue-specific in effect or elicits differentiation along two different lineages.

regulation of differentiation of both tissues, or in separate loci that are located close together within chromosome region 13q14. A direct test of this hypothesis was made possible by the isolation of the target for these mutations, the retinoblastoma susceptibility gene (Friend *et al.* 1986). Detailed studies of the defects in this gene over the past few years have shown a variety of mutations that share the effect of elimination of the *RBl* gene product; such mutations become homozygous through the mechanisms described above (Newsham and Cavenee 1990). Perhaps most important in the development of this model is the demonstration of diminished growth or tumorigenicity of retinoblastoma and osteosarcoma cells upon retroviral transfer of a wild-type copy of the *RBl* gene (Huang *et al.* 1988).

The information derived from these studies raises two points relevant to familial predisposition to cancer. Chromosomal mechanisms capable of unmasking predisposing recessive mutations occur in more than one tumour and, at least for chromosome 13, clinically associated tumours share this mechanism of pathogenesis. This latter point suggests that these loci have pleiotropic tissue specificity; however, this pleiotropy appears to be restricted to a small number of tissue types. Finally, since the experiments that allowed detection of genetic alterations which occurred early enough to comprise monoclonal characteristics of the tumour are conceptually and experimentally fairly simple, they have been extended to a variety of human tumour types. Table 15.3 documents the remarkable success of the approach (Lasko *et al.* 1991).

Virtually every type of human cancer examined has shown regions of the tumour genome which have become homozygous. These include common and rare cancers, diseases which affect primarily children or adults, and neoplasia of almost every ontogeny from which human cancer can arise. The data in most of these cases can be fitted to the model outlined in Fig. 15.1 such that predispositions that are somatically recessive first undergo homozygosis and then are compounded by genotoxic damage which leads to increasingly malignant phenotypes. The concerted losses of alleles at

Table 15.3. Losses of heterozygosity in human tumours.

Chromosome region	Tumours
1p	Neuroblastoma, melanoma, breast carcinoma, pheochromocytoma, medullary thyroid carcinoma
2	Uveal melanoma
3p	Renal cell carcinoma, testicular, small cell lung carcinoma, uterine carcinoma
5q	Colon carcinoma
10	Glioblastoma
11p	Wilms' tumour, embryonal, rhabdomyosarcoma, hepatoblastoma, adrenal carcinoma, breast carcinoma, transitional cell bladder carcinoma, hepatocellular carcinoma, testicular tumours
11q	Insulinomas, parathyroid tumours
13q	Retinoblastoma, osteosarcoma, small cell lung carcinoma, breast carcinoma, hepatocellular carcinoma
14q	Neuroblastoma
17p	Colon carcinoma, breast carcinoma, osteosarcoma, small cell lung carcinoma, astrocytoma
18q	Colon carcinoma
22	Acoustic neuroma, meningioma

loci on several chromosomes in some tumours (e.g. breast carcinoma, colon carcinoma, and small cell lung carcinoma) suggest that losses of heterozygosity probably play a role in later stages of malignant progression as well.

Genetics of cancer progression

The foregoing section suggests that the attainment of complete defectiveness at a 'tumour locus' is one means of initiating the pathway of tumorigenesis. There is no *a priori* reason why similar mechanisms should be excluded from comprising at least some more distal events as well. In order to test this hypothesis we searched for a disease system characterized by the increasing acquisition of histologically defined malignant criteria. We chose the glial tumours for our first efforts at utilizing genotypic analyses to place tumours into various stages of malignant progression. Gliomas, as a class, are the most common primary neoplasms of the central nervous system. Tumours of this type can be subclassified according to their cellular differentiation, displaying either astrocytic, oligodendrocytic, ependymal, or mixed composition, with astroyctic tumours occurring most frequently. Prognoses for individuals having astrocytoma vary according to the histopathologically assessed malignancy grade of the tumour; however, all adult malignancy grades (grades II–IV) of astrocytoma respond poorly to radiation and/or chemotherapy and the 5-year survival rate for individuals with the most malignant form, glioblastoma (astrocytoma grade IV), is less than 5 per cent. The propensity of low malignancy grade astrocytomas to relapse with recurrent tumours that often display a malignant progression accentuates the severity of the disease. Several cytogenetic analyses have described frequent chromosome aberrations in direct preparations and short-term cultures of astrocytomas (Bigner *et al.* 1989). In contrast, studies involving astrocytic tumours of low malignancy grade have consistently demonstrated cells with normal karyotypes. This could be due to the analysed mitoses not being representative of the tumour cells or, alternatively, genetic information could be lost following chromosomal mechanisms not detectable by cytogenetic analysis such as those described in Fig. 15.1 and the preceding section.

To determine whether astrocytomas and glioblastomas share a progressional lineage, and whether specific losses of heterozygosity were preferentially associated with some of the stages of the pathway, we initially compared constitutional and tumour genotypes at loci on each human chromosome for samples from a series of adult cases of astrocytoma representing each malignancy grade. Data obtained for the loci on chromosome 17p are shown in Table 15.4. This subset of tumours with astrocytic differentiation lost heterozygosity in a clonal fashion through the loss and duplication of mitotic recombination mechanisms described in Fig. 15.1, regardless of malignancy stage (James *et al.* 1989). We cannot now be sure that such events comprise the initial steps; they may occur somewhat later in the progression. They do, however, indicate shared insults among all the stages and, as such, in all likelihood represent early events which appear to confer selective advantages used in the outgrowth of the clone. Clearly, it would be desirable to uncover events which occurred at progressively later stages.

Cytogenetic analyses have provided important clues to the gross chromosomal changes taking place in these tumours. For example, monosomy for chromosome 10 has been detected in about one-third of grade IV astrocytomas. We therefore compared constitutional and tumour genotypes at loci on chromosome 10 for the cases of astrocytoma analysed for chromosome 17 above. Allelic combinations were determined with probes homologous to three different chromosome 10 loci: D10S1, D10S4, and PLAU. Representative data obtained with samples of various histological grades are shown in Table 15.4. Each of the grade IV (glioblastoma; GB) tumours examined showed loss of constitutional heterozygosity at one or more of the chromosome 10 loci and these losses appeared to be elicited by nondisjunction resulting in monosomy. In sharp distinction, none of the tumours of lower malignancy grades showed loss of alleles at any of the chromosome 10 loci examined (James *et al.* 1988).

These data have two major implications. The first is the demonstration of a clonal origin of the cells comprising these tumours. The cellular pleomorphism of malignant astrocytomas and karyotypic heterogeneity of *in vitro* derived cell sub-

Table 15.4. Loss of heterozygosity in stages of glioma malignancy

Histological grade	Patient	Tissue[a]	Alleles present at locus							Mechanism	
			Chromosome 17			Chromosome 10					
			GH1	D17S71	D17S5	D10S1	D10S4	PLAU	Chr. 17	Chr. 10
AII	A3	N	**1,2**[b]	**1,2**	**1,2**	1,1	**1,2**	—[c]		
		T	1,2	1,2	1,1	1,1	1,2		Mitotic recombination	None
	A5	N	**1,2**	**1,2**	**1,2**	2,2	**1,2**	**1,2**		
		T	1,2	1,1	2,2	2,2	1,2	1,2	Mitotic recombination	None
AIII	G2	N	1,1	**1,2**	**1,2**	**1,2**	**1,2**	**1,2**		
		T	1,1	1,2	1,1	1,2	1,2	1,2	Mitotic recombination	None
	G25	N	**1,2**	**1,2**	**1,2**	**1,2**	**1,2**	1,1		
		T	1,2	1,1	1,1	1,2	1,2	1,1	Mitotic recombination	None
AIV	G14	N	**1,2**	**1,2**	**1,2**	**1,2**	**1,2**	2,2		
		T	1,1	2,2	1,1	1	2	2	Isodisomy	Monosomy
	G21	N	**1,2**	**1,2**	**1,2**	1,1	**1,2**	1,1		
		T	1,2	2,2	1,1	1	1	1	Mitotic recombination	Monosomy

[a]N, normal; T, tumour.
[b]Alleles designated in bold type are combinations that were heterozygous in constitutional tissue.
[c]Not determined.

populations arising from primary tumours has complicated attempts to determine the nature of any relationship between the cells that constitute this type of neoplasm. These data show that grade IV astrocytomas arise from the expansion of cells deficient in all or part of chromosomes 10 and 17. The second issue involves the histopathological evidence that astrocytomas progress and become more malignant with time. Because the losses of heterozygosity for chromosome 10 loci were restricted to tumours of the highest malignancy grade, it may be that this aberration is an event of tumour progression, rather than of initiation. Conversely, it is possible that the aetiologies and ontogenies of astrocytomas exhibiting low and high degrees of cellular differentiation have no interrelating molecular pathways. However, the post-operative, post-therapeutic recurrence and histological progression of astrocytoma is well documented, providing clinical support for the ontogenic relationship suggested by the shared chromosome 17p aberrations.

In order to provide a direct test of the imputed descendancy relationship between low and high grade tumours, we took advantage of two clinical observations. The first, as mentioned above, are instances of patients with initial low grade disease who relapsed with high grade tumours. We were able to test a small number of such paired samples and could show that the low grade tumours had alterations of chromosome 17 which were identically maintained in the high grade recurrences. The latter had compounded the problem with chromosome 10 changes of the type described above for unrelated cases. These data were in substantial support of the hypothesis but the interpretations were limited due to limitations in the numbers of tissues available and the proportion of these which were informative for the analysis.

The second clinical situation alluded to above arises from the well-described heterogeneity in the histological characteristics of gliomas. This can result in the microscopic appearance of areas resembling high grade tumours adjacent to areas more similar to lower grade disease. One interpretation of these observations is that these tissues represent the transition state between two tumour grades before the lower is displaced by the higher. In order to test this notion, the tissues must be clearly dissected and analysed with a technique that is sufficiently sensitive for the small amount of tissue available. Although the genetic marker analyses are limited in capability with regard to the latter consideration, they focused attention on a specific region of the genome, namely the distal end of the short arm of chromosome 17 (James et al. 1989). One gene which is resident in this region and which is mutated in many types of human cancer including glioblastoma is the gene encoding the p53 protein.

We therefore determined the nucleotide sequence of the p53 gene in tissues of recurrent and histologically heterogeneous disease. In either instance, mutations of the p53 gene of low grade tumours were precisely maintained in their higher grade counterparts and, in fact, these mutations were often compounded by other mutations in the same gene. Finally, the identification of these mutations allowed the design of oligonucleotides specific to the mutation. The analysis of cDNAs from low grade and high grade tumours with these as probes allowed a direct experimental visualization of the evolution of the population towards those with more mutations.

Conclusions

The original observations of behaviour of malignancy in cell hybrids (Harris 1969) suggested that some transformed traits are recessive to their normal counterparts. The analysis of human cancer supports this conclusion and allows a powerful paradigm for the involvement of genetic lesions in the initiation and progression of neoplasia. The utility of this approach in providing reagents of diagnostic and prognostic significance is only now being realized but should over the next decade assume some prominence in efforts to subdue this disease.

References

Bigner, S. H., Mark, J., Mahaley, M. S., and Bigner, D. D. (1989). Patterns of the early gross chromosomal changes in malignant human gliomas. *Hereditas*, **101**, 103–13.

Cavenee, W. K., Dryja, T. P., Phillips, R. A., Benedict, W. F. Godbout, R., Murphree, A. L., *et al.* (1983). Expression of recessive alleles by chromosomal mechanisms in retinoblastoma. *Nature*, **305**, 770–84.

Cavenee, W. K., Hansen, M. F., Kock, E., Nordenskjöld, M., Maumenee, I., Squire, J. A., *et al.* (1985). Genetic origins of mutations predisposing to retinoblastoma. *Science*, **228**, 501–3.

Francke, U. (1976). Retinoblastoma and chromosome 13. *Cytogenet. Cell Genet.*, **16**, 131–4.

Friend, S. H., Bernards, R., Rogelj, S., Weinberg, R. A., Rapoport, J. M., Albert, D. M., *et al.* (1986). A human DNA segment with properties of the gene that predisposes to retinoblastoma and osteosarcoma. *Nature*, **323**, 654–6.

Harris, H., Miller, O. J., Klein, G., Worst, P., and Tachibana, T. (1969). Suppression of malignancy by cell fusion. *Nature*, **223**, 363–8.

Huang, H.-J. S., Yee, J.-K., Chen, P.-L., Chen, R., Bookstein, R., Friedmann, T., *et al.* (1988). Suppression of the neoplastic phenotype by replacement of the Rb gene in human cancer cells. *Science*, **242**, 1563–6.

James, C. D., Carlbom, E., Dumanski, J. P., Hansen, M. F., Nordenskjöld, M., Collins, V. P., *et al.* (1988). Clonal genomic alterations in glioma malignancy stages. *Cancer Res.*, **48**, 5546–51.

James, C. D., Carlbom, E., Nordenskjöld, M., Collins, V. P., and Cavenee, W. K. (1989). Mitotic recombination mapping of chromosome 17 in astrocytomas. *Proc. Natl Acad. Sci. USA*, **86**, 2858–62.

Knudson, A. G., Jr (1971). Mutation and cancer: statistical study of retinoblastoma. *Proc. Natl Acad. Sci. USA*, **68**, 820–3.

Lasko, D., Cavenee, W. K., and Nordenskjöld, M. (1991). Loss of constitutional heterozygosity in human cancer. *Annu. Rev. Genet.*, **25**, 281–314.

Mittelman, F. (1988). *Catalog of chromosome aberrations in cancer*, (3rd edn). Liss, New York.

Mulvihill, J. J. (1977). Genetic repertory of human neoplasia. In *Genetics of human cancer*, (ed. J. J. Mulvihill, R. W. Miller, and J. F. Fraumeni), pp. 137–43. Raven, New York.

Newsham, I. F. and Cavenee, W. K. (1990). Molecular mutagenesis in the pathology of retinoblastoma: a review. *Brain Pathol.*, **1**, 25–32.

Sidransky, D., Mikklesen, T., Schwechheimer, K. Rosenblum, M. L., Cavenee, W. K., and Vogelstein, B. (1992). Clonal expansion of p53 mutant cells leads to brain tumor progression. *Nature*, **355**, 846–7.

16. Cancer suppression by the retinoblastoma gene

Wen-Hwa Lee, David Goodrich, and Eva Lee

A concentrated effort toward the molecular cloning of cancer suppressor genes had an important recent success: a candidate retinoblastoma (RB) gene was cloned, identified, and extensively characterized by us and others (Friend *et al.* 1986; Fung *et al.* 1987; Lee *et al.* 1987*a*). Our demonstration of the tumour suppression activity of this single gene when transferred into tumour cells containing inactivated endogenous RB genes constituted a critical milestone for the theory of cancer suppression (Huang *et al.* 1988). Since several other cancers are known to be familial, are associated with chromosomal deletions, or exhibit loss of heterozygosity, an extension of this concept to other genes and to other cancers is widely anticipated (Malkin *et al.* 1990).

Tumour-suppressing activity of the retinoblastoma gene

The existence of inactivating mutations of RB in some natural human tumours supports the idea that RB normally functions to prevent tumour formation in susceptible precursor cells (Dryja *et al.* 1986; Friend *et al.* 1987; Harbour *et al.* 1988; Lee *et al.* 1988; T'Ang *et al.* 1988; Toguchida *et al.* 1988; Yokota *et al.* 1988; Horowitz *et al.* 1989, 1990; Reissmann *et al.* 1989; Shew *et al.* 1989, 1990*b*; Bookstein *et al.* 1990*b*; Cheng *et al.* 1990). To test whether complementation with a wild-type gene could revert neoplastic cells to more normal behaviour, we used retrovirus-mediated gene transfer to introduce exogenous RB expression into cultured cells. Transcription of RB RNA would be initiated by the Moloney leukaemia virus promoter; in addition the vector expressed the selectable marker gene neomycin phosphotransferase under the control of the Rous sarcoma virus promoter. For comparison, an identical vector containing the luciferase cDNA in place of the RB cDNA was made. Replication-defective viral stocks were obtained by transfecting vector DNA into murine retroviral packaging cell lines and the murine virus was used to infect an amphotropic retrovirus packaging cell line to produce viral stocks that could infect human cells.

We attempted to express the normal RB gene product in the retinoblastoma cell line Weri-27 and in the osteosarcoma cell line Saos-2, both of which contain partial deletions of RB (Huang *et al.* 1988). After infection with the RB-containing virus and selection, bulk populations of cells expressed an apparently full-length RB protein that localized to the nucleus. Cells infected with either RB or the luciferase virus were analysed for morphology, growth rate, and soft agar colony formation. Initially, expression of RB characteristically retarded growth rate and increased the average cell size. Infection of the same cells with the control luciferase-containing virus produced no change in morphology or growth rate.

After isolation of single-cell clones, RB-expressing cells have growth rates and morphology similar to those of uninfected cells, however (Chen *et al.* 1992). This implies that cells can adapt to the presence of RB and achieve growth rates similar to that of cells not expressing RB. Also, RB is not generally toxic to cells. The tumorigenicity of cells containing wild-type RB gene has been compared with that of cells infected with the luciferase virus by injection of cells into the flanks of nude mice. Both single-cell clones and bulk-infected Weri-27 cell lines have been tested (Huang *et al.* 1988; Chen *et al.* 1992). Nude mice were injected with 2×10^7 viable RB- or luciferase-containing cells in opposite flanks. After 2 months, luciferase-containing cells had produced large subcutaneous tumours while little or no tumour cell growth was observed for RB-containing cells (Huang *et al.* 1988; Chen *et al.* 1992). Similar results were obtained by Sumegi *et al.* (1990). As an alternative assay for tumorigenicity, retinoblastoma cell lines with and without exogenous RB expression are injected intraocularly. The results of such assays are in general agreement with the results of subcutaneous injection in nude mice; expression of wild-type RB correlates with reduced frequency and size of tumours (Madreperla *et al.* 1991; Xu *et al.* 1991). In sum, these results provide evidence that RB has the ability to inhibit tumour growth. Furthermore, because RB is able to suppress tumorigenicity in osteosarcoma, the involvement of the gene in inhibition of tumorigenesis may extend beyond retinoblastoma.

To examine such a possibility, the effect of wild-type RB expression on the neoplastic properties of other human tumour cell lines was examined. Using the previously described retroviral RB expression vector, the ability of wild-type RB to inhibit tumorigenicity in one prostate cell line (Bookstein *et al.* 1990*b*) and three bladder carcinoma cell lines (Goodrich *et al.* 1992) was tested. In the prostate carcinoma cell line DU145 and bladder carcinoma cell lines 5637 and HT1376, cells that expressed exogenous RB had a reduced ability to form tumours in nude mice, although their growth rate in culture and soft agar colony-forming ability were largely unaltered relative to cells lacking normal RB expression. In a third bladder carcinoma cell line, J82, both growth rate in culture and soft agar colony formation were reduced upon expression of RB. Transfection was also used to create stable bladder carcinoma cell lines expressing exogenous, wild-type RB (Takahashi *et al.* 1991). Again, expression of RB was associated with a reduced frequency or size of tumour formation after subcutaneous injection of the cells into nude mice. Preliminary results suggest that re-expression of RB in the breast carcinoma cell line MDA-MB468 suppresses some of the neoplastic properties of these cells (Wang *et al.* 1993). In sum, these observations indicate that RB can affect the tumorigenicity of a wide variety of tumours and that RB inactivation is also a biologically significant alteration in adult neoplasms.

Mouse model for the dosage effect of RB on growth and development

The above experiments were done to show that reconstitution with the wild-type RB gene in RB⁻ tumour cells suppressed their tumorigenic potential. To move one step further, it is important to test whether it is possible to suppress tumour formation in animals. Since no such naturally occurring RB⁻ tumour is available in animals, it is, for several reasons, very important to build such a model. Recent successes in constructing such mice using embryonic stem cells indicated that an endeavour of this magnitude was feasible. Mice heterozygous for a null allele of RB have been created (Lee *et al.* 1992). Surprisingly, no retinoblastoma was found in more than 100 heterozygous mice examined at an early age. Continuous observation has led to the finding of brain tumours in older (8–12 months) mice. The general phenotype appears to be loss of body weight, sickness, and infertility. The incidence of brain tumour was unusually high; all out of the 22 mice examined had brain tumours only. By analysis of the tumour DNA by Southern blotting, we have determined that the wild-type allele of the RB gene was lost in every case examined but not in other normal tissues. This result strongly suggested that inactivation of the RB gene contributes a significant role in the genesis of brain tumour. Although the tumour type in mice was not retinoblastoma, the availability of these mice with brain tumours provides an excellent model for testing tumour suppression by replacement with the wild-type RB gene in animals.

Upon crossing two heterozygous RB null mice, we observed that homozygous RB null embryos die before the 16th day of gestation. Thus complete inactivation of RB causes embyonic lethality. Examination of embryos prior to day 16 reveals multiple defects. The haematopoietic system is abnormal; there is a significant increase in the number of immature nucleated erythrocytes. In the nervous system, ectopic mitoses and massive cell death are found, particularly in the hindbrain. All spinal ganglion cells die. These result indicate that RB is essential for normal mouse development. Interestingly, transgenic mice carrying extra copies of the human RB transgene driven by the human RB promoter were smaller and lighter than normal. The degree of dwarfism roughly correlated with the copy number of the transgene and its corresponding level of expression (Bignon *et al.*, 1993). Transferring the transgene to homozygous RB-deficient mice produced normal healthy mice (Bignon *et al.* 1993). These results suggest that the human RB gene can functionally complement the mouse homologue and that the effect of RB protein on overall mouse development is closely dependent upon its dosage (Chang *et al.* 1993).

RB protein and cell cycle regulation

To begin to understand how the RB protein exerts it tumour suppressive influence on cells, it is necessary to have a basic knowledge of the protein and its gene. The 200 kb gene encoding RB contains 27 exons and is transcribed into a mature mRNA of about 4.7 kb (Lee *et al.* 1987*a*; Bookstein *et al.* 1988; Wiggs *et al.* 1988; Hong *et al.* 1989; McGee *et al.* 1989). The promoter region has features analogous to the control regions of constitutively expressed housekeeping genes (Hong *et al.* 1989). The predicted RB gene product has 928 amino acids and a molecular weight of 106 kDa (Lee *et al.* 1987*b*). The gene is evolutionarily conserved and is ubiquitously expressed in vertebrates suggesting that RB may play an important role in cellular function.

The first hint concerning the cellular function of RB came with the demonstration that RB is a nuclear phosphoprotein (Lee *et al.* 1987*b*). The 110 kDa RB protein is phosphorylated on serine and threonine residues and multiple electrophoretic forms of the protein are accounted for on the basis of varying amounts of phosphorylation (Ludlow *et al.* 1989; Shew *et al.* 1989), with the fastest-migrating band being the least modified. Oscillation of RB protein phosphorylation with the cell division cycle hints that RB may effect cell growth by regulating the cell division cycle. The G_1 phase of the eukaryotic cell division cycle contains important checkpoints for continuation or exit from the cell cycle (Pardee 1989; Lewin 1990). Genetic studies on yeasts have defined a control point just prior to S phase, termed 'start', where the cell makes its decision to proceed with the next round of cell division. For example, exposure of the cell to mating pheromone before start inhibits proliferation. The start checkpoint is also required to prevent entry into the subsequent cycle before the cell has reached sufficient size and accumulated enough macromolecules. Animal cells have similar control mechanisms operating in the G_1 phase of the cell cycle. Withdrawal of growth factors before a 'restriction point' in G_1 prevents entry into the following S phase and places the cell in a state of quiescence, often referred to as G_0 phase. In G_0, the cell continues to metabolize, but it has exited from the cell division cycle. This G_0 state defined in culture is believed to be analogous to the state of a terminally differentiated cell that has exited the cell division cycle. Conceivably, RB may play some role at a restriction point in the G_1 phase, either to allow exit from the cell cycle or to delay entry into S phase until a suitable growth status is achieved.

To test this hypothesis, purified RB proteins, either full-length or a truncated form containing the SV40 T antigen binding region, were injected into cells and the effect on entry into S phase determined (Goodrich *et al.* 1991). Synchronized cells injected early in G_1 with either type of RB protein inhibits progression into S phase. The effect is specific, reversible, and can be antagonized by the presence of SV40 T antigen. Injection of RB protein into cells arrested at the G_1–S boundary or 6–10 h before the end of G_1 has no effect on S phase entry, suggesting that RB protein inhibits G_1 progression rather than DNA synthesis. Therefore, it is likely that RB may regulate cell growth by affecting progression through the G_1 phase of the cell cycle. Consistent with this idea, transfection of RB genes designed to express either full-length or N-terminally truncated forms of RB protein also result in growth inhibition (Templeton *et al.* 1991; Qin *et al.* 1992).

Phosphorylation appears to be the major method for regulation of the RB protein. While the levels of expression remain rather constant in dividing and arrested cells, the phosphorylation status of the RB protein fluctuates in a cell cycle-dependent manner (Buchkovich et al. 1989; Chen et al. 1989; Mihara et al. 1989; Ludlow et al. 1990; Maheswaran et al. 1991; Burke et al. 1992; DeCaprio et al. 1992; Zhang et al. 1992). The most likely candidate for an RB kinase is the Cdc2 complex comprised of the products of the *CDC2* gene and one of the cyclin genes. Significantly, the cyclical activity of the Cdc2 complex is concurrent with the phosphorylation pattern of the RB protein; each demonstrates peaks of activity occurring at the G_1–S and G_2–M transitions (Furukawa et al. 1990). In addition, the RB protein contains several consensus Cdc2 kinase target sites, B-S/T-P-X-B (where X is a polar residue and B is generally a basic residue) (Shenoy et al. 1989). By analysis of RB phosphopeptides, it was shown that S249, T252, T373, S807, and S811 are phosphorylated *in vivo*, and in each case these sites closely match the Cdc2 consensus sequence (Taya et al. 1989; Lees et al. 1991; Lin et al. 1991). This study also suggested that other Cdc2 sites may be phosphorylated *in vivo*, especially T5, T356, S612 and S788. A partial analysis of the requirement of some of the Cdc2 target sites in RB's ability to bind T antigen was performed by mutating groups of two or four S or T residues in adjacent Cdc2 target sites (Hamel et al. 1990). The results suggested that at least one of the residues phosphorylated *in vivo* occurs in one of four C-terminal Cdc2 sites. However, a systematic study of how phosphorylation may modulate RB biochemical activity requires a full understanding of its functional role in cellular physiology.

Biochemical properties of RB protein

There are two known biochemical functions of RB protein; one is its ability to bind DNA (Lee et al. 1987b; Wang et al. 1990) and the other is to interact with several oncoproteins of DNA tumour viruses (DeCaprio et al. 1988; Whyte et al. 1988; Dyson et al. 1989; Ludlow et al. 1989). One of the earliest properties of purified preparations of RB protein to be discovered was its ability to associate with DNA. Using trpE–RB fusion proteins, we have demonstrated that this non-specific DNA binding activity is intrinsic to the C-terminal region of the RB protein (Wang et al. 1990). Fusion proteins containing RB exons 19–27 bound DNA with high affinity while proteins containing as little as exons 19–22 or exons 23–27 had reduced DNA binding affinity. An attempt to identify specific sequences recognized by RB directly has been unsuccessful; it is likely that RB binds to DNA in a non-specific manner.

Another interesting biochemical property of the RB gene product is its ability to form specific complexes with the transforming proteins of several DNA tumour viruses including SV40 large T, adenovirus E1A, and human papillomavirus E7 (DeCaprio et al. 1988; Whyte et al. 1988; Dyson et al. 1989; Munger et al. 1989). The regions of the transforming proteins required for complex formation are similar to those required for transformation. The amino acids of the RB protein that are required for binding to SV40 T antigen have also been determined (Hu et al. 1990; Huang et al. 1990). Two distinct regions of the RB protein, amino acids 394–571 and

649–773, are required for high-affinity complex formation. These regions correspond to regions frequently mutated in tumour cells; in all cases analysed to date mutated RB protein from human tumour cells has been unable to form complexes with T antigen. A naturally occurring point mutant at one (amino acid 706) of these cysteines has been detected in a small cell lung carcinoma (Bignon *et al.* 1990). The correlation between the T antigen binding domains and the naturally occurring mutants suggest that these regions may constitute an important functional domain of the RB protein.

Analysis of naturally mutated RB proteins showed that both phosphorylation and binding to the larger T antigen (T) of SV40 are, in all cases, disturbed even by as little as a single amino acid substitution (Horowitz *et al.* 1989, 1990; Bignon *et al.* 1990; Bookstein *et al.* 1990a,b; Kaye *et al.* 1990; Shew *et al.* 1990a,b), whereas DNA binding is retained in several mutated RB proteins (Lee *et al.* 1987b; Wang *et al.* 1990). In order to explore further the significance of the interaction of RB with T antigen, an extensive series of artificially mutated RB constructs containing small in-frame deletions or insertions was made, expressed, and assayed for T-binding. Two separate regions that together are required for T-binding were mapped (Hu *et al.* 1990; Huang *et al.* 1990). Strikingly, all naturally mutated RB proteins in tumour cells are structurally altered within these regions. Therefore, the regions of RB important for T-binding are also likely to be important for RB's putative tumour suppression activity. A reasonable speculation is that these regions contribute to the tertiary structure of the protein that is normally recognized by cellular RB kinase(s) (Bignon *et al.* 1990; Hamel *et al.* 1990). These studies suggested that cellular proteins mimicking T's ability to interact with RB could be important for the physiological function of RB.

There probably are cellular proteins that, similarly to SV40 T antigen, antagonize the activity of RB protein. The c-Myc protein seems a good candidate for a cellular antagonist of RB since it is localized in the nucleus and can physically associate with RB protein *in vitro* (Rustgi *et al.* 1991), and overexpression of the c-*myc* gene promotes cell proliferation and blocks differentiation (Land *et al.* 1983; Copolla and Cole 1986; Dmitrovsky *et al.* 1986; Prochownik and Kukowska 1986; Bar-Ner *et al.* 1992). This possibility has been confirmed by findings that co-injection of c-Myc protein but not EJ-Ras, c-Fos, or c-Jun, inhibits the ability of RB protein to arrest the cell cycle (Goodrich *et al.* 1991). Conversely, c-Myc does not inhibit the activity of another tumour suppressor, p53 (Levine *et al.* 1991). Since injection of c-Myc is dominant to injection of RB, c-Myc may function downstream of RB in the same or an overlapping regulatory pathway. Consistent with this notion, tumour growth factor-β1 (TGF-β1) down-regulation of c-*myc* transcription in keratinocytes is blocked by viral transforming proteins that inactivate RB (Pietenpol *et al.* 1990, 1991). RB may therefore be an intermediary in regulation of c-*myc* by TGF-β1.

These observations have inspired a search for other cellular proteins that bind RB protein in a manner similar to the DNA tumour virus transforming proteins. One approach to identifying such proteins is to purify them from cell extracts with RB protein affinity columns (Huang *et al.* 1991; Kaelin *et al.* 1991). SV40 T antigen or a T antigen derived peptide can compete for binding to RB protein with some of these cellular proteins. One of these RB binding proteins has been purified and partially sequenced, and the gene encoding the protein has been molecularly cloned (Qian *et al.* 1993). The gene encodes a protein that has homology to the *msi1* gene of *Saccharomyces cerevisia*. The *msi1* gene suppresses *ira1* and *ira2* mutations, and is therefore involved in *ras1* and *ras2* signal transduction. The human gene encoding the RB binding protein may also function in signal transduction since it can complement *msi1* mutations in yeast.

Another approach is to test candidate proteins directly for their ability to bind RB protein *in vitro*. For example c-Myc, N-Myc, E2F, ATF-2, and Cdc2 proteins are capable of binding RB protein *in vitro* (Bandara and LaThangue 1991; Chellappan *et al.* 1991; Chittenden *et al.* 1991; Rustgi *et al.* 1991; Hu *et al.* 1992; Kim *et al.* 1992*a*). All of these proteins require a C-terminal region of RB that includes the domain required for binding to the oncoproteins of DNA tumour viruses. However, these proteins seem to require additional C-terminal sequences for binding that are not required by SV40 T antigen.

A third approach to identifying cellular, RB-associated proteins is to screen phage λ protein expression libraries using purified RB protein as a probe. Several novel genes of unknown function have been identified by this method. One of these genes, however, encodes a protein with properties similar to the transcription factor E2F (Helin *et al.* 1992; Kaelin *et al.* 1992; Shan *et al.* 1992). In addition, previously cloned genes encoding nuclear lamin A and several previously unknown genes with nuclear matrix properties have been isolated by this method.

Another method used to isolate RB-associated proteins involves a yeast two-hybrid screening system (Fields and Song 1989; Chien *et al.* 1991). Briefly, the Gal4 DNA binding domain is fused to parts of the RB protein. This fusion protein can bind DNA, but cannot activate a Gal4-dependent promoter since the Gal4 activation domain has been replaced by RB protein. An expression library is constructed that fuses random cDNAs to the transactivation domain of Gal4. If the random cDNA encodes a protein that associates with RB protein, then the Gal4 transactivation domain will be brought into proximity with the DNA binding domain and the Gal4-dependent promoter will be activated. Activation of the Gal4-dependent promoter serves as the basis for selection and/or screening of the cDNA clones. Several RB associated proteins have been isolated by this method including phosphatase type 1A (T. Durfee *et al.* 1993).

The range of proteins that associate with RB *in vitro* is quite large and includes transcription factors, proto-oncogenes, structural proteins, possible signal transducing proteins, cyclins, kinases, and phosphatases. It is difficult to find clues to the mechanism of RB protein action from such a varied list. The best characterized RB-associated protein, however, is the transcription factor E2F (Bandara and LaThangue 1991; Chellappan *et al.* 1991; Chittenden *et al.* 1991). Hence the possible role of RB in regulation of transcription factors has been explored in detail.

E2F, originally discovered through its role in the activation of the adenovirus E2 promoter (Kovesdi *et al.* 1986; Yee *et al.* 1989), may regulate genes important for the cell division cycle since the E2F binding site is found in the promoters of genes such as DNA polymerase α, ribonucleotide reductase, c-*myc*, thymidylate synthase, and c-

myb (Blake and Azizkhan 1989; Hiebert *et al.*, 1989, 1991; Mudryj *et al.* 1990). A related factor, DRTF1, binds the same DNA site (LaThangue *et al.* 1990; Bandara and LaThangue 1991). The RB–E2F complex is found predominantly in the G_1 and S phases of the cell cycle (Chellappan *et al.* 1991; Mudryj *et al.* 1991), corresponding roughly to the transition in RB protein phosphorylation. The RB–E2F complex can be disrupted by adenovirus E1a, SV40 T antigen, and human papillomavirus E7 protein (Chellappan *et al.* 1992). In the presence of E1a, which can free E2F from RB-containing protein complexes, the E2F DNA binding sites become transcriptional activators. E2F sites act as transcriptional activators in the absence of E1a in cell lines lacking RB, and co-transfection with RB-expressing plasmids switches the E2F sites to transcriptional inhibitors in RB⁻ cell lines. These results suggest that binding of RB converts the E2F factor from an activator to an inhibitor of transcription.

RB has also been shown to activate transcription from the TGF-β2 promoter and the fourth promoter of the insulin-like growth factor II gene (Kim *et al.* 1992a,b). Activation of the TGF-β2 promoter requires intact ATF DNA binding sites. ATF2 has high affinity for the TGF-β2 ATF DNA binding site and can mediate activation by RB from an artificial promoter. Hence RB can apparently positively and negatively regulate transcription.

Although provocative, these conclusions are based on indirect, co-transfection assays. The effects of RB on these various promoters may be a consequence of more global effects of RB overexpression on cell growth. Since RB can dramatically inhibit progression through the cell cycle, inhibition or activation of transcription from certain promoters may be due to the global effects of cell cycle arrest rather than to specific interaction with RB protein. In fact, little change in expression of growth regulatory genes such as c-*fos*, c-*myc*, or N-*myc* have been observed upon stable expression of physiological levels of RB (Chen *et al.* 1992). In this regard, it would be interesting to test the effects of another gene known to arrest cell cycle progression, such as p53, on transcription. Such a control would establish if the interaction between RB expression and transcriptional regulation from certain promoters is specific.

RB protein can bind to types of proteins other than transcription factors (Shan *et al.* 1992). Several of these proteins, lamin C for example, seem to be associated with or are components of the nuclear matrix. In fact, RB protein itself can bind to an insoluble, nuclear compartment (Templeton, 1992). This 'tethering' to a nuclear structure is reduced by phosphorylation or mutation of the RB protein (Mittnacht and Weinberg 1991; Templeton *et al.* 1991). A fraction of the unphosphorylated RB protein survives extractions designed to reveal the nuclear matrix (M. Mancini and W.-H. Lee unpublished). Hence, association of RB protein with a structural component of the nucleus may be important for its function.

Summary

The molecular isolation of the RB gene is an important achievement in research on cancer. For the first time, it has become possible to examine, at the molecular level, genes which inhibit the growth of tumour cells. Although the exact function of the RB gene in controlling cell growth remains to be elucidated, continued study of this gene should yield insight into mechanisms of oncogenesis, gene regulation, and cellular differentiation complementary to the knowledge which has long been accumulating from the study of oncogenes.

References

Bandara, L. R. and LaThangue, N. B. (1991). Adenovirus E1a prevents the retinoblastoma gene product from complexing with a cellular transcription factor. *Nature*, **351**, 494–7.

Bar-Ner, M., Messing, L. T., Cultraro, C. M., Birrer, M. J., and Segal, S. (1992). Regions within the c-myc protein that are necessary for transformation are also required for inhibition of differentiation of murine erythroleukemia cells. *Cell Growth Differ.*, **3**, 183–90.

Bignon, Y.-J., Chen, Y., Chang, C.-Y., Riley, D. J., Windle, J. J., Melon, P. L., Lee, W.-H. (1993) Expression of a retinoblastoma transgene results in dwarf mice. *Gene & Develop.* **7**, 1654–62.

Bignon, Y.-J., Shew, J.-Y., Rappolee, D., Naylor, S. L., Lee, E. Y.-H. P., Schnier, J., *et al.* (1990). A single Cys706 to Phe substitution in the retinoblastoma protein causes the loss of binding to SV40 T antigen. *Cell Growth Differ*, **1**, 647–51.

Blake, M. C. and Azizkhan, J. C. (1989). Transcription factor E2F is required for efficient expression of the hamster dihydrofolate reductase gene *in vitro* and *in vivo*. *Mol. Cell. Biol.*, **9**, 4994–5002.

Bookstein, R., Lee, E. Y.-H. P., To, H., Young, L.-J., Sery, T., Hayes, R. C., *et al.* (1988). Human retinoblastoma susceptibility gene: genomic organization and analysis of heterozygous intragenic deletion mutants. *Proc. Natl Acad. Sci. USA*, **85**, 2210–14.

Bookstein, R., Rio, P., Madreperla, S., Hong, F., Allred, C., Grizzle, W. E., *et al.* (1990*a*). Promoter deletion and loss of retinoblastoma gene expression in human prostate carcinoma. *Proc. Natl Acad. Sci. USA*, **87**, 7762–6.

Bookstein, R., Shew, J.-Y., Chen, P.-L., Scully, P., and Lee, W.-H. (1990b). Suppression of tumorigenicity of human prostate carcinoma cells by replacing a mutated *RB* gene. *Science*, **247**, 712–5.

Buchkovich, K., Duffy, L. A., and Harlow, E. (1989). The retinoblastoma protein is phosphorylated during specific phases of the cell cycle. *Cell*, **58**, 1097–105.

Burke, L. C., Bybee, A., and Thomas, N. S. (1992). The retinoblastoma protein is partially phosphorylated during early G_1 in cycling cells but not in G_1 cells arrested with α-interferon. *Oncogene*, **7**, 783–8.

Chang, C.-Y., Riley, D. J., Lee, E. Y.-H. P., Lee, W.-H. (1993). Quantitative effects of the retinoblastoma gene on mouse development and tissue-specific tumorigenesis. *Cell Growth Differ*. **4**, 1057–64.

Chellappan, S., Hiebert, S., Mudryj, M., Horowitz, J. M., and Nevins, J. R. (1991). The E2F transcription factor is a cellular target for the RB protein. *Cell*, **65**, 1053–61.

Chellappan, S., Kraus, V. B., Kroger, B., Munger, K., Howley, P. M., Phelps, W. C., *et al.* (1992). Adenovirus E1A, simian virus 40 tumor antigen, and human papillomavirus E7 protein share the capacity to disrupt the interaction between transcription factor E2F and the retinoblastoma gene product. *Proc. Natl Acad. Sci. USA*, **89**, 4549–53.

Chen, P.-L., Scully, P., Shew, J.-Y., Wang, J. Y.-J., and Lee, W.-H. (1989). Phosphorylation of the retinoblastoma gene product is modulated during the cell cycle and cellular differentiation. *Cell*, **58**, 1193–8.

Chen, P.-L., Chen, Y., Shan, B. Bookstein, R., and Lee, W.-H. (1992). Stability of RB expression determines the tumorigenicity of reconstituted retinoblastoma cells. *Cell Growth Differ.*, **3**, 119–25.

Cheng, J., Scully, P., Shew, J.-Y., Lee, W.-H., Vila, V., and Haas, M. (1990). Homozygous deletion of the retinoblastoma gene in an acute lymphoblastic leukemia (T) cell line. *Blood*, **75**, 730–35.

Chien, C.-T., Bartel, P. L., Sternglanz, R., and Fields, S. (1991). The two-hybrid system: a method to identify and clone genes for proteins that interact with a protein of interest. *Proc. Natl Acad. Sci. USA*, **88**, 9578–82.

Chittenden, T., Livingston, D. M., and Kaelin, W. G. Jr (1991). The T/E1A-binding domain of the retinoblastoma product can interact selectively with a sequence-specific DNA-binding protein. *Cell*, **65**, 1071–82.

Copolla, J. S. and Cole, M. D. (1986). Constitutive c-*myc* oncogene expression blocks mouse erythroleukemia cell differentiation but not commitment. *Nature*, **320**, 760–3.

DeCaprio, J. A., Ludlow, J. W., Figge, J., Shew, J.-Y., Huang, C.-M., Lee, W.-H., *et al.* (1988). SV40 large tumor antigen forms a specific complex with the product of the retinoblastoma susceptibility gene. *Cell*, **54**, 275–83.

DeCaprio, J. A., Furukawa, Y., Ajchenbaum, F., Griffin, J. D., and Livingston, D. M. (1992). The retinoblastoma-susceptibility gene product becomes phosphorylated in multiple stages during cell cycle entry and progression. *Proc. Natl Acad. Sci. USA*, **89**, 1795–8.

Dmitrovsky, E., Kuehl, W. M., Hollis, G. F., Kirsch, I. R., Bender, T. P., and Segal, S. (1986). Expression of a transfected human c-*myc* oncogene inhibits differentiation of a mouse erythroleukemia cell line. *Nature*, **322**, 748–50.

Dryja, T. P., Rapaport, J. M., Joyce, J. M., and Petersen, R. A. (1986). Molecular detection of deletions involving band q14 of chromosome 13 in retinoblastomas. *Proc. Natl Acad. Sci. USA*, **83**, 7391–4.

Durfee, T., Becherer, K., Chen, P.-L., Yeh, S.-H., Yang, Y., Kilburn, A. E., *et al.* (1993). The retinoblastoma protein associates with the protein phosphatase type 1 catalytic subunit. *Genes Devel.*, **7**, 555–69.

Dyson, N., Howley, P. M., Munger, K., and Harlow, E. (1989). The human papilloma virus-16 E7 oncoprotein is able to bind to the retinoblastoma gene product. *Science*, **243**, 934–7.

Fields, S. and Song, O.-K. (1989). A novel genetic system to detect protein–protein interactions. *Nature*, **340**, 245–6.

Friend, S. H., Bernards, R., Rogelj, S., Weinberg, R. A., Rapaport, J. M., Albert, D. M., *et al.* (1986). A human DNA segment with properties of the gene that predisposes to retinoblastoma and osteosarcoma. *Nature*, **323**, 643–6.

Friend, S. H., Horowitz, J. M., Gerber, M. R., Wang, X.-F., Bogenman, E., Li, F. P., *et al.* (1987). Deletions of a DNA sequence in retinoblastomas and mesenchymal tumors: organization of the sequence and its encoded protein. *Proc. Natl Acad. Sci. USA*, **84**, 9059–63.

Fung, Y. K. T., Murphree, A. L., T'Ang, A., Qian, J., Hinrichs, S. H., and Benedict, W. F. (1987). Structural evidence for the authenticity of the human retinoblastoma gene. *Science*, **236**, 1657–61.

Furukawa, Y., Piwnica-Worms, H., Ernst, T. J., Kanakura, Y., and Griffen, J. D. (1990). *cdc2* gene expression at the G_1 to S transition in human T lymphocytes. *Science*, **250**, 805–8.

Goodrich, D. W., Wang, N. P., Qian, Y.-W., Lee, E. Y-H. P., and Lee, W.-H. (1991). The retinoblastoma gene product regulates progression through the G_1 phase of the cell cycle. *Cell*, **67**, 293–302.

Goodrich, D. W., Chen, Y., Scully, P., and Lee, W.-H. (1992). Expression of the retinoblastoma gene product in bladder carcinoma cells associates with a low frequency of tumor formation. *Cancer Res.*, **52**, 1968–73.

Hamel, P. A., Cohen, B. L., Sorce, L. M., Gallie, B. L., and Phillips, R. A. (1990). Hyperphosphorylation of the retinoblastoma gene product is determined by domains outside the simian virus 40 large-T-antigen-binding regions. *Mol. Cell. Biol.*, **10**, 6586–95.

Harbour, J. W., Lai, S.-H., Whang-Peng, J., Gazdar, A. F., Minna, J. D., and Kaye, F. J. (1988). Abnormalities in structure and expression of the human retinoblastoma gene in SCLC. *Science*, **241**, 353–7.

Helin, K., Lees, J. A., Vidal, M., Dyson, N., Harlow, E., and Fattaey, A. (1992). A cDNA encoding a pRB-binding protein with properties of the transcription factor E2F. *Cell*, **70**, 337–50.

Hensel, C. H., Hsieh, C. L., Gazdar, A. F., Johnson, B. E., Sakaguchi, A. Y., Naylor, S. L., *et al.* (1990). Altered structure and expression of the human retinoblastoma susceptibility gene in small cell lung cancer. *Cancer Res.*, **50**, 3067–72.

Hiebert, S. W., Lipp, M., and Nevins, J. R. (1989). E1a-dependent *trans*-activation of the human *myc* promoter is mediated by the E2F factor. *Proc. Natl Acad. Sci. USA*, **86**, 3594–8.

Hiebert, S. W., Blake, M., Azizkhan, J., and Nevins, J. R. (1991). Role of E2F transcription factor in E1a-mediated *trans*-activation of cellular genes. *J. Virol.*, **65**, 3547–52.

Hong, F. D., Huang, H.-J. S., To, H., Young, L.-J. S., Oro, A., Bookstein, R., *et al.* (1989). Structure of the human retinoblastoma gene. *Proc. Natl Acad. Sci. USA*, **86**, 5502–6.

Horowitz, J. M., Yandell, D. W. Park, S. H. Canning, S., Whyte, P., Buchkovich, K., *et al.* (1989). Point mutational inactivation of the retinoblastoma antioncogene. *Science*, **243**, 937–40.

Horowitz, J. M., Park, S.-H., Bogenmann, E., Cheng, J.-C., Yandell, D. W., Kaye, F. J., *et al.* (1990). Frequent inactivation of the retinoblastoma anti-oncogene is restricted to a subset of human tumor cells. *Proc. Natl Acad. Sci. USA*, **87**, 2775–9.

Hu, Q., Dyson, N., and Harlow, E. (1990). The regions of the retinoblastoma protein needed for binding to adenovirus E1A or SV40 T antigen are common sites for mutations. *EMBO J.*, **9**, 1147–55.

Hu, Q. J., Lees, J. A., Buchkovich, K. J., and Harlow, E. (1992). The retinoblastoma protein physically associates with the human cdc2 kinase. *Mol. Cell. Biol.*, **12**, 971–80.

Huang, H.-J. S., Yee, J.-K., Shew, J.-Y., Chen, P.-L. Bookstein, R., Friedmann, T., *et al.* (1988). Suppression of the neoplastic phenotype by replacement of the retinoblastoma gene product in human cancer cells. *Science*, **242**, 1563–6.

Huang, S., Wang, N.-P., Tseng, B. Y., Lee, W.-H., and Lee, E. Y.-H. P. (1990). Two distinct and frequently mutated regions of retinoblastoma protein are required for binding to SV40 T antigen. *EMBO J.*, **9**, 1815–22.

Huang, S., Lee, W.-H., and Lee, E. Y.-H. P. (1991). Identification of a cellular protein that competes with SV40 T antigen for binding to the retinoblastoma gene product. *Nature*, 160–62.

Kaelin, W. G. J., Pallas, D. C., DeCaprio, J. A., Kaye, F. J., and Livingston, D. M. (1991). Identification of cellular proteins that can interact specifically with the T/E1A-binding region of the retinoblastoma gene product. *Cell*, **64**, 521–32.

Kaelin, W. G., Krek, W., Sellers, W. R., DeCaprio, J. A., Ajchenbaum, F., Fuchs, C. S., *et al.* (1992). Expression cloning of a cDNA encoding a retinoblastoma-binding protein with E2F-like properties. *Cell*, **70**, 351–64.

Kaye, F. J., Kratzke, R. A., Gerster, J. L., and Horowitz, J. M. (1990). A single amino acid substitution results in a retinoblastoma protein defective in phosphorylation and oncoprotein binding. *Proc. Natl Acad. Sci. USA*, **87**, 6922–6.

Kim, S.-J., Wagner, S., Liu, F., O'Reilly, M. A., Robbins, P. D., and Green, M. R. (1992*a*). Retinoblastoma gene product activates expression of the human TGF-$\beta 2$ gene through transcription factor ATF-2. *Nature*, **358**, 331–4.

Kim, S. J., Onwuta, U. S., Lee, Y. I., Li, R., Botchan, M. R., and Robbins, P. D. (1992b). The retinoblastoma gene product regulates Sp1-mediated transcription. *Mol. Cell. Biol.*, **12**, 2455–63.

Kovesdi, I., Reichel, R., and Nevins, J. R. (1986). Identification of a cellular transcription factor involved in E1a *trans*-activation. *Cell*, **45**, 219–28.

Land, H., Parada, L. F., and Weinberg, R. A. (1983). Tumorigenic conversion of primary embryo fibroblasts requires at least two cooperating oncogenes. *Nature*, **304**, 596–602.

LaThangue, N. B., Thimmappaya, B., and Rigby, P. W. J. (1990). The embryonal carcinoma stem cell E1a-like activity involves a differentiation-regulated transcription factor. *Nucleic Acids Res.*, **18**, 2929–38.

Lee, W.-H., Bookstein, R., Hong, F., Young, L.-J., Shew, J.-Y., and Lee, E. Y.-H. P. (1987a). Human retinoblastoma susceptibility gene: cloning, identification, and sequence. *Science*, **235**, 1394–9.

Lee, W.-H., Shew, J.-Y., Hong, F., Sery, T., Donoso, L. A., Young, L. J., et al. (1987b). The retinoblastoma susceptibility gene product is a nuclear phosphoprotein associated with DNA binding activity. *Nature*, **329**, 642–5.

Lee, E. Y.-H. P., To, H., Shew, J.-Y., Bookstein, R., Scully, P., and Lee, W.-H. (1988). Inactivation of the retinoblastoma susceptibility gene in human breast cancers. *Science*, **241**, 218–21.

Lee, E. Y.-H. P., Chang, C.-Y., Hu, N., Wang, Y.-C., J., Herrup, K., Lee, W.-H., et al. (1992). Mice deficient for RB are nonviable and show defects in neurogenesis and hematopoiesis. *Nature*, **359**, 288–94.

Lees, J. A., Buchkovich, K. J., Marshak, D. R., Anderson, C. W., and Harlow, E. (1991). The retinoblastoma protein is phosphorylated on multiple sites by human cdc2. *EMBO J.*, **10**, 4279–90.

Levine, A. J., Momand, J., and Finlay, C. A. (1991). The p53 tumour suppressor gene. *Nature*, **351**, 453–6.

Lewin, B. (1990). Driving the cell cycle: M phase kinase, its partners, and substrates. *Cell*, **61**, 743–52.

Lin, B. T., Gruenwald, S., Morla, A. O., Lee, W.-H., and Wang, J. Y. (1991). Retinoblastoma cancer suppressor gene product is a substrate of the cell cycle regulator cdc2 kinase. *EMBO J.*, **10**, 857–64.

Ludlow, J. W., deCaprio, J. A., Huang, C.-M., Lee, W.-H., Paucha, E., and Livingston, D. M. (1989). SV40 large T antigen binds preferentially to an underphosphorylated member of the retinoblastoma susceptibility gene family. *Cell*, **56**, 57–65.

Ludlow, J. W., Shon, J., Pipas, J. M., Livingston, D. M., and DeCaprio, J. A. (1990). The retinoblastoma susceptibility gene product undergoes cell cycle-dependent dephosphorylation and binding to and release from SV40 large T. *Cell*, **60**, 387–96.

McGee, T. L., Yandell, D. W., and Dryja, T. P. (1989). Structure and partial genomic sequence of the human retinoblastoma susceptibility gene. *Gene*, **80**, 119–28.

Madreperla, S. A., Whittum-Hudson, J. A., Prendergast, R. A., Chen, P. L., and Lee, W. H. (1991). Intraocular tumor suppression of retinoblastoma gene-reconstituted retinoblastoma cells. *Cancer Res.*, **51**, 6381–4.

Maheswaran, S., McCormack, J. E., and Sonenshein, G. E. (1991). Changes in phosphorylation of *myc* oncogene and RB antioncogene protein products during growth arrest of the murine lymphoma WEHI 231 cell line. *Oncogene*, **6**, 1965–71.

Malkin, D., Li, F. P., Strong, L., Fraumeni, J. F., Nelson, C., Kim, D., et al. (1990). Germ line p53 mutations in a familial syndrome of breast cancer, sarcomas, and other neoplasms. *Science*, **250**, 1233–8.

Mihara, K., Cao, X.-R., Yen, A., Chandler, S., Driscoll, B., Murphree, A. L., et al., (1989). Cell cycle-dependent regulation of phosphorylation of the human retinoblastoma gene product. *Science*, **246**, 1300–3.

Mittnacht, S. and Weinberg, R. A. (1991). G_1/S phosphorylation of the retinoblastoma protein is associated with an altered affinity for the nuclear compartment. *Cell*, **65**, 381–93.

Mudryj, M., Hiebert, S. W., and Nevins, J. R. (1990). A role for the adenovirus inducible E2F transcription factor in a proliferation dependent signal transduction pathway. *EMBO J.*, **9**, 2179–84.

Mudryj, M., Devoto, S. H., Hiebert, S. W., Hunter, T., Pines, J. and Nevins, J. R. (1991). Cell cycle regulation of the E2F transcription factor involves interaction with cyclin A. *Cell*, **65**,1243–53.

Pardee, A. B. (1989). G_1 events and regulation of cell proliferation. *Science*, **246**, 603–8.

Pietenpol, J. A., Stein, R. W., Moran, E., Yaciuk, P., Schlegel, R., Lyons, R. M., et al. (1990). TGF-beta 1 inhibition of c-*myc* transcription and growth in keratinocytes is abrogated by viral transforming proteins with pRB binding domains. *Cell*, **61**, 777–85.

Pietenpol, J. A., Munger, K., Howley, P. M., Stein, R. W., and Moses, H. L. (1991). Factor-binding element in the human c-*myc* promoter involved in transcriptional regulation by transforming growth factor beta1 and by the retinoblastoma gene product. *Proc. Natl Acad. Sci. USA*, **88**, 10227–31.

Prochownik, E. V. and Kukowska, J. (1986). Deregulated expression of c-*myc* by murine erythroleukemia cells prevents differentiation. *Nature*, **322**, 848–50.

Qian, Y.-W., Wong, Y.-C., J., Hollingsworth, R. E., Jr, Jones, D., Ling, N., and Lee, E. Y.-H. P. (1993). A cDNA encoding a retinoblastoma-binding protein, RbAp48, has properties related to MSI1, a negative regulator of *ras* in yeast. *Nature*. **364**, 648–52.

Qin, X.-Q., Chittenden, T., Livingston, D. M. and Kaelin, W. G. (1992). Identification of a growth suppression domain within the retinoblastoma gene product. *Genes Dev.*, **6**, 953–64.

Reissmann, P. T., Simon, M. A., Lee, W.-H., and Slamon, D. J. (1989). Studies of the retinoblastoma gene in human sarcomas. *Oncogene*, **4**, 839–43.

Rustgi, A. K., Dyson, N., and Bernards, R. (1991). Amino-terminal domains of c-myc and N-myc proteins mediate binding to the retinoblastoma gene product. *Nature*, **352**, 541–4.

Shan, B., Zhu, X., Chen, P.-L., Durfee, T., Yang, Y., Sharp, D., and Lee, W.-H. (1992). Molecular cloning of cellular genes encoding retinoblastoma-associated proteins: identification of a gene with properties of the transcription factor E2F. *Mol. Cell. Biol.*, **12**, 5620–31.

Shenoy, S., Choi, J.-K., Bagrodia, S., Copeland, T. D., Maller, J. L., and Shalloway, D. (1989). Purified maturation promoting factor phosphorylates pp60$^{c\text{-}src}$ at the sites phosphorylated during fibroblast mitosis. *Cell*, **57**, 763–74.

Shew, J.-Y., Ling, N., Yang, X., Fodstad, O., and Lee, W.-H. (1989). Antibodies detecting abnormalities of the retinoblastoma susceptibility gene product (pp110RB) in osteosarcomas and synovial sarcomas. *Oncogene Res.*, **1**, 205–14.

Shew, J.-Y., Chen, P.-L., Bookstein, R., Lee, E. Y.-H. P., and Lee, W.-H. (1990*a*). Deletion of a splice donor site ablates expression of the following exon and produces an unphosphorylated RB protein unable to bind SV40 T antigen. *Cell Growth Differ.*, **1**, 17–25.

Shew, J.-Y., Lin, B., Chen, P.-L., Tseng, B. Y., Yang-Feng, T. L., and Lee, W.-H. (1990*b*). C-terminal truncation of the RB protein leads to functional inactivation. *Proc. Natl Acad. Sci. USA*, **87**, 6–10.

Sumegi, J., Uzvolgyi, E., and Klein, G. (1990). Expression of the RB gene under the control of MuLV-LTR suppresses tumorigenicity of WERI-Rb-27 retinoblastoma cells in immunodefective mice. *Cell Growth Differ.*, **1**, 247–50.

T'Ang, A., Varley, J. M., Chakraborty, S., Murphree, A. L., and Fung, Y.-K. T. (1988). Structural rearrangement of the retinoblastoma gene in human breast carcinoma. *Science*, **242**, 263–6.

Takahashi, R., Hashimoto, T., Hong-Ji, X., Hu, S.-X., Matsui, T., Miki, T., *et al.* (1991). The retinoblastoma gene functions as a growth and tumor suppressor in human bladder carcinoma cells. *Proc. Natl Acad. Sci. USA*, **88**, 5257–61.

Taya, Y., Yasuda, H., Kamijo, M., Nakaya, K., Nakamura, Y., Ohba, Y., *et al.* (1989). *In vitro* phosphorylation of the tumor suppressor gene RB protein by mitosis-specific histone H1 kinase. *Biochem. Biophys. Res. Commun.*, **164**, 580–6.

Templeton, D. J. (1992). Nuclear binding of purified retinoblastoma gene product is determined by cell cycle-regulated phosphorylation. *Mol. Cell. Biol.*, **12**, 435–43.

Templeton, D. J., Park, S. H., Lanier, L., and Weinberg, R. A. (1991). Nonfunctional mutants of the retinoblastoma protein are characterized by defects in phosphorylation, viral oncoprotein association, and nuclear tethering. *Proc. Natl Acad. Sci. USA*, **88**, 3033–7.

Toguchida, J., Ishizaki, K., Sasaki, M. S., Ikenaga, M., Sugimoto, M., Kotoura, Y., *et al.* (1988). Chromosomal reorganization for the expression of recessive mutation of retinoblastoma susceptibility gene in the development of osteosarcoma. *Cancer Res.*, **48**, 3939–43.

Wang, N.-P., Chen, P.-L., Huang, S., Donoso, L. A., Lee, W.-H., and Lee, E. Y.-H. P. (1990). DNA-binding activity of retinoblastoma protein is intrinsic to its carboxyl-terminal region. *Cell Growth Diff.*, **1**, 233–9.

Wang, N.-P., To, H., Lee, W.-H., and Lee, E. Y.-H. P. (1993). Tumor suppressor activity of RB and p53 genes in human breast carcinoma cells. *Oncogene*, **8**, 279–88.

Whyte, P., Buchkovich, K. J., Horowitz, J. M., Friend, S. H., Raybuck, M., Weinberg, R. A. *et al.* (1988). Association between an oncogene and an anti-oncogene: the adenovirus E1A proteins bind to the retinoblastoma gene product. *Nature*, **334**, 124–9.

Wiggs, J., Nordenskjold, M., Yandell, D., Rapaport, J., Grondin, V., Janson, M., *et al.* (1988). Prediction of the risk of hereditary retinoblastoma using DNA polymorphisms within the retinoblastoma gene. *N. Engl. J. Med.*, **318**, 151–7.

Xu, H. J., Sumegi, J., Hu, S. X., Banerjee, A., Uzvolgyi, E., Klein, G., *et al.* (1991). Intraocular tumor formation of RB reconstituted retinoblastoma cells. *Cancer Res.*, **51**, 4481–5.

Yee, A. S., Raychaudhuri, P., Jakoi, L., and Nevins, J. R. (1989). The adenovirus-inducible factor E2F stimulates transcription after specific DNA binding. *Mol. Cell. Biol.*, **9**, 578–85.

Yokota, J., Akiyama, T., Fung, Y.-K. T., Benedict, W. F., Namba, Y., Hanaoka, M., *et al.* (1988). Altered expression of the retinoblastoma (RB) gene in small-cell carcinoma of the lung. *Oncogene*, **3**, 471–5.

Zhang, W., Hittelman, W., Van, N., Andreeff, M., and Deisseroth, A. (1992). The phosphorylation of retinoblastoma gene product in human myeloid leukemia cells during the cell cycle. *Biochem. Biophys. Res. Commun.*, **184**, 212–16.

17. The p53 pathway, past and future
David P. Lane

Abstract

Mutation of the p53 tumour suppressor gene and the accumulation of the p53 protein are amongst the most common specific changes found in human cancer. An understanding of the precise biochemical and biological affects of these changes will provide new insights into the fundamental nature of the neoplastic process. New results show that the p53 protein is a critical part of a damage-induced cell cycle checkpoint control that maintains genetic stability. The p53 protein provides an attractive target for the development of new diagnostic and therapeutic agents that will hopefully reduce the impact of cancer on society.

Introduction and discussion

The revolution in methods brought about by the development of modern molecular biology is allowing enormous progress to be made in the analysis of human cancer. It is clear that many separate genetic changes are required before a normal cell will behave as a malignant cell. Principal among these changes are the mutational activation of the growth-promoting functions of the proto-oncogenes and the mutational inactivation of the function of the tumour suppressor genes whose existence was first deduced by the cell fusion studies of Henry Harris (see Klein (1987) and Bishop (1991) for reviews). Tumours in different tissues are typically associated with certain specific changes to defined oncogenes and suppressor genes. One gene, however, the p53 gene, stands out because of its almost universal involvement in the development of the common tumours in man (Lane and Benchimol 1990; Vogelstein 1990; Levine *et al.* 1991; Vogelstein and Kinzler 1992). The existence of a common step in the development of cancer compels us to determine the precise function(s) of p53 because that knowledge should deepen our understanding of the whole neoplastic process.

The properties of the p53 protein

The p53 protein was first discovered through its interaction with large T antigen, the oncogene product of the SV40 small DNA tumour virus (Lane and Crawford 1979; Linzer and Levine 1979). Subsequent work has shown that p53 also forms physical complexes with the oncogene products of two other groups of DNA tumour virus, the E1b product of adenovirus 5 (Sarnow *et al.* 1982) and the E6 product of human papilloma virus 16 (Werness *et al.* 1990). The p53 gene is on chromosome 17p and consists of 11 exons (Soussi *et al.* 1990). It has been conserved in evolution among the vertebrates but has not so far been found in invertebrates. The gene, in man, encodes a 393 amino acid phosphoprotein that is commonly located in the cell nucleus. The p53 protein binds specifically to double-stranded DNA and can act as a regulator of transcription both to promote the transcription of some genes, for example the muscle creatine kinase gene, but also to reduce the transcription of several other genes, for example the *myc* gene. High levels of the protein can inhibit both viral and cellular DNA synthesis *in vivo*. The p53 protein is normally present at very low levels in the tissues of the adult organism due to its very short half-life (reviewed in Lane and Benchimol (1990), Vogelstein (1990), Levine *et al.* (1991), Vogelstein and Kinzler (1992)). The protein accumulates to high levels in cells exposed to DNA damaging agents (Maltzman and Czyzyk 1984; Kastan *et al.* 1991; Hall *et al.* 1992; Lu *et al.* 1992) which suggests, as will be discussed later, that p53 may act as a checkpoint control to prevent DNA synthesis in cells with damaged template. The breakdown of this control pathway, brought about by mutations in p53 or other components of the pathway, may be a key step in the progression of cancer, as it will permit the proliferation of cells that contain aberrant DNA and give rise to a pool of variants from which malignant or drug resistant clones of cancer cells will arise (Livingstone *et al.* 1992; Yin *et al.* 1992).

p53 tumour suppressor gene and oncogene?

Mice in which both alleles of the p53 gene have been inactivated develop a wide variety of tumours at a very young age (Donehower *et al.* 1992). This establishes that loss of p53 function is sufficient to predispose to the development of neoplasia. In humans the germ-line inheritance of one mutant allele of the p53 gene predisposes the affected individual to a wide range of neoplasia sometimes manifest as the Li–Fraumeni syndrome (Malkin *et al.* 1990; Srivastava *et al.* 1990; Santibanez-Koref *et al.* 1991). The mutant p53 proteins found in human tumour cells have lost their growth-suppressive function; however, several lines of evidence suggest that they might also have gained a growth-promoting function. Transfection of mutant p53 with an activated *ras* gene will transform primary rodent cells. This function depends on the conformation and continued expression of the p53 protein since temperature-sensitive point mutants in p53 have been identified that will act as transforming genes at 39 °C in this assay but will suppress transformation and growth of the same cells at 32 °C (Michalovitz *et al.* 1990). Part of the mechanism of this dominant transforming activity resides in the capacity of mutant p53 proteins to bind to and inactivate the function of the wild-type protein. This classic dominant negative mutant behaviour is not sufficient, however, to explain all the characteristics of the p53 system. The mutant proteins may also act on other targets to promote growth since at least in some systems introduction of a mutant p53 gene into cells that have lost both wild-type alleles enhances their neoplastic properties (Wolf *et al.* 1985). In human tumours it is very apparent that there is a strong selection for loss of the normal allele and retention of the mutant allele. There also seems to be a selection for high level expression of the mutant protein in the tumour.

Expression of p53 protein in human tumours

The production of recombinant p53 protein in bacteria has allowed the isolation of polyclonal and monoclonal antibodies that efficiently detect the protein in routine histopathology samples (Midgley *et al.* 1992; Vojtesek *et al.* 1992). This has confirmed earlier work that identified high levels of p53 protein in 20 per cent of breast cancers (Crawford *et al.* 1984; Cattoretti *et al.* 1988) but has also extended the observation to many other tumour types including for example lung, colon, stomach, brain, bladder, skin, cervical, and ovarian cancer (Bartek *et al.* 1990*a,b*, 1991; Iggo *et al.* 1990; Rodrigues *et al.* 1990; Bennett *et al.* 1991; Gusterson *et al.* 1991; Pignatelli *et al.* 1991; Wrede *et al.* 1991; Wright *et al.* 1991; Eccles *et al.* 1992; Hanski *et al.* 1992). High levels of p53 are frequently found in tumours of all major cell lineages including sarcomas, carcinomas, glioblastomas, and haemopoetic malignancies. The molecular basis for the accumulation of high levels of p53 specifically in tumours seems to be an increase in the post-translational stability of the protein as no consistent increase in mRNA levels is found. Direct sequencing of the p53 gene in immunohistochemically strongly positive tumours has shown that the protein is usually the product of a p53 gene carrying a point mutation within the coding region. The tumours are typically homozygous at the p53 locus having lost the wild-type allele. High levels of the p53 protein are clearly associated with a poor prognosis in breast (Isola *et al.* 1992), lung, and gastric cancers (Martin *et al.* 1992). In the case of breast cancer p53 levels are fast emerging as a key indicator of disease-free interval and may soon be used to identify a subset of women requiring differential post-operative therapy (D. Barnes, personal communication). The patterns of expression of p53 seen by immunohistochemistry are of great interest. The most striking, which seems to be closely associated with point mutation of the protein, is an intense staining of all cells within the tumour. This pattern is seen in about 20 per cent of primary breast cancers and in 50–70 per cent of colon, gastric, oesophageal, and lung cancers (Bennett *et al.* 1991, 1992; Midgley *et al.* 1992). In these tumours the events that lead to the morphological appearance of malignancy must be closely coincident with the events that lead to the high levels of p53. In other cases more focal expression of p53 immunoreactivity is apparent suggesting that the appearance of immunoreactivity may mark a stage in tumour progression. Molecular studies and the high levels of p53 apparent in the metastases of heterogeneously positive primary tumours support this view. In other cases just a few scattered positive cells are seen and finally some tumours show no enhancement of immunoreactivity over background (Midgley *et al.* 1992).

Regulation of p53 protein levels

The association between high levels of p53 protein and poor prognosis, and the variable expression patterns of immunoreactivity seen in individual tumours have prompted a detailed investigation of the molecular basis of the accumulation of p53 protein in human cancer. A series of initial studies from my group and others established that in a series of cell lines and primary tumours high levels of p53 were associated with point missense mutations in the p53 gene (Bartek *et al.* 1990; Iggo *et al.* 1990; Rodrigues *et al.* 1990). It is high levels of mutant protein that are being detected. Many of the point mutations in p53 change the conformation of the protein and this conformational change can be detected with monoclonal antibodies (Gannon *et al.* 1990; Stephen and Lane 1992). Several studies have shown that this conformational change renders the protein more stable metabolically and may in part account for its accumulation. However, this effect whilst important is not sufficient to explain all cases. Mutation in the p53 protein is not the only way in which it can become stable in tumour cells. We recently found a cancer-prone family (Barnes *et al.* 1992) in which the proband and her affected mother showed a high level of p53 constitutively expressed in normal epithelial, endothelial, and stromal cells. The p53 open reading frame was normal in sequence in these individuals suggesting that mutation in another gene was causing both the high level of p53 and the associated increased risk of neoplasia. In families with germ-line p53 mutations, by contrast, high levels of p53 are not found in normal tissues though they are often readily apparent in the tumours of these individuals. This suggests that other events may be required to stabilize even mutant p53 proteins. Certainly wild-type p53 can be stabilized dramatically by treating cells with DNA damaging agents, and by the action of SV40 large T and adenovirus E1b protein. In some tumours containing high levels of p53 the gene appears to be wild-type. In a mouse model of prostrate cancer (Lu *et al.* 1992) we found high levels of wild-type p53 in tumour

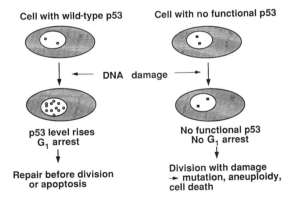

Fig. 17.1. A model for the function of p53. In normal cells where p53 is wild-type and the p53 pathway is intact, DNA damage—and perhaps other stresses—lead to the stabilization of p53 which accumulates to high levels. This induces a cell cycle arrest. In some cells (stem cells?) this arrest will lead to apoptosis, in others the growth arrest is transitory and once the inducing signal is gone p53 levels return to normal and the cell can go back into cycle. In tumour cells the p53 pathway is often inactivated by mutation in p53 itself or in other genes that act upstream or downstream of p53 and affect its function. When these cells are exposed to DNA damaging agents, DNA synthesis continues leading to genetic errors and instability. This may result in cell death but also give rise to aneuploidy. From this pool of damaged cells malignant clones will arise at elevated frequencies.

cells derived from precursors expressing an activated *ras* and an activated *myc* gene. The pattern of expression was sporadic and was clearly associated with cells within the population that had nuclear abnormalities consistent with cell cycle failures. It is thus possible that the stabilization of both mutant and wild-type p53 in tumours is a direct result of their genetic instability which is itself caused by a breakdown in the functioning of the p53 pathway (Fig. 17.1). Failure in the p53 pathway could be due to mutations in p53 itself or in the gene products responsible for inducing the increase in p53 levels following DNA damage or in the gene products required for the cellular response to high levels of p53. To understand this in detail we will need to identify these other genes in the p53 pathway. This process is already under way: we have found human cells that do not produce stable p53 in the presence of UV radiation or SV40 T antigen (X. Lu and D. P. Lane, 1993). Their p53 gene is, however, wild-type in sequence and will respond to UV when the defective cells are fused to responsive mouse cells. Similarly Kastan's laboratory have shown that the p53 response to γ radiation is defective in some cases of ataxia telangiectasia (Kastan *et al.* 1992). Intense efforts are now under way to determine how p53 is degraded and what blocks this process following DNA damage. Identification of the gene products involved in this pathway will allow an investigation of their function in normal and tumour cells and may identify new cancer susceptibility syndromes.

Regulation of p53 protein function

If p53 can act as a potent suppressor of cell growth in cells exposed to DNA damaging agents, or transfected with extra copies of the p53 gene, how is it possible for normal growth and development to take place? Three models could explain this, first that normal levels of p53 in undamaged cells are too low to restrict growth, second that p53 function is normally ablated by the activity of other gene products, and third that p53 is normally produced in a latent form and only becomes activated in the presence of signals induced by DNA damage. All three models are probably correct, leading to a very tight control on the growth-suppressive function of p53. The level of p53 is very low in normal cells and a gene product, MDM2, that binds and inactivates p53 has recently been discovered (Momand *et al.* 1992). Work in my own laboratory has shown that p53 is normally present in cells in a form that cannot bind to double-stranded DNA with sequence specificity. To do so the latent protein must be activated by conformational modifications at the C-terminus of the protein. These modifications can be brought about by phosphorylation with case kinase II, by the action of members of the hsp70 heat shock protein family, and finally by the binding of specific antibodies (Hupp *et al.* 1992). This allosteric regulation of p53 function is exciting because it suggests that small molecules may be discovered or designed that will modulate p53 function.

Conclusions

We are now at an immensely exciting stage in cancer research. We have identified a common step in the molecular progression of the disease and novel findings about the structure and function of p53 are being made at breakneck speed as the full resources of modern molecular and cellular biology are brought to bear on this one protein. One can only be optimistic for the future. I am convinced that novel therapeutics targeted to the p53 pathway will be discovered soon. They will exploit the fundamental difference in genetic stability and cell cycle control found in cells that lack the p53 pathway. They will exploit the tumour-specific accumulation and conformational alteration in p53 protein. They will therefore strike selectively at the fundamental differences between the normal cell and the tumour cell. We can hope that they will be more effective and kinder to the patient than any current agent. If that dream comes true then all our work will have been worthwhile.

Acknowledgement

My work is currently supported by the Cancer Research Campaign.

References

Barnes, D. M., Hanby, A. M., Gillett, C. E., Mohammed, S., Hodgson, S., Bobrow, L. G., *et al.* (1992). Abnormal expression of wild type p53 protein in normal cells of a cancer family patient. *Lancet*, **340**, 259–63.

Bartek, J., Bartkova, J., Vojtesek, B., Staskova, Z., Rejthar, A., Kovarik, J., *et al.* (1990*a*). Patterns of expression of the p53 tumour suppressor in human breast tissues and tumours *in situ* and *in vitro*. *Int. J. Cancer*, **46**, 839–44.

Bartek, J., Iggo, R. Gannon, J., and Lane, D. P. (1990*b*). Genetic and immunochemical analysis of mutant p53 in human breast cancer cell lines. *Oncogene*, **5**, 893–9.

Bartek, J., Bartkova, J., Vojtesek, B., Staskova, Z., Lukas, J., Rejthar, A., *et al.* (1991). Abberant expression of the p53 oncoprotein is a common feature of a wide spectrum of human malignancies. *Oncogene*, **6**, 1699–703.

Bennett, W. P., Hollstein, M. C., He, A., Zhu, S. M., Resau, J., Trump, B. F., *et al.* (1991). Archival analysis of p53 genetic and protein alterations in Chinese esophageal cancer. *Oncogene*, **6**, 1779–84.

Bishop, J. M. (1991). Molecular themes in oncogenesis, *Cell*, **64**, 235–48.

Catoretti, G., Rilke, F., Andreola, S., D'Amato, L., and Delia, D. (1988). p53 expression in breast cancer. *Int. J. Cancer*, **41**, 178–83.

Crawford, L. V., Pim, D. C., and Lamb, P. (1984). The cellular protein p53 in human tumors. *Mol. Biol. Med.*, **2**, 261–72.

Donehower, L. A., Harvey, M., Slagle, B. L., McArthur, M. J., Montgomery, C. A., Butel, J. S., *et al.* (1992). Mice deficient for p53 are developmentally normal but susceptible to spontaneous tumours. *Nature*, **356**, 215–21.

Eccles, D. M., Brett, L., Lessells, A., Gruber, L., Lane, D., Steel, C. M., *et al.* (1992). Overexpression of the p53 protein and allele loss at 17p13 in ovarian carcinoma. *Br. J. Cancer*, **65**, 40–44.

Gannon, J. V., Greaves, R., Iggo, R., and Lane, D. P. (1990). Activating mutations in p53 produce a common conformational effect. A monoclonal antibody specific for the mutant form. *EMBO J.*, **9**, 1595–602.

Gusterson, B. A., Anbazhagan, R., Warren, W., Midgely, C., Lane, D. P., O'Hare, M., *et al.* (1991). Expression of p53 in premalignant and malignant squamous epithelium. *Oncogene*, **6**, 1785–9.

Hall, P. A., McKee, P. H., Menage, H. D., Dover, R., and Lane, D. P. (1992). High levels of p53 protein in UV irradiated human skin. *Oncogene*, **8**, 203–7.

Hanski, C., Bornhoeft, G., Shimoda, T., Hanski, M. L., Lane, D. P., Stein, H., *et al.* (1992). Expression of p53 protein in invasive colorectal carcinomas of different histological type. *Cancer*, **70**, 2772–77.

Hupp, T. R., Meek, D. W., Midgley, C. A., and Lane, D. P. (1992). Regulation of the specific DNA binding function of p53. *Cell*, **71**, 875–886.

Iggo, R., Gatter, K., Bartek, J., Lane, D., and Harris, A. L. (1990). Increased expression of mutant forms of p53 oncogene in primary lung cancer. *Lancet*, **335**, 675–9.

Isola, J., Visakorpi, T., Holli, K., and Kallioniemi, O.-P. (1992). Association of over-expression of tumour suppressor protein p53 with rapid cell proliferation and poor prognosis in mode-negative breast cancer patients. J. Natl *Cancer Inst.*, **84**, 1109–14.

Kastan, M. B., Onyekwere, O., Sidransky, D., Vogelstein, B., and Craig, R. W. (1991). Participation of p53 protein in the cellular response to DNA damage. *Cancer Res.*, **51**, 6304–11.

Kastan, M. B., Zhan, Q., El-Deiry, W. K., Carrier, F., Jacks, T., Walsh, W. V., *et al.* (1992). A mammalian cell cycle checkpoint pathway utilizing p53 and gadd45 is defective in ataxia-telangiectasia. *Cell*, **71**, 587–97.

Klein, G. (1987). The approaching era of the tumour suppressor genes. *Science*, **238**, 1539–44.

Lane, D. and Benchimol, S. (1990). p53: oncogene or anti-oncogene. *Genes Dev.*, **4**, 1–8.

Lane, D. P. and Crawford, L. V. (1979). T-antigen is bound to host protein in SV40-transformed cells. *Nature*, **278**, 261–3.

Levine, A. J., Momand, J., and Finlay, C.A. (1991). The p53 tumour suppressor gene. *Nature*, **351**, 453–6.

Linzer, D. I. H. and Levine, A. J. (1979).Characterization of a 54K dalton cellular SV40 tumour antigen present in SV40 transformed cells and uninfected embryonal carcinoma cells. *Cell*, **17**, 43–52.

Livingstone, L. R., White, A., Sprouse, J., Livanos, E., Jacks, T., and Tisty, T. D. (1992). Altered cell cycle arrest and gene amplification potential accompany loss of wild type p53. *Cell*, **70**, 923–35.

Lu, X. and Lane, D. P. (1993). Differential induction of transcriptionally active p53 following UV or ionising radiation: defects in chromosome instability syndromes? *Cell*, **75**, 765–78.

Lu, X., Park, S. H., Thompson, T. C., and Lane, D. P. (1992). *ras*-induced hyperplasia occurs with mutation of p53, but an activated *ras* and *myc* together can induce carcinoma without p53 mutation. *Cell*, **70**, 153–61.

Malkin, D., Li, F. P., Strong, L. C., Fraumeni, J. F. J., Nelson, C. E., Kim, D. H., *et al.* (1990). Germ-line p53 mutations in a familial syndrome of breast cancer, sarcomas, and other neoplasms. *Science*, **250**, 1233–8.

Maltzman, W. and Czyzyk, L. (1984). UV irradiation stimulates levels of p53 cellular tumour anitgen in nontransformed mouse cells. *Mol. Cell. Biol.*, **4**, 1689–94.

Martin, H. M., Filipe, M. I., Morris, R. W., Lane, D. P., and Silvestre, F. (1992). p53 expression and prognosis in gastric carcinoma. *Int. J. Cancer*, **50**, 859–62.

Michalovitz, D., Halvey, O., and Oren, M. (1990). Conditional inhibition of transformation and of cell proliferation by a temperature sensitive mutant of p53. *Cell*, **62**, 671–80.

Midgley, C. A., Fisher, C. J., Bartek, J., Vojtesek, B., Lane, D. P., and Barnes, D. M. (1992). Analysis of p53 expression in human tumours: an antibody raised against human p53 expressed in *E.coli. J. Cell. Sci.*, **101**, 183–9.

Momand, J., Zambetti, G. P., Olson, D. C., George, D., and Levine, A. J. (1992). The *mdm*-2 oncogene product forms a complex with the p53 protein and inhibits p53-mediated transactivation. *Cell*, **69**, 1237–45.

Pignatelli, M., Stamp, G. W. H., Kafiri, G., Lane, D. P., and Bodmer, W. F. (1991). Overexpression of p53 nuclear oncoprotein in colorectal adenomas. *Int. J. Cancer*, **50**, 683–8.

Rodrigues, N. R., Rowan, A., Smith, M. E. F., Kerr, I. B., Bodmer, W. F., Gannon, J., *et al.* (1990). p53 mutations in colorectal cancer. *Proc. Natl Acad. Sci. USA*, **87**, 7555–9.

Santibanez-Koref, M. F., Birch, J. M., Hartley, A. L., Morris Jones, P. H., Craft, A. W., Eden, T., *et al.* (1991). p53 germ-line mutations in Li-Fraumeni syndrome. *Lancet*, **338**, 1490–91.

Sarnow, P., Ho, Y. S., Williams, J., and Levine, A. J. (1982). Adenovirus E1b-58 kd tumour antigen and sv40 large tumour antigen are physically associated with the same 54 kd cellular protein in transformed cells. *Cell*, **28**, 387–94.

Soussi, T., Caron de Fromentel, C., and May, P. (1990). Structural aspects of the p53 protein in relation to gene evolution. *Oncogene*, **5**, 945–52.

Srivastava, S., Zou, Z., Pirollo, K., Blattner, W., and Chang, E. H. (1990). Germ-line transmission of a mutated p53 gene in a cancer prone family with Li-Fraumeni syndrome. *Nature*, **348**, 747–9.

Stephen, C. W. and Lane, D. P. (1992). Mutant conformation of p53: precise epitope mapping using a filamentous phage epitope library. *J. Mol. Biol.*, **225**, 577–83.

Vogelstein, B. (1990). A deadly inheritance. *Nature*, **348**, 681–2.

Vogelstein, B. and Kinzler, K. W. (1992). p53 function and dysfunction. *Cell*, **70**, 523–6.

Vojtesek, B., Bartek, J., Midgley, C. A., and Lane, D. P. (1992). An immunochemical analysis of human p53: new monoclonal antibodies and epitope mapping using recombinant p53. *J. Immunol. Methods*, **151**, 237–44.

Werness, B. A., Levine, A. J., and Howley, P. M. (1990). Association of human papillomavirus types 16 and 18 E6 proteins with p53. *Science*, **248**, 76–79.

Wolf, D., Harris, N., Goldfinger, N., and Rotter, V. (1985). Reconstitution of p53 expression in nonproducer Ab-MuLV-transformed cell line by transfection of a functional p53 gene. *Cell*, **38**, 119–26.

Wrede, D., Tidy, J. A., Crook, T., Lane, D., and Vousden, K. H. (1991). Expression of abnormal p53 and RB protein detectable only in HPV negative cervical carcinoma cell lines. *Mol. Carcinogen.* **4**, 171–5.

Wright, C., Mellon, K., Johnston, P., Lane, D. P., Harris, A. L., Wilson Horne, C. H., *et al.* (1991). Expression of mutant p53, c-erbB-2 and the epidermal growth factor receptor in transitional cell carcinoma of the human urinary bladder *Br. J. Cancer*, **63**, 967–70

Yin, Y., Tainsky, M. A., Bischoff, F. Z., Strong, L. C., and Wahl, G. M. (1992). Wild-type p53 restores cell cycle control and inhibits gene amplification in cells with mutant p53 alleles. *Cell*, **70**, 937–48.

18. Somatic cell genetics and the search for colon cancer genes

H.J.W. Thomas, A-M. Frischauf, and E. Solomon

Introduction

As this volume is in honour of Professor Sir Henry Harris, it seems appropriate, under the general topic of genetics of colon cancer, to focus specifically on the role of somatic cell genetics in the isolation of genes responsible for this disease. Two rather different types of somatic cell genetics have been used in the identification of the genes mutated in colon cancer. The first type might be called tumour somatic cell genetics and has developed from the fact that loss of genetic material from tumour cells can be detected either microscopically, as whole or partial chromosome loss or by loss of heterozygosity (LOH) (for reviews see Goddard and Solomon, (1993) and Lasko *et al.*, (1991)). This genetic loss in tumours has come to be recognized as the hallmark of tumour suppressor genes. The second type of somatic cell genetics is the methodological one, in which the segregation of human chromosomes from interspecific hybrids, can be used both for mapping purposes and for the production of probes. Part of this methodology has evolved directly from experiments of Goss and Harris (1975) in which breakage of chromosomes through irradiation was used as a statistical mapping method. We and others have used both types of somatic cell genetics in various phases in the search for the adenomalous polyposis coli (APC) gene, as well as other colon cancer genes.

Inherited cancer syndromes

One of the major indications that cancer is a genetic disease comes from those rare families in which susceptibility to a particular cancer is inherited, usually in a dominant manner. Although these families are of interest in themselves, it was not initially clear what bearing they and their particular mutations, would have on the far more common sporadic forms of the disease. The search for cancer susceptibility genes became compelling, however, with the first evidence in the case of familial and sporadic retinoblastoma (Cavenee *et al.* 1983; Godbout *et al.*, 1983), that Knudson had been correct and that the same genes causing inherited susceptibility were likely to be mutated in the sporadic tumours as well (Knudson, 1985, 1991). If this were true, as predicted by Knudson, for the extremely common solid tumours such as breast and colon, these families would become of major importance.

Through linkage analysis the genes could be localized, thereby providing an extremely powerful tool towards the cloning and identification of cancer genes.

From statistical considerations, Knudson also predicted that these genes would act recessively at the cellular level and that both alleles would sustain mutations resulting in loss of activity (Knudson 1971). These tumour suppressor genes were believed to function in the regulation of normal growth and cell division. Knudson predicted that the early genetic events leading to sporadic tumour growth would be loss of both copies of a tumour suppressor gene within the somatic cells of the tissue involved. The same gene would be mutated in the germ-line of affected individuals in familial cases, and in these individuals loss of the second copy would occur within a cell of the particular tissue. If this were the case, the search for cancer genes would have an enormous advantage over the search for other types of inherited disorders. Somatic cells of the tumours would be expected to have mutations in the same region as that identified by linkage in the families. Mutational mechanisms which could lead to loss of activity could include not only localized stop codons, for example, within the gene, but relatively large events such as interstitial deletions in the region of the gene and loss of the entire chromosome. Highly informative polymorphic DNA markers in the region of interest would enable the detection of such events. Comparison of tumour DNA with normal matched DNA for loss of heterozygosity (LOH) would reveal those regions in which tumour material was lost.

Colon cancer

Each year 25 000 new cases of colon cancer are diagnosed in the UK and 19 000 deaths occur due to this disease. The majority of these cases are sporadic, that is, family history shows no indication of a genetic factor leading to this disease. A small percentage however, (approximately 1 per cent), are due to a genetic disease known as familial adenomatous polyposis (FAP) in which a dominantly inherited susceptibility to colon cancer leads to successive generations within a family being affected with colon cancer. The gene responsible for this disease is known as the adenomatous polyposis coli gene, or APC. The disease is characterized by the appearance in the late teens or early twenties of hundreds to thousands of benign adenomatous polyps in the colon. Unless removed, these inevitably progress to carcinomas (Knudson 1971). The usual treatment is prophylactic colectomy, that is, removal of the colon before progression to carcinoma occurs. The histopathology of the individual adenomas and carcinomas from FAP patients is not distinguishable from that of sporadic tumours suggesting that similar molecular or biochemical events occur in both types.

To determine whether Knudson's model would apply to colon cancer, it would ultimately be necessary to compare the germ-line gene mutations causing FAP with the same gene in sporadic tumours. However, knowing approximately where the APC gene was located in the genome would allow that region to be examined in sporadic tumours for loss of genetic material (LOH) as was done for retinoblastoma. Observation of such loss in sporadic colon tumours in the same region as the APC gene would be good evidence that the model was correct.

Genetic linkage and loss of heterozygosity in FAP

The location of a tumour suppressor gene can be determined in a number of ways. In familial cases, the mutations will segregate as any dominant gene and linkage analysis will reveal its location. Sporadic tumours will show loss of heterozygosity at polymorphic loci when compared with paired normal tissue. If this loss is localized by interstitial deletion, it too will point to the location of a tumour suppressor gene. Without initial clues, however, both of these approaches have been extremely labour intensive. With the current developments of oligonucleotide repeat polymorphisms (Weber and May 1989) at small genomic intervals and automated technology (Weissenbach *et al.* 1992) the task has become somewhat easier. Fortunately, in the case of FAP, an individual was identified with a constitutional interstitial deletion in 5q15-q22, with multiple physical anomalies and mental retardation and FAP (Herrera *et al.* 1986). That this was the clue as to the location of the FAP gene and linkage to this region in families (Bodmer *et al.* 1987; Leppert *et al.* 1987) as well as LOH in sporadic tumours (Solomon *et al.* 1987) was immediately demonstrated. Although the deletion covered a large region of 10–15 Mb, it led to the demonstration that the tumour suppressor hypothesis was likely to be correct for colon cancer.

Somatic cell genetic approaches

As with any positional cloning effort, in order to isolate the APC gene we needed to isolate a large number of probes within the region defined by flanking genetic markers and the deletion. To do this we first produced a mapping panel from individuals with deletions in the region and then used a variety of approaches to generate probes. Ultimately we were able to use these to produce a 10 Mb physical map of the region (Solomon and Frishcauf 1993; Ward *et al.* 1993).

Hybrid mapping panel

Somatic cell genetics and the production of interspecific hybrids allowed us to produce a mapping panel with which to localize probes in the APC region. Three individuals were identified with deletions on chromosome 5. The first, M, was a girl of 10 years with mild dysmorphic features and mild developmental delay, with no family history of FAP. The deletion was described as 5q15-q22 (Varesco *et al.* 1989). The second, P, was one of two brothers, both with FAP, mental retardation, and dysmorphic features whose mother had died of colorectal cancer. Again, the deletion was described as 5q15-q22 (Hockey *et al.* 1989). The third case, S, with FAP, dysmorphic features, and mental retardation had a 5q deletion described as 5q22.1-q23.3. This was believed to be due to an intrachromosomal insertion in which the distal boundary of the deletion appeared considerably more telomeric than the others (Cross *et al.* 1992).

Chromosome 5 encodes the gene for leucyl-tRNA synthetase (Giles *et al.* 1980). A Chinese hamster line, temperature sensitive for leu tRNA synthetase (Thompson

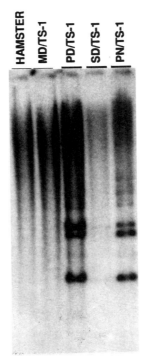

Figure 18.1. DNA from hamster line tsH1, hybrids containing normal chromosome 5 (PN/TS-1), and hybrids with chromosome 5 deletions (MD/TS-1, PD/TS-1, and SD/TS-1). DNA is digested with *Bgl*II and probed with 12.25*Hin*d3.1, derived from an irradiation hybrid. Adapted from Solomon and Frischauf (1993).

et al. 1973) was fused with lymphoid lines from M, P, and S and selected at the non-permissive temperature. Resulting hybrids contained human chromosome 5 on a Chinese hamster background. We obtained a hybrid PN/TS-1 containing a normal 5 and no other human material (Varesco, *et al.* 1989). Each of the other 5 del chromosomes was also selected in the individual hybrids, PD/TS-1, MD/TS-1, and SD/TS-1, in which little or no other human material was retained. These four hybrids were the basis of a mapping panel onto which probes from the region could be mapped (Thomas 1991). Figure 18.1 shows an example of an irradiation hybrid clone (see below) which is deleted in MD/TS-1 and SD/TS-1 but not PD/ST-1, showing that the PD deletion is smaller at one boundary. In fact, this is the only clone that distinguishes the PD and SD distal breakpoints. Although the SD breakpoint was described as being cytologically further distal, it is clearly very close to that of PD. Using the MD/TS-1 hybrid we were able to show that genetic markers flanking APC were both deleted, indicating that M is therefore at high risk of developing FAP. This provided the first molecular pre-symptomatic diagnosis of this disease.

Irradiation hybrids

To generate markers within the FAP region we chose the approach of producing irradiation-reduced hybrids (Cox et al. 1990). A single chromosome 5-containing hybrid, PN/TS-1, was irradiated with 50 000 rad and fused to the Chinese hamster line A23, selecting for TK. We chose a high dose expecting that the fragments would be small and specific for the deleted region (Benham et al. 1989). On average approximately 10 per cent of the chromosome was retained in the 150 colonies examined. This figure is considerably lower than that obtained in other studies with lower doses of irradiation (Cox et al. 1990; Richard et al. 1991). DNA from these hybrids analysed on pulsed field gel electrophoresis (PFGE) however did not give bands in common with a hybrid containing a normal 5. This suggested that although 10 per cent of markers were retained, that these were not from a contiguous piece of chromosomes but rather in fragments which we could estimate to be <1 MB.

Nine hybrids which were positive for a marker within the FAP deletions (ECB27) were used to produce phage libraries. These were then mapped onto the panel of deletion hybrids. Of 201 phage tested, 10 (5 per cent) fell within the deletions. This is approximately the number that would be expected to map within the deletion using clones from a whole chromosome 5 library and mapping them at random. No enrichment was therefore obtained although a number of useful probes were isolated. *In situ* hybridization of several hybrids with total human DNA had indicated that the chromosome 5 material was present in one or two blocks within the hamster chromosomes. However, *in situ* hybridization, to normal human chromosomes, with a pool of phage clones from three libraries showed a scattered signal along all of chromosome 5, indicating the chromosome 5 material, originated from multiple regions of the irradiated chromosome. The hybrids could be estimated to contain as many as 20 small fragments. Ordering of markers was unsuccessful using these hybrids as markers were not associated in pairwise combinations at frequencies greater than random even when very close (Thomas 1991). Lower doses of radiation would have been necessary for this strategy to have been more successful.

Other physical mapping methods

One of the markers first linked to FAP, C11P11, we knew from PFGE to be in a region of large *Bss*HII fragments and relatively few CpG-rich islands (Bird 1987; Varesco et al. 1989). We therefore took the approach of isolating and cloning the ends of large *Bss*HII fragments (Michiels et al. 1987) from a hybrid containing only human chromosomes 4 and 5 (MacDonald et al. 1987) and mapping the end clones back to the deletions. In this way we obtained two clones within the deletions, one of which played a key role in further physical mapping of the region.

As an additional source of probes from the deleted region we used the approach of directly cutting out band 5q22 from metaphase chromosomes and generating a library of microclones from the APC region (Ludecke et al. 1989, 1990). These were then mapped back to the deletion hybrid panel (Hampton et al. 1991a,b). Those mapping within the deletions were then used directly as probes on PFGE blots or hybridized to a chromosome 5 gridded cosmid library (L. Devean, Los Alamos). The isolated cosmids were again checked on this panel. The microclones were well distributed over the region and provided an excellent source of markers.

While this work was in progress, MCC, a candidate gene for FAP showing mutations in sporadic colorectal tumours was isolated (Kinzler *et al.* 1991). Using cDNA from this gene we were able to isolate several YACs (yeast artificial chromosomes) from the region which were converted to cosmid clones (Hampton *et al.* 1992). One of these YACs contained the MCC and APC genes and the probes isolated from this YAC were used in constructing the surrounding physical map.

Using irradiation hybrid phage clones, microclones, endclones, and YACs we were able to construct a long range physical map of the APC region, covering 10 Mb (Solomon and Frischauf 1993; Ward *et al.* 1993).

The APC gene

In the course of this work the APC gene was isolated simultaneously by two groups, through a combination of physical mapping strategies, isolation of candidate cDNAs from the region, and use of two germ-line deletions in FAP patients (Groden *et al.* 1991; Joslyn *et al.* 1991; Kinzler *et al.* 1991; Nishisho *et al.* 1991). LOH studies had indicated a region of very high allele loss in colorectal tumours, suggesting a candidate region of approximately 10–15 Mb surrounding these probes (Ashton-Rickardt *et al.* 1989). Using PFGE analysis on DNA from FAP patients, shifts were seen in two cases, using probes from the region of high allele loss. These were due to germ-line deletions, estimated to be 260 and 100 kb, which proved to be of enormous help in defining the region in which to look for candidate cDNAs (Joslyn, *et al.* 1991). Somatic cell hybrids were used to segregate the chromosomes 5 from these two deletion cases. The patient with the 260 kb deletion appeared to have only the abnormal band and not a normal band on PFGE. Separation of each of the chromosomes 5 from this patient in hybrids confirmed that one indeed had an abnormally small band and that the other had not been detected on the PFGE because it was missing a *Not*I site and therefore had a larger *Not*I fragment than could be resolved on the gel. Three candidate genes mapped within the 100 kb deletion and one of these, APC, was truncated at the deletion breakpoint. RNase protection and SSCP (single stranded conformational polymorphism) (Orita *et al.* 1989) analysis on this gene revealed it to be mutated in both the germ-line of FAP patients and in sporadic tumours.

Extensive analysis has now been done on the APC gene with respect to the types of mutations and the positions of mutations within this gene in the germ-line and sporadic tumours (Groden, *et al.* 1991; Nishisho, *et al.* 1991; Cottrell *et al.* 1992; Fodde *et al.* 1992; Miyoshi *et al.* 1992*a,b*). It is unequivocal that mutations in this gene are responsible for FAP and for sporadic colon tumours. Examination of very small pre-malignant adenomas of sporadic cases shows a frequency of APC mutations very similar to that found in the carcinomas, suggesting that these mutations are very early events (Powell *et al.* 1992). The failure to find LOH in adenomas of

FAP patients, as well as the low level of LOH in small sporadic adenomas would indicate that loss of activity from one allele is sufficient to promote adenoma formation (Bodmer et al. 1987) and that subsequent loss of the second allele gives further growth advantage. The positions of the mutations in the gene are essentially identical in both the germ-line and sporadic cases and occur towards the 5′-portion of the gene. Most striking is the observation that all of the mutations either generate stop codons directly or are small insertions or deletions leading to frame shifts and early termination. Additionally, as observed in the FAP cases with whom the initial mapping was done, deletion of the whole gene can result in the FAP phenotype. On the whole missense mutations are not seen. The protein products of these mutations are truncations toward the N-terminus and are believed to represent completely inactive products rather than ones capable of acting in a dominant negative fashion. Western blot analysis on lymphocytes from an APC patient reveals these truncated proteins (Smith et al. 1993). The normal gene codes for a large mRNA of approximately 8.5 kb and a protein of approximately 2843 amino acids. The deduced amino acid sequence does not have homology with any known proteins and is not predictive of the function of this protein. The nature of the mutations in APC and the finding of the same mutations in sporadic and germ-line tumours has made a convincing case for a tumour suppressor gene.

Other colon cancer genes

The power of detecting LOH in somatic tumour cells is well illustrated with two other examples of genes involved in colon cancer. Consistent high LOH on chromosome 17p and specifically at 17p12-p13.3, led to the testing of p53 as a candidate gene in colorectal carcinogenesis (Baker et al. 1989). p53 was found to be mutated not only in colon cancer, but in almost every type of tumour and is probably the most common genetic event in tumour progression (Malkin et al. 1990; Srivastava et al. 1990). Additionally, p53 germ-line mutations have been found in the Li-Fraumeni syndrome, characterized by early onset of multiple tumours including soft tissue sarcomas, breast tumours, and brain tumours (Malkin et al. 1990; Srivastava et al. 1990).

Although the initial indication that p53 is involved in colon tumour growth came from LOH studies, the mode of action of this gene is considerably more complicated than that suggested by simple loss of activity (for reviews see Vogelstein and Kinzler (1992), Lane (1992), and Lane (Chapter 17, this volume). The most common mutations in the p53 gene are missense mutations, which lead to loss of function as a transcriptional activator. These mutant proteins, however, act in a dominant negative fashion by forming heterotetramers with the wild-type protein. The result is reduction of the level of normal tetramers significantly below the 2-fold reduction that would occur with simple loss or truncation of one allele, as occurs with APC. Twofold reduction of p53 is apparently insufficient to promote growth for many types of tumours. In others, however, such as lung and oesophogeal tumours, loss or truncation of one allele seems to be a more frequent event. With either type of mutation, however, loss of the second allele, with concomitant complete loss of p53 activity, is common. These are relatively late events in tumour formation and are believed to be related to the fact that p53 is only induced under conditions of stress or instability at which time loss of its activity would result in further uncontrolled growth.

An additional locus where LOH is seen with high frequency in colon cancer is 18q, in the region of 18q21-qter. Again, this observation led to the cloning of the DCC (deleted in colon cancer) gene (Fearon et al. 1990). This gene undergoes LOH in 71 per cent of colorectal carcinomas. It also undergoes somatic mutations including deletions, insertions, and point mutations in 13 per cent of colorectal carcinomas. DCC mutations are later events than the mutations in APC, but clearly also occur in a high percentage of sporadic colon cancers. DCC encodes a putative 190 kDa transmembrane phosphoprotein with homology to neural cell adhesion molecules (NCAMs) (Fearon et al. 1990) and may therefore play a role in cell–cell or cell–matrix interactions. To date no germ-line mutations have been published in DCC, nor linkage demonstrated to any of the other hereditary colon cancer syndromes.

Conclusions

Somatic genetic changes in tumours are now known to indicate the position of tumour suppressor genes. Additionally, dominantly inherited familial cancers involve genes which are also mutated in sporadic tumours. The combined power of these families and linkage analysis, with the phenomenal technical advances in molecular genetics, including highly informative polymorphic markers, YAC cloning, and a variety of methods for identifying coding sequences within genomic DNA has led or is leading to the cloning of tumour suppressor genes from most of the major types of solid tumours. The immediate and obvious benefits accrue to those families in whom predictive testing can be done, so that carriers are screened and non-carriers are reassured. In colon cancer, the APC gene is the earliest change we know about. The development of screening methods for early detection of the truncated protein and mutated gene will by now be under way.

The model of familial and sporadic tumours will no doubt reveal other interesting aspects of cancer genetics. In some solid tumours such as lung and renal cell carcinoma, families with inherited disease are almost never found (for a review see Goddard and Solomon (1993)). In both of these cases however, LOH and deletions in the sporadic tumours point to regions of chromosome 3p in which tumour suppressor genes must surely lie. When such genes are isolated it will be of interest to find from their cell biology and, perhaps, transgenic models, whether germ-line mutations either do not survive or do not express this phenotype. In other instances it is becoming clear that while sporadic and familial forms of a disease may indeed involve mutations in the same gene, they need not be tumour suppressor genes. In these cases LOH is not seen and the search for these genes is not aided by somatic genetic events in the tumours. Lynch syndrome II, or hereditary non-polyposis colon cancer (HNPCC), is a dominant disorder predisposing to multiple types of cancer, including colorectal cancer (Lynch et al. 1985). The disease has recently been linked to a locus on chromosome 2, at which LOH is not seen (Peltomaki et al. 1993). Tumours from

HNPCC patients as well as 13 per cent of apparent sporadic tumours show instability of dinucleotide repeats throughout the genome, suggesting a gene involve in DNA replication repair may be involved. (Aaltonen *et al.* 1993; Ionov *et al.* 1993; Thibodeau *et al.* 1993). As yet, there remain several regions of high LOH in colon cancers, including 8p and 22 (Okamoto *et al.* 1988; Vogelstein *et al.* 1989). In addition, a number of hereditary colon cancer syndromes have not yet been mapped, either by linkage or LOH. Cloning of the regions of high LOH may eventually lead to the isolation of other genes contribution to the progression of colon cancer.

References

Aaltonen, L. A., Peltomaki, P., Leach, F. S., Sistonen, P., Pylkklanen, L., Mecklin, J.-P., *et al.* (1993). Clues to the pathogenesis of familial colorectal cancer. *Science*, **260**, 812–16.

Ashton-Rickardt, P. G., Dunlop, M. G., Nakamura, Y., Morris, R. G., Purdie, C. A., Steel, C. M., *et al.* (1989). High frequency of APC loss in sporadic colorectal carcinoma due to breaks clustered in 5q21-22. *Oncogene*, **4**, 1169–74.

Baker, S. J., Fearon, E. R., Nigro, J. M., Hamilton, S. R., Preisinger, A. C., Jessup, J. M., *et al.* (1989). Chromosome 17 deletions and p53 gene mutations in colorectal carcinomas. *Science*, **244**, 217–21.

Benham, F., Hart, K., Crolla, J., Bobrow, M., Francavilla, M., and Goodfellow, P.N. (1989). A method for generating hybrids containing nonselected fragments of human chromosomes. *Genomics*, **4**, 509–17.

Bird, A. P. (1987). CpG islands as gene markers in the vertebrate nucleus. *Trends Genet.*, **3**, 342–7.

Bodmer, W. F., Bailey, C. J., Bodmer, J., Bussey, H. J. R., Ellis, A., Gorman, P., *et al.* (1987). Localization of the gene for familial adenomatous polyposis on chromosome 5. *Nature*, **328**, 614–16.

Cavenee, W. K., Dryja, T. P., Philips, R. A., Benedict, R., Gallie, B. L., Murphree, A. L., *et al.* (1983). Expression of recessive alleles by chromosomal mechanisms in retinoblastoma. *Nature*, **305**, 779–84.

Cottrell, S., Bicknell, D., Kaklamanis, L., and Bodmer, W. F. (1992). Molecular analysis of APC mutations in familial adenomatous polyposis and sporadic colon carcinomas. *Lancet*, **340**, 626–30.

Cox, D. R., Burmeister, M., Price, E. R., Kim, S., and Myers, R. M. (1990). Radiation hybrid mapping: a somatic cell genetic method for constructing high-resolution maps of mammalian chromosomes. *Science*, **250**, 245–50.

Cross, I., Delhanty, J., Chapman, P., Bowles, L. V., Griffin, D., Wolstenholme, J., *et al.* (1992). An intrachromosomal insertion causing 5q22 deletion and familial adenomatous polyposis coli in two generations. *J. Med. Genet.*, **29**, 175–9.

Fearon, E. R., Cho, K. R., Nigro, J. M., Kern, S. E., Simons, J. W., Ruppert, J. M., *et al.* (1990). Identification of a chromosome 18q gene that is altered in colorectal cancers. *Science*, **247**, 49–56.

Fodde, R., van der Luijt, R., Wijnen, J., Tops, C., van der Klift, H., van Leeuwen-Cornelisse, I., *et al.* (1992). Eight novel inactivating germ line mutations at the APC gene identified by denaturing gradient gel electrophoresis. *Genomics*, **13**, 1162–8.

Giles, R. E., Shimizu, N. and Ruddle, F. H. (1980). Assignment of a human genetic locus to chromosome 5 which corrects the heat sensitive lesion associated with reduced leucyl-tRNA synthetase activity in ts025Cl Chinese hamster cells. *Somat. Cell Genet.*, **6**, 667–87.

Godbout, R., Dryja, T. P., Squire, J., Gallie, B. L. and Phillips, R. A. (1983). Somatic inactivation of genes on chromosome 13 is a common event in retinoblastoma. *Nature*, **304**, 451–3.

Goddard, A. D. and Solomon E. (1993). Genetic aspects of cancer. *Adv. Hum. Genet.* **21**, 321–76.

Goss, S. J. and Harris, H. (1975). New method for mapping genes in human chromosomes. *Nature*, **255**, 680–3.

Groden, J., Thliveris, A., Samowitz, W., Carlson, M., Gelbert, L., Albertsen, H., *et al.* (1991). Identification and characterization of the familial adenomatous polyposis coli gene. *Cell*, **66**, 589–600.

Hampton, G., Leuteritz, G., Ludecke, H. J., Senger, G., Trautmann, U., Thomas, H., *et al.* (1991*a*). Characterization and mapping of microdissected genomic clones from the adenomatous polyposis coli (APC) region. *Genomics*, **11**, 247–51.

Hampton, G. M., Howe, C., Leuteritz, G., Thomas, H., Bodmer, W. F., Solomon, E., *et al.* (1991*b*). Regional mapping of 22 microclones around the adenomatous polyposis coli (APC) locus on chromosome 5q. *Hum. Genet.* **88**, 112–14.

Hampton, G. M., Ward, J. R. T., Cottrell, S., Howe, K., Thomas, H. J. W., Ballhausen, W. G., *et al.* (1992). Yeast artificial chromosomes for the molecular analysis of the familial polyposis APC gene region. *Proc. Natl Acad. Sci. USA*, **89**, 8249–53.

Herrera, L., Katati, S., Gibas, L., Pietrzak, E., and Sandberg, A. A. (1986). Gardner syndrome in a man with an interstitial deletion of 5q. *Am. J. Med. Genet.* **25**, 473–6.

Hockey, K. A., Mulcahy, M. T., Montgomery, P., and Levitt, S. (1989). Deletion of chromosome 5q and familial adenomatous polyposis. *J. Med. Genet.*, **26**, 61–2.

Ionov, Y., Peinado, M. A., Malkhosyan, S., Shibata, D., and Perucho, M. (1993). Ubiquitous somatic mutations in simple repeated sequences reveal a new mechanism for colonic carcinogenesis. *Nature*, **363**, 558–61.

Joslyn, G., Carlson, M., Thliveris, A., Albertsen, H., Gelbert, L., Samovitz, W. *et al.* (1991). Identification of deletion mutations and three new genes at the familial polyposis locus. *Cell*, **66**, 601–13.

Kinzler, K. W., Nilbert, M. C., Su, L.-K., Vogelstein, B., Bryan, T. M., Levy, D. B., *et al.* (1991). Identification of FAP locus genes from chromosome 5q21. *Science*, **253**, 661–5.

Knudson, A.G. (1971). Mutation and cancer: statistical study of retinoblastoma. *Proc. Natl Acad. Sci. USA*, **68**, 820–3.

Knudson, A. J. (1985). Hereditary cancer, oncogenes, and antioncogenes, *Cancer Res.* **45**, 1437–43.

Knudson, A. J. (1991). Overview: genes that predispose to cancer. *Mutat. Res.* **247**, 185–90.

Lane, D. P. (1992). p53, guardian of the genome. *Nature*, **358**, 15–16.

Lasko, D., Cavenee, W.and Nordenskjold, M. (1991). Loss of constitutional heterozygosity in human cancer. *Annu. Rev. Genet.* **25**, 281–314.

Leppert, M., Dobbs, M., Scambler, P., O'Connell, P., Nakamura, Y., Stauffer, *et al.* (1987). The gene for familial polyposis coli maps to the long arm of chromosome 5. *Science*, **238**, 1411–13.

Ludecke, H. J., Senger, G., Claussen, U., and Horsthemke, B. (1989). Cloning defined regions of the human genome by microdissection of banded chromosomes and enzymatic amplification. *Nature*, **338**, 348–50.

Ludecke, H. J., Senger, G., Claussen, U., and Horsthemke, B. (1990). Construction and characterization of band-specific DNA libraries. *Hum. Genet.*, **84**, 512–16.

Lynch, H. T., Kimberling, W., Albano, W. A., Lynch, J. F., Biscone, K., Scheulke, G. S. *et al.* (1985). Hereditary nonpolyposis colorectal cancer (Lynch syndromes 1 and 11). *Cancer*, **56**, 934–8.

MacDonald, M. E., Anderson, M. A., Gilliam, T. C., Traneblaerg, L., Carpenter, N. J., Magenis, *et al.* (1987). A somatic cell hybrid panel for localizing DNA segments near the Huntington's disease gene. *Genomics*, **1**, 29–34.

Malkin, D., Li, F. P., Strong, L. C., Fraumeni, J. F., Nelson, C. E., Kim, D. H., *et al.* (1990). Germ line p53 mutations in a familial syndrome of breast cancer, sarcomas, and other neoplasms. *Science*, **250**, 1233–8.

Michiels, F., Burmeister, M., and Lehrach, H. (1987). Derivation of clones close to the cystic fibrosis marker met by preparative field inversion gel electrophoresis. *Science*, **236**, 1305–7.

Miyoshi, Y., Ando, H., Nagase, H., Nishisho, I., Horii, A., Miki, Y., *et al.* (1992*a*). Germline mutations of the APC gene in 53 familial adenomatous polyposis patients. *Proc. Natl Acad. Sci. USA*, **89**, 4452–6.

Miyoshi, Y., Nagase, H., Ando, H., Horii, A., Ichii, S., Nakatsuru, S., *et al.* (1992*b*). Somatic mutations of the APC gene in colorectal tumors: mutatin cluster region in the APC gene. *Hum. Mol. Genet.*, **1**, 229–33.

Nishisho, I., Nakamura, Y., Miyoshi, Y., Miki, Y., Ando, H., Horii, A., *et al.* (1991). Mutations of chromosome 5q21 genes in FAP and colorectal cancer patients. *Science*, **253**, 665–9.

Okamoto, M., Sasaki, M., Sugio, K., Sato, C., Iwama, T., Ikeuchi, T., *et al.* (1988). Loss of constitutional heterozygosity in colon carcinoma from patients with familial polyposis coli. *Nature*, **331**, 273–7.

Orita, M., Iwahana, H., Kanazawa, H., Hayashi, K., and Sekiya, T. (1989). Detection of polymorphisms of human DNA by gel electrophoresis as single-strand conformation polymorphisms. *Proc. Natl Acad. Sci. USA*. **86**, 2766–70.

Peltomaki, P., Aaltonen, L. A., Sistonen, P., Pylkkanen, L., Mecklin, J.-P., Jarvinen, H., *et al.* (1993). Genetic mapping of a locus predisposing to human colerectal cancer. *Science*, **260**, 810–12.

Powell, S. M., Zilz, N., Beazer-Barclay, Y., Bryan, T. M., Hamilton, S. R., Thibodeau, S. N., *et al.* (1992). APC mutations occur early during colorectal tumorigenesis. *Nature*, **359**, 235–7.

Richard, C. W. I., Withers, D. A., Meeker, T. C., Maurer, S., Evans, G. A., Myers, T. M., *et al.* (1991). A radiation hybrid map of the proximal long arm of chromosome 11 containing the multiple endocrine neoplasia type I (MEN-I) and bcl-1 disease loci. *Am. J. Hum. Genet.* **49**, 1189–96.

Smith, K. J., Johnson, K. A., Bryan, T. M., Hill, D. E., Markowitz, S., Willson, J. K. V., *et al.* (1993). The APC gene product in normal and tumor cells. *Proc. Natl Acad. Sci. USA*, **90**, 2846–50.

Solomon, E. and Frischauf, A.-M. (1993). Construction of a 10-Mb physical map in the adenomatous polyposis region of chromosome 5. *Genome Anal.*, **5**, 89–105.

Solomon, E., Voss, R., Hall, V., Bodmer, W. F., Jass, J. R., Jeffreys, A. J., *et al.* (1987). Chromosome 5 allele loss in human colorectal carcinomas. *Nature*, **328**, 616–19.

Srivastava, S., Zou, Z., Pirollo, K., Blattner, W., and Chang, E. H. (1990). Germ-line transmission of a mutated p53 gene in a cancer-prone family with Li-Fraumeni syndrome. *Nature*, **348**, 747–9.

Thibodeau, S. N., Bren, G., and Schaid, D. (1993). Microsatellite instability in cancer of the proximal colon. *Science*, **260**, 816–19.

Thomas, H. J. W. (1991). Molecular analysis of the adenomatous polyposis coli gene region. Ph.D. thesis. University College, London.

Thompson, L. H., Harkins, J. L., and Stanners, C.P. (1973). A mammalian cell mutant with a temperature-sensitive leucyl-transfer RNA synthetase. *Proc. Natl Acad. Sci. USA*, **70**, 3094–8.

Varesco, V., Thomas, H. J. W., Cottrell, S. V. M., Fennell, S. J., Williams, S., Searle, S., *et al.* (1989). CpG island clones from a deletion encompassing the gene for adenomatous polyposis coli. *Proc. Natl Acad. Sci. USA*, **86**, 10118–22.

Vogelstein, B., Fearon, E. R., Kern, S. E., Hamilton, S. R., Preisinger, A. C., Nakamura, Y., *et al.* (1989). Allelotypes of colorectal carcinomas, *Science*, **244**, 207–11.

Vogelstein, B. and Kinzler, K. W. (1992). p53 function and dysfunction. *Cell*, **70**, 523–6.

Ward, J. R. T., Cottrell, S., Thomas, H. J. W., Jones, T. A., Howe, K., Hampton, G. M., *et al.* (1993). A long range restriction map of human chromosome 5q21-23. *Genomics*, **17**, 15–24 (1993).

Weber, J. L. and May, P. E. (1989). Abundant class of human DNA polymorphisms which can be typed using the polymerase chain reaction. *Am. J. Hum. Genet.*, **44**, 388–96.

Weissenbach, J., Gyapay, G., Dib., C., Vignal, A., Morissette, J., Milasseau, G., *et al.* (1992). A second-generation linkage map of the human genome. *Nature*, **359**, 795–801.

19. Roles of the *myc* gene in cell proliferation and differentiation, as deduced from its role in tumorigenesis

George Klein

Myc, *myb*, *fos*, *rel*, and *ski* are oncogenes that code for DNA binding nuclear phosphoproteins that can *trans*-activate other genes and can stimulate DNA replication, directly or indirectly. They may contribute to tumour development and/or progression, after having been activated by structural and/or regulatory changes (Eisenmann 1989). The best known gene of this category, c-*myc* (Penn et al. 1990), is also most widely implicated in spontaneous tumour development *in vivo*. It is the cellular counterpart of v-*myc*, which has been isolated from four different acute avian leukaemia viruses. c-*myc* and v-*myc* can induce a broad spectrum of tumours, such as myelocytomatosis, endotheliomas, liver and kidney carcinomas, and erythroblastosis. One strain can also transform fibroblasts.

The c-*myc* proto-oncogene encodes a short-lived nuclear phosphoprotein that is well conserved in vertebrates. It is a member of a family of structurally and functionally closely related genes. Next to c-*myc*, N-*myc* and L-*myc* are the best known members of the family. The c-*myc* gene can be constitutively activated by retroviral insertion on either side, or by translocation to a highly active chromosome region, such as an immunoglobulin locus in B-cells. It can also contribute to tumour progression by amplification, as will be discussed below. The extraordinary frequency of c-*myc* activation by these mechanisms in different tumours, and in the apparent absence of any structural changes in the protein, indicates that c-*myc* can turn into an oncogene by purely regulatory changes that lead to the constitutive expression of the gene and make it refractory to normal regulatory factors.

The role of the pathologically activated, constitutively expressed *myc* gene in the oncogenic process may be seen in relation to the normal expression of the gene (Klein and Klein 1986). Resting fibroblasts and lymphocytes do not express any detectable *myc* message but activated cells do. Growth factor or mitogen stimulation induces transcription of c-*fos* and c-*myc*. The induction of the 'competence genes', involved in preparing the entry of resting cells into the mitotic cycle, is among the earliest activation associated changes.

High constitutive *myc* expression can abrogate the growth factor requirements of normal cells. Interleukin 2 (IL2) or IL3 dependent haemopoietic cell lines infected with v-*myc*-carrying retroviruses acquired the ability to grow in the absence of the stimulatory interleukin and became tumorigenic in mice (Rapp *et al.* 1985). This was not accompanied by constitutive IL3 or IL2 production and was therefore not due to an autocrine mechanism. It would appear that the uninterrupted expression of *myc* obviates the need of the cell for an external growth stimulatory signal.

When continuously proliferating leukaemia, teratocarcinoma, neuroblastoma, or histiocytoma cells are induced to differentiate into a quiescent (G0) cell, *myc* expression is down-regulated (Penn *et al.* 1990). This can be prevented by the introduction of a constitutively expressed *myc* construct that fails to respond to normal regulatory signals. As long as *myc* expression is maintained, differentiation is inhibited (Coppola and Cole 1986; Prochownik and Kukowsa 1986). Contrariwise, inhibition of c-*myc* expression by antisense RNA can lead to accelerated differentiation (Griep and Westphal 1988; Prochownik *et al.* 1988). These experiments are consistent with the postulate that the pathological activation of *myc* expression by retroviral insertion or chromosomal translocation can prevent the cell from leaving the cycling compartment in accordance with its normal program. Such cells have an increased self-renewal potential *in vivo*, and can readily grow as immortalized lines *in vitro*.

The role of *myc* in oncogene complementation

Following the discovery that established strains of rodent fibroblasts can be transformed by DNA from human tumours, Land, Parada, and Weinberg have shown that even normal diploid rodent fibroblasts can be transformed, but only if two activated oncogenes are combined (Land *et al.* 1983). The first successful combination was *myc* and *ras*. Constitutively activated but otherwise normal *myc* genes immortalized the diploid fibroblasts without any sign of morphological transformation. Mutationally activated *ras* genes changed the social relationships between the cells. The normal flat, two-dimensional monolayer transformed into disorderly and partly three-dimensional growth. This was due to the lack of contact inhibition between the cells and was correlated with increased agarose clonability.

The oncogene complementation system was also helpful for the classification of other oncogenes as immortalizing or '*myc*-related' and transforming or '*ras*-related' (Weinberg 1989). The former category includes, in addition to the other members of the *myc* family, also *jun*, *myb*, *p53*, *ski,* and *fos*. This category also includes several DNA virus-encoded transforming proteins, such as SV40 and polyoma large T, adenoviral E1A, and the human papillomavirus-encoded E7 proteins. It is noteworthy that all immortalizing gene products in this group are *nuclear* proteins. In contrast, oncogenes of the *ras*-like group encode membrane-associated proteins. In addition to other members of the *ras* family, the group includes polyoma middle T, a membrane-associated, virally encoded transforming protein, and the cell-derived oncogenes *src*, *erbB*, *ros*, *fms*, *fos*, *mil*, *mos*, and *abl*.

Weinberg (1989) has suggested that the nuclear proteins of the Myc group may immortalize diploid fibroblasts by emancipating them from their requirement for exogenous signals. Cells with constitutively expressed nuclear oncogenes can divide in the absence of growth factors secreted by other cells. Membrane proteins of the activated Ras type liberate the cells from their own programmed, inner growth limitations. They can achieve this either by faulty signal transmission as in the case of Ras, or by permanent damage to a membrane receptor, as in the case of v-ErbB and v-Fms, or by increased kinase activity which changes the normal membrane structure, as in the case of Src and Abl.

The activities of the nuclear and the cytoplasmic oncogenes are complementary rather than additive. Certain cancer traits are induced more effectively by nuclear proteins while others are achieved more readily by gene products that act in the cytoplasm, depending on the cell type and its actual position in the chain of tumour progression (Weinberg 1985).

Myc activation by chromosomal translocation

A 'natural' mode of cellular oncogene activation was discovered by the analysis of tumour-associated chromosomal translocations (Klein 1983). Burkitt's lymphoma, a B-cell-derived, immunoglobulin-producing tumour, carries one of three alternative translocations in virtually 100 per cent of cases. The highly endemic, largely Epstein–Barr virus (EBV)-carrying and the sporadic, mainly EBV-negative form contain the same translocations. They arise by a break in chromosome 8 at band q24 (the location of the c-*myc* gene) and a break of one of the three immunoglobulin gene carrying chromosomes, followed by a reciprocal exchange of the terminal segments. In the most frequent or 'typical' translocation, chromosome 14 breaks at the site of the IgH locus. In the two alternative 'variant' translocations, 2;8 and 8;22, chromosome 2 breaks at the site of the κ gene and chromosome 22 at the site of the λ gene respectively. This juxtaposition of c-*myc* to the immunoglobulin sequences subordinates them to the control of the constitutively active immunoglobulin loci. The latter are not programmed to switch off at any time during the continued lifespan of the B-cell. The c-*myc* gene is regularly expressed in proliferating cells but not in resting ones (Penn et al. 1990). The *myc* gene juxtaposed to the immunoglobulin sequences is constitutively expressed like an immunoglobulin locus. In contrast to the normal, non-translocated *myc* gene, it does not obey the normal down-regulation program of *myc* at the time when activated B-cells turn into long-lived resting cells in G_0. This happens on at least two different occasions during the life cycle of the B-cell: following the normal immunoglobulin gene rearrangement when a pre-B cell turns into a virgin B-cell, and on the cessation of antigen-induced B-cell proliferation when a stimulated blast turns into a memory cell.

Myc was first discovered as the transforming sequence of an avian leukaemia virus (Penn et al. 1990). The v-*myc* gene can transform a wide range of cells, including haemopoietic, mesenchymal, and epithelial targets. Activation of its cellular counterpart (c-*myc*) can participate in the development and/or progression of many different tumours. Its activation by retroviral insertion is a regular event in the initiation of chicken bursal lymphoma. It is activated by chromosomal translocation to an immunoglobulin locus, not only in Burkitt's lymphoma, but also in mouse and rat plasmacytoma. Moreover, the progression of several human tumours to more malignant forms can be favoured by the amplification of c-*myc* or its close relatives, N- or L-*myc*, as discussed in the section on oncogene amplification below. The extraordinary frequency of *myc* activation by these three different mechanisms and in so many different tumours probably reflects the fact that the normal proto-oncogene can turn into an oncogene by purely regulatory changes that lead to higher or more constitutive expression, without necessarily requiring any structural changes.

The dysregulation of c-*myc* by its juxtaposition to an immunoglobulin locus is the clearest known example of tissue-specific oncogene activation by a purely cellular mechanism. The molecular anatomy of the translocation is closely similar in the corresponding human, murine, and rat tumours (Klein 1989). The breakpoints in and around the c-*myc* gene vary widely in different tumours, suggesting that the breakage is a random accident of cell division. The coding sequences always retain their integrity, indicating that the constitutively expressed Myc protein needs to be intact in order to provide the translocation-carrying cells with a selective advantage. The regular occurrence of the translocations indicates that they represent an essential rate-limiting step in the tumorigenic process. This conclusion could be confirmed by 'facsimile' experiments. Transgenic mice that carry a *myc* gene under the control of an immunoglobulin enhancer develop pre-B and B-cell-derived lymphomas in 90 per cent or more (Adams *et al.* 1985). Mice infected with a retrovirally activated v-*myc* gene developed pristane oil induced plasmacytomas much faster and in a higher frequency than uninfected, pristane-treated controls (Potter *et al.* 1987). The plasmacytomas that expressed the experimentally introduced v-*myc* gene did not have any immunoglobulin/*myc* translocations, in contrast to those that did not express v-*myc* and had either a typical or a variant translocation. These experiments confirmed that activated *myc* genes were tumorigenic for B-cells and supported the conclusion that the immunoglobulin/*myc* translocations trigger the development of the tumour by the constitutive activation of *myc*. However, they also showed that additional changes were required for the development of fully autonomous neoplasia, since all facsimile tumours were mono- or oligoclonal. Activation of additional oncogene(s) and/or loss of tumour suppressor genes may be required. This is in line with our current knowledge concerning the multi-step development of most cancers (Klein and Klein 1985).

The study of the numerous translocations found in B- and T-cell lymphomas has been particularly rewarding, since these translocations involve the normally rearranging immunoglobulin or T-cell receptor sequences, respectively. The attached sequence that has become translocated from another chromosome could be identified by departing from the breakpoint (Haluska *et al.* 1988). The discovery of the *bcl-2* oncogene, the chromosome 18-derived, IgH-juxtaposed sequence in the follicular lymphoma associated 14;18 translocation has proved particularly important. Facsimile experiments with transgenic mice have shown that constitutive activation of the *bcl-2* gene by a linked IgH enhancer (Emu) increases the life span of resting B-cells and prevents their programmed death by apoptosis. The corresponding Emu–*myc* transgenics have gone one step further: they have no resting B-cells at all, all their B-cells having transformed into proliferating immunoblasts (Adams 1988).

The immunoglobulin/*myc* translocation does not involve any specific sequences in or around the *myc* locus. In contrast, the IgH/*bcl-2* (14;18) translocation takes place by recombination between J (joining) sequences of the IgH locus on chromosome 14 and corresponding homologous sequences in the neighbourhood of the *bcl-2* locus. The 14;18 translocation leads to the activation of the normal *bcl-2* protein, lending further support to the notion that the major mode of oncogene activation in lymphoid malignancies is due to the action of juxtaposed (*cis*) regulatory sequences. It must be noted, however, that the oncogenic effect of different displaced oncogenes can be quite different. This can be best seen in the contrast between *myc* and *bcl-2*. The *myc*/immunoglobulin translocation leads to highly malignant lymphomas whereas the *bcl-2*/IgH translocation causes low grade malignancy (Haluska *et al.* 1988). This can be seen in relation to the different effects of the two genes in the transgenic mice, mentioned above. Occasionally a *bcl-2*/IgH-carrying low grade lymphoma may progress to high malignancy by a second immunoglobulin/*myc* translocation. This is a good example of tumour progression by the sequential activation of different oncogenes. Progression of Ph1 positive chronic myelogenous leukaemia (CML) to acute leukaemia provides another example (De Klein *et al.* 1988). Blast crisis can proceed through one of several alternative pathways. Duplication of the Ph1 chromosome, appearance of an extra chromosome no. 8 and morphological change of a chromosome no. 17 are the most common 'major routes'.

Myc and other oncogene translocations have also been identified in T-cell leukaemia and lymphomas (Haluska *et al.* 1988). Instead of the immunoglobulin loci, the T-cell translocations involve one of the T-cell receptor (TCR) genes. The architecture of T-cell translocations resembles the variant (light chain), rather than the typical (heavy chain) translocations in the B-cells. The most common translocation disrupts J-segments of the TCR-α locus on chromosome 14q11. The TCR-β locus on chromosome 7 is frequently involved.

Myc amplification

A wide variety of solid and haemopoietic tumours have been found to harbour amplified oncogenes (Cooper *et al.* 1984; Sümegi *et al.* 1985; Alitalo and Schwab 1986; Slamon *et al.* 1987; Collins and Groudine 1988). Genes of the *myc* family were amplified most frequently. In small cell lung carcinoma (SCLC), amplification of *myc* was correlated with a 'variant tumour cell phenotype' (Little *et al.* 1977), characterized by shorter doubling time, higher cloning efficiency, increased radioresistance, increased invasiveness, increased tumorigenicity in nude mice, high metastatic ability, and morphological and enzymatic changes associated with poor prognosis. Interestingly, the three known *myc*-genes, C-, N-, and L-*myc*, were alternatively amplified, suggesting a close functional relationship. N-*myc* was originally discovered due to its prognostically unfavourable amplification in neuroblastoma (Brodeur et al. 1984) and L-myc through its amplification in some SCLC variants (Little et al. 1977). Their postulated functional relationship is consistent with the similar transforming potential of c- and N-myc, when cotransfected with activated ras genes into rat embryo fibroblasts (Schwab et al. 1985). This does not exclude minor functional differences, as suggested by the finding that SCLC variants with an N-myc amplification were less aggressive than their c-myc amplified counterparts.

Apparently, myc amplification confers a growth advantage on the lung carcinoma cell, particularly after it has been displaced to a foreign tissue environment. In the neuroblastomas, Brodeur et al. (1984) found a significant correlation between N-myc amplification and the more aggressive stage III–IV forms of the disease. There was a close correlation between patient survival and the N-myc copy number in the tumours. In human B plasmacytomas, we have found a relationship between c-myc amplification and the appearance of plasma cell leukaemia (Sümegi et al. 1985), the most malignant form of the disease.

Amplification of the neu oncogene in about 30 per cent of breast carcinomas was correlated with poor prognosis (Slamon et al. 1987; Borg et al. 1990) and particularly in patients with extensive axillary lymph node involvement. For node-positive patients, neu amplification was a significant predictor of early relapse and death. It is not known how amplified oncogenes contribute to tumour progression, but their persistence during tumour growth and after metastasis suggests that the presence of the amplified copies is needed for the maintenance of highly malignant growth.

Amplified oncogenes may act by increasing the concentration of the relevant oncoprotein. An increased level of Myc protein in the more invasive and metastatic forms of SCLC could act by increasing the proliferative drive of the cells, so that they become more apt to overcome local growth-restricting forces in distant tissue compartments. Alternatively, gene amplification may not be accompanied by a parallel increase of the corresponding oncoprotein, but could act by titrating out regulatory factors that would normally inhibit the gene. A delicate balance may exist between the transcription of regulatory factors and their target DNA sequences in or near the gene. Amplification of the target sequences may tilt the balance in favour of cell division, at the expense of differentiation (Penn et al. 1990). It should be noted that strong differentiation-inducing signals elicited by TPA or DMSO could down-regulate the 40- to 60-fold amplified, highly expressed myc genes in HL60, a promyelocytic leukaemia line, concomitantly with terminal differentiation. Similarly, the highly expressed N-myc was down-regulated when neuroblastoma cells were induced to differentiate by retinoic acid exposure (Thiele et al. 1985).

Concluding remarks

What was first a hypothesis based on the chromosomal localization of the translocations associated with Burkitt's lymphoma and mouse plasmacytoma (Klein 1981) has been fully proven as a reality. Chromosomal translocations associated with a given tumour type can contribute to the carcinogenic process in a decisive way, by bringing a proto-oncogene (or a functionally analogous gene) under the influence of a highly active chromatin region. Alternatively, as in the case of the (9;22) *bcr/abl* translocation, the rearrangement acts by generating a new fusion protein, composed by sequences on both sides of the breakpoint. The regular occurrence of the same translocation (or its variants) in tumours of the same type is the best evidence that it plays an essential role in the tumorigenic process. These 'experiments of nature' represent the most regular oncogene activation events in the course of spontaneous tumour development in *any* system.

Quite in analogy with the two other fields concerned with oncogene activation in tumour development, namely retroviral insertion and experimental transformation by DNA transfection, the analysis of the tumour-associated chromosomal translocations has led to the discovery of previously unknown proto-oncogenes. The *bcl-2* gene, which is transposed from chromosome 18 to the IgH region of chromosome 14 in follicular lymphomas, is the most notable example. Further study of tumour-associated translocations may reveal a gamut of new proto-oncogenes.

The current scenario of Burkitt's lymphoma development illustrates the complex interaction between viral transformation and cellular oncogene activation during a specific carcinogenic process. It also provides the best analysed example of escape from immune surveillance so far. Finally, lymphomas carrying the double (*IgH/bcl-2* and immunoglobulin/*myc*) translocation represent a beautiful example of tumour progression.

References

Adams, J. M. (1988). Consequences of constitutional activation of oncogenes within transgenic mice. In *Cellular oncogene activation*, (ed. G. Klein), pp. 365–88. Marcel Dekker, New York.

Adams, R. J., Harris, A. W., Pinkert, C. A., Corcoran, L. M., Alexander, W. A., Cory, S., *et al.* (1985). The c-*myc* oncogene driven by immunoglobulin enhancers induces lymphoid malignancy in transgenic mice. *Nature*, **318**, 533–8.

Alitalo, K. and Schwab, M. (1986). Oncogene amplification in tumor cells. *Adv. Cancer Res.*, **47**, 235–82.

Borg, A., Tandon, A. K., Sigurdsson, H., Clark, G. M., Fernö, M., Fuqua, S. A. W., *et al.* (1990). HER-2/*neu* amplification predicts poor survival in node-positive breast cancer. *Cancer Res.*, **50**, 4332–7.

Brodeur, G., Seeger, R., Schwab, M., Varmus, H., and Bishop, J. M. (1984). Amplification of N-myc in untreated human neuroblastomas correlates with advanced disease stage. *Science*, **224**, 1121–4.

Collins, S. J. and Groudine, M. (1988). Oncogene amplification and tumor progression. In *Cellular oncogene activation*, (ed. G. Klein), pp. 313–4. Marcel Dekker, New York.

Cooper, C. S., Park, M., Blair, D. G., Tainsky, M. A., Huebner, K., Croce, C. M., *et al.* (1984). Molecular cloning of a new transforming gene from a chemically transformed human cell line. *Nature*, **311**, 29–33.

Coppola, J. A. and Cole, M. D. (1986). Constitutive c-*myc* oncogene expression blocks mouse erythroleukemia cell differentiation but not commitment. *Nature*, **320**, 760–63.

De Klein, A., Hermans, A., Heisterkamp, N., Groffen, J., and Grosveld, G. (1988). The Philadelphia chromosome and the bcr/c-abl translocation in chronic myelogenous and acute leukemia. In *Cellular oncogene activation*, (ed. G. Klein), pp. 295–302. Marcel Dekker, New York.

Eisenman, R. N. (1989). Nuclear oncogenes. In *Oncogenes and molecular origins of cancer*, (ed. R. Weinberg), pp. 175–222, Cold Spring Harbor Laboratory Press, Cold Spring Harbor, NY.

Griep, A. E. and Westphal, H. (1988). Anti-sense myc sequences induced differentiation of F9 cells. *Proc. Natl Acad. Sci. USA*, **85**, 6806–10.

Haluska, F. G., Tsujimoto, Y., and Croce, C. M. (1988). Oncogene activation in human T-cell and non-Burkitt B-cell malignancies. In *Cellular oncogene activation*, (ed. G. Klein), pp. 275–94. Marcel Dekker, New York.

Klein, G. (1981). The role of gene dosage and genetic transpositions in carcinogenesis. *Nature*, **294**, 313–18.

Klein, G. (1983). Specific chromosomal translocations and the genesis of B-cell-derived tumors in mice and men. *Minireview Cell*, **32**, 311–15.

Klein, G. (1989). Multiple phenotypic consequences of the Ig/myc translocation in B-cell derived tumors. *Genes, Chromosomes Cancer*, **1**, 3–8.

Klein, G. and Klein, E. (1985). Evolution of tumours and the impact of molecular oncology. *Nature*, **315**, 190–95.

Klein, G. and Klein, E. (1986). Conditioned tumorigenicity of activated oncogenes. *Cancer Res.*, **46**, 3211–24.

Land, H., Parada, L. F., and Weinberg, R. A. (1983). Tumorigenic conversion of pulmonary embryo fibroblasts requires at least two cooperating oncogenes. *Nature*, **304**, 596–602.

Little, C. D., Nau, M. M., Carcney, D. N., Gazdar, A. F., and Minna, J. D. (1977). Amplification and expression of the c-*myc* oncogene in human lung cancer cell lines. *Nature*, **306**, 194–6.

Penn, L. J. Z., Laufer, E. M., and Land, H. (1990). C-*myc*: evidence for multiple regulatory functions. *Cancer Biol.* **1**, 69–80.

Potter, M., Mushinski, J. F., Mushinski, E. B., Brust, S., Wax, J. S., Wiener, F., *et al.* (1987). Avian v-*myc* replaces chromosomal translocation in murine plasmacytomagenesis. *Science*, **235**, 787–9.

Prochownik, E. V. and Kukowsa, J. (1986). Deregulated expression of c-*myc* by murine erythroleukemia cells prevents differentiation. *Nature*, **322**, 848–50.

Prochownik, E. V., Kukowsa, J., and Rodgers, C. (1988) c-*myc* antisense transcripts accelerate differentiation and inhibit G1 expression in murine erythroleukemia cells. *Mol. Cell. Biol.*, **8**, 3683–95.

Rapp, U., Cleveland, J., Brightman, K., Scott, A., and Ihle, J. (1985). Abbrogation of IL-3 and IL-2 dependence by recombinant murine retroviruses expressing v-*myc* oncogene. *Nature*, **317**, 434–8.

Schwab, M., Varmus, H., and Bishop, J. M. (1985). Human N-*myc* gene contributes to neoplastic transformation of mammalian cells in culture. *Nature*, **316**, 160–62.

Slamon, D. J., Clark, G. M., Wong, S. G., Levin, W. J., Ulrich, A., and McGuire, W. L. (1987). Human breast cancer: correlation of relapse and survival with amplification of the HER-2/*neu* oncogene. *Science*, **235**, 177–82.

Sümegi, J., Hedberg, T., Björkholm, M., Godal T., Mellstedt, H., Nilsson, M.-G., *et al.* (1985). Amplification of the c-myc oncogene in human plasma-cell leukemia. *Int. J. Cancer*, **36**, 367–71.

Thiele, C., Reynolds, C., and Israel, M. (1985) Decreased expression of N-*myc* precedes retinoic acid-induced morphological differentiation of human neuroblastoma. *Nature*, **313**, 404–6.

Weinberg, R. A. (1985). The action of oncogenes in the cytoplasm and nucleus. *Science*, **230**, 770–76.
Weinberg, R. A. (1989) Oncogenes, antioncogenes, and the molecular bases of multistep carcinogenesis. *Cancer Res.*, **49**, 3713–21.

20. Genes that suppress the action of mutated *ras* genes

Makoto Noda

Introduction

Three major types of approach have been taken to address the question of how activated *ras* oncogenes induce malignant cell transformation: enzymology, genetics, and cell biology. Ras proteins have intrinsic guanine nucleotide-binding activity as well as GTP-hydrolysing (GTPase) activity (for review see Barbacid (1987)). Therefore, cellular factors which affect these activities in the test tube can be sought. The GTPase activating protein (GAP) and the guanine nucleotide-exchange factors (GNEFs) have been identified in this fashion (McCormick 1989; Downward *et al.* 1990; Huang *et al.* 1990; Wolfman and Macara 1990). Oncogenically mutated ('activated') Ras proteins show either decreased intrinsic GTPase activity or poor susceptibility to GAP (reviewed in Barbacid (1987) and Krengel *et al.* (1990)) and, therefore, GAP probably serves as a negative regulator and GNEFs as positive regulators of Ras. Although enzymology can yield precise results, it has its limitations. For instance, interaction between purified proteins does not necessarily reflect the physiological functions of the proteins *in vivo*. Also, the applicability of this technique is confined to the molecules directly interacting with Ras proteins.

The structures of Ras proteins are extremely well conserved across species, and this allows the genetic dissection of the Ras-related signal transduction pathways using experimental organisms such as yeast, fruit-fly, and nematode. In particular, a set of new proteins involved in the Ras-mediated signalling pathways have recently been discovered through the study of two relatively simple systems: eye development in fruit-fly (Rubin 1989) and vulva development in the nematode *Caenorhabditis elegans* (Horvitz and Sternberg 1991). The studies so far have shown links between homologues of epidermal growth factor (EGF) (Hill and Sternberg 1992), receptor-type protein tyrosine kinase (Hafen *et al.* 1987; Aroian *et al.* 1990), Src-homology-2 (SH2)-containing protein (Clark *et al.* 1992), GNEF, Ras (Simon *et al.* 1991), and GAP (Gaul *et al.* 1992). Genetics is powerful in discovering functional links, direct or indirect, between proteins, but it is rather poor in providing evidence for direct interaction between the proteins and for actual biochemical activities of individual proteins. In addition, there is always the possibility that information obtained in these simple organisms cannot be directly extrapolated to mammalian systems.

One of the systems which can serve to compensate the problems in the above two approaches would be cultured mammalian cells. Although it is time-consuming and subject to some technical limitations, one can study both the biochemistry and the genetics of mammalian cells by using tissue culture systems. For instance, we have previously shown that the activated *ras* genes induce neuronal differentiation, rather than malignant cell growth, in the rat pheochromocytoma cell line PC12 (Noda *et al.* 1985). This system has recently been used to show the link between Ras, Raf, and MAP kinase in the transduction of the nerve growth factor-mediated signal (Thomas *et al.* 1992; Wood *et al.* 1992). In the following sections, another set of examples for the study of Ras function using mammalian cell culture, namely the genetic approach, will be discussed.

Study of 'flat' revertants

Common strategies in genetics include isolation and characterization of mutants which cannot express certain phenotype. If the mutation is of a loss-of-function type, one may speculate that the wild-type gene is likely to be necessary for the expression of that phenotype. If the mutation is of a gain-of-function type, the wild-type gene may be involved in the negative regulation of the phenotype. To apply this principle to the phenotype of malignant cell transformation, a simple experiment is:

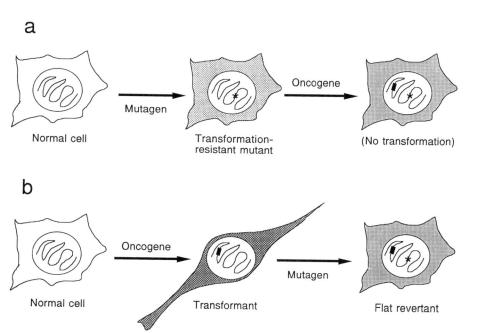

Fig. 20.1. Transformation-resistant mutant (a) compared with flat revertant (b).

to take normal cells and make mutants which can no longer be transformed (Inoue et al. 1983) (Fig. 20.1(a)), or alternatively, to take transformed cells and make mutants which look normal (non-transformed or 'flat' revertants) (Fig. 20.1(b)). The latter type of experiment has the advantage that the mutant phenotype can be directly detected, although it is usually more difficult to detect revertants than transformants. In the second type of experiments, one can also design selection techniques based on the differences between the properties of normal cells and transformed cells.

These are the aim and rationale for most of the studies using flat revertants to be discussed below. Throughout this chapter, I use the term 'revertant' to indicate the non-transformed variant derived from transformed cells, and 'transformation' to indicate malignant cell transformation induced by the activated *ras* oncogenes unless otherwise stated.

Revertants induced by chemical mutagens

The study of flat revertants was initiated long before the discovery of oncogenes. In early studies, emphasis was on the demonstration that single genes were responsible for the malignant phenotypes exhibited by virally transformed cell lines (Ozer and Jha 1977; Bassin and Noda 1987). Naturally, most mutations represented inactivation of viral transforming genes. There are a few cases which apparently represented mutations in host genes (e.g. Stephenson *et al.* 1973), but little has been done to elucidate the nature of the mutated genes.

In 1981, we initiated a series of experiments designed to isolate flat revertants caused by mutations in cellular genes rather than viral oncogene, and soon found that in this type of experiment the following two tactics were imperative:

1. use of a parental cell line carrying multiple copies of retrovirally transduced oncogene to minimize the chances of getting revertants due to inactivation or elimination of the oncogene;
2. use of techniques which efficiently and selectively kill or purge the transformed cells from the mutagenized population.

Our parental cell line, named DT, was hypoxanthine-guanine-phosphoribosyltransferase (HGPRT)-deficient NIH3T3 carrying two copies of the Kirsten murine sarcoma virus provirus stably integrated into the genome. We chose to use this cell line because NIH3T3 was known to be transformed in a single step by the sarcoma virus and because the marker HGPRT$^-$ would facilitate genetic characterization of the resulting revertants.

We mutagenized DT cells with 5'-azacytidine and selected with a cardiac glycoside, ouabain, which we found is highly toxic to DT cells and only mildly so to NIH3T3 cells (Bassin *et al.* 1984). Two independent revertants were isolated in this fashion (Noda *et al.* 1983). Both revertants were found to retain and express intact v-Ki-*ras* genes, and yet were flat and non-tumorigenic in nude mice.

Cell fusion experiments between the revertants and various partner cells (Noda *et al.* 1983) revealed two interesting properties of these revertants:

1. The non-transformed phenotypes of both revertants were dominant in two types of cell hybrid (revertant × NIH3T3) and (revertant × v-*ras*/NIH3T3), suggesting that a gene or genes whose product(s) inhibit the oncogenic function of Ras may be activated in these revertants.
2. This dominant transformation-suppressing activity was effective on multiple oncogenes including v-*ras*, v-*src*, and v-*fec*, but not on others such as v-*mos*, v-*fms*, v-*sis*, and polyoma virus.

The latter finding provided the first indication that there may be a functional link between the tyrosine kinase oncoproteins (Src and Fes) and Ras proteins. This possibility was later substantiated by microinjection experiments with a Ras-neutralizing antibody (Smith *et al.* 1986) and more recently by genetic experiments using fruit-fly and *C. elegans* mentioned above. Biochemical analysis of the revertants revealed changes in the expression patterns of certain proteins such as tropomyosin (Cooper *et al.* 1985) and fibronectin (Noda and Ikawa 1987). Interestingly, the revertants, despite their non-malignant phenotype, produce transforming

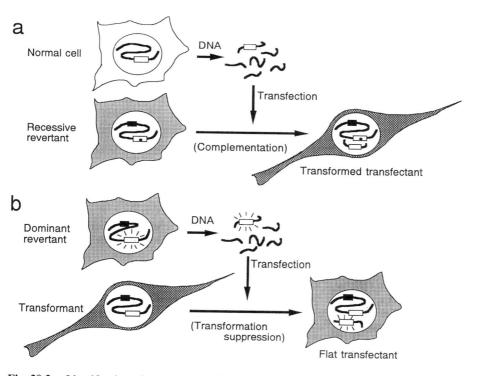

Fig. 20.2. Identification of genes responsible for recessive (a) or dominant (b) reversion via DNA transfection.

growth factors, as do the parental DT cells (Salomon et al. 1984), indicating that not all of the v-ras-associated phenotypes are suppressed in the revertants. Attempts to identify the genes primarily responsible for the reversion, however, have been unsuccessful so far.

To date, several other revertants caused by chemically induced mutations probably in cellular genes have been reported (Norton et al. 1984; Ryan et al. 1985; Bauer et al. 1987; Zarbl et al. 1987; Haynes and Downing 1988; Sincar et al. 1988; Kuzumaki et al. 1989; Yanagihara et al. 1990). Some of them were shown to be the gain-of-function type (Kuzumaki et al. 1989) while others were found to be the loss-of-function type (Bauer et al. 1987; Zarbl et al. 1987; Haynes and Downing 1988). One way to identify the mutated genes would be to use the DNA-mediated gene transfer technique. In the case of recessive mutants, the identification would theoretically be relatively easy because one can transfect DNA from normal cells into the revertants and look for rare transformed cells (Fig. 20.2(a)). In the case of dominant mutants, however, it would be more difficult because one would have to transfect revertant DNA into the transformed cells and look for rare flat revertants again (Fig. 20.2(b)). So far, this type of approach has been successful in only one case (i.e. rsp-1) as discussed below.

Revertants induced by DNA transfection

Normal human fibroblasts are known to be able to suppress malignant growth properties of cell lines harbouring activated ras genes, including DT, upon cell fusion (Craig and Sager 1985; Geiser et al. 1986; our unpublished observation) (Fig. 20.3(a)). This is reminiscent of the behaviour of the dominant revertants discussed above, and suggests that NIH3T3 may lack the activity of one or more genes normally expressed in human fibroblasts. Therefore, we reasoned that we might be able to induce at least partial reversion in DT cells by transfecting them with DNA from normal human fibroblasts (Fig. 20.3(b)). For several technical reasons (Noda 1990), we chose to use a cDNA expression library for transfection.

Seven flat revertants exhibiting greatly reduced tumorigenicity in nude mice, but still expressing the viral oncogene, were isolated from the library-transfected DT cells following several different selection protocols (Noda et al. 1989). Independence of each clone was confirmed by the patterns of integrated plasmids detected by Southern blot analysis using a vector probe. The dominant nature of all the revertants was demonstrated by cell fusion with NIH3T3 and v-ras/NIH3T3. The result was consistent with the assumption that the expression of exogenously introduced cDNA was responsible for the revertant phenotype. By virtue of the flanking vector sequences, cDNA clones could easily be recovered by introducing cut and circularized revertant DNA directly into Escherichia coli (Fig. 20.4). At this time, individual cDNA clones were tested for their transformation suppression activity by transfection into DT cells. cDNA clones (Krev series) possessing the ability to induce revertants

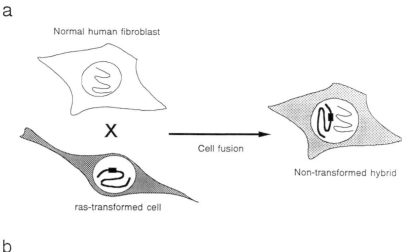

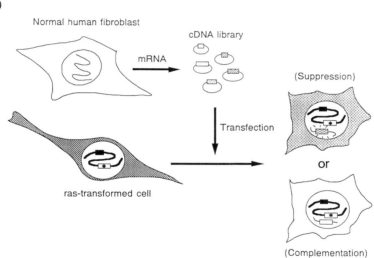

Fig. 20.3. Suppression of *ras*-associated transformed phenotype after cell fusion with normal human fibroblast (a) or transfection with normal human fibroblast cDNA library (b).

at certain frequencies could indeed be isolated in this fashion. The value of such an approach has also been demonstrated by Cutler *et al.* (1992).

Attempts to find tumour/transformation suppressor genes by using transfection of genomic DNA have also been reported (Padmanabhan *et al.* 1987; Schaefer *et al.* 1988). However, the nature of the products of the candidate genes isolated in these studies has not been reported.

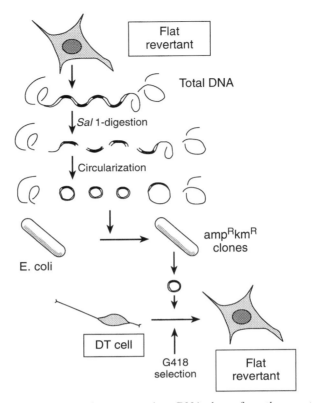

Fig. 20.4. Recovery of transformation-suppressing cDNA clones form the revertants.

The transformation suppressor genes

Krev-1

The K*rev*-1-encoded protein shares around 50 per cent amino acid identity with Ras proteins (Fig. 20.5(a)) (Kitayama et al. 1989) and was found to be identical to Rap la (Pizon et al. 1988) and *smg*p21a (Kawata et al. 1988), independently discovered through different approaches (Noda, 1993a). Besides the conserved motifs commonly found in GTP binding proteins, Krev-1 and Ras share an identical amino acid sequence in the region known as the effector domain. C-terminal region of the Krev-1 protein contains two characteristic structures playing roles in membrane association: lysine repeats and a CAAL box. The latter serves as the signal for geranylgeranylation on the cysteine residue (Kawata et al. 1990; Der and Cox 1991). The serine residue adjacent to the CAAL box has been shown to serve as a target for protein kinase A (PKA) (E. Lapetina, personal communication).

Two types of protein modulating the biochemical activities of the Krev-1 protein (i.e. GDP/GTP-binding and GTPase) have been found: GTPase activating proteins

a Krev-1

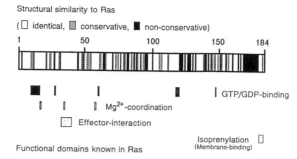

b Krev-3

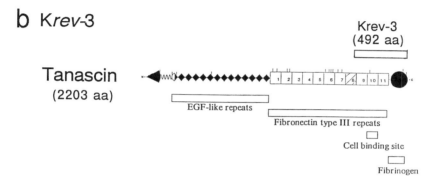

c rsp-1
(Cutler et al.)

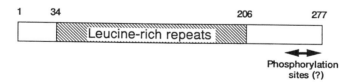

Fig. 20.5. Structural features of the proteins encoded by Krev-1 (a), Krev-3 (b), and rsp-1 (c).

(rap1 GAPs) (Kikuchi *et al.* 1989; Polakis *et al.* 1991; Rubinfeld *et al.* 1991; Zhang *et al.* 1991) and GDP dissociation stimulator (GDS) (Yamamoto *et al.* 1990; Kaibuchi *et al.* 1991) (Fig. 20.6). An additional activity of GDS is to remove smg p21 from the membrane and this process is accelerated after the serine-phosphorylation of smg p21 by PKA. The Krev-1 gene is expressed ubiquitously in various tissues (Kitayama *et al.* 1989) and is especially abundant in neurones (Kim

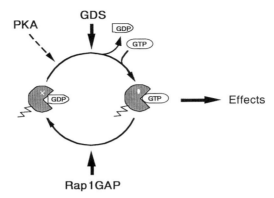

Fig. 20.6. Regulatory proteins for K*rev*-1.

et al. 1990), platelets (Nagata *et al.* 1989), and neutrophils (Quilliam *et al.* 1991). Association of the Krev-1 protein with the Golgi apparatus in fibroblasts has been demonstrated (Beranger *et al.* 1991).

When a K*rev*-1 expression plasmid is transfected into DT cells, 2–3 per cent of the transfectant colonies show flat morphology, and only subtle morphological changes can be found in the rest of the transfectants (Kitayama *et al.* 1989). Part of the reason for this low efficiency of revertant-induction may be explained by the model in which the K*rev*-1 protein is switched on in response to an unknown upstream signal. Although certain point mutations in the Krev-1 gene, that would be expected to impair the intrinsic GTPase activity of its product ('activating' mutations), increased the efficiency of revertant-induction up to 5-fold, more than 80 per cent of the transfectants still remain transformed (Kitayama *et al.* 1990*a*). These findings may suggest the existence of additional regulatory mechanisms for the Krev-1 protein, besides the GDP/GTP exchange reaction, and/or clonal heterogeneity among the recipient DT cells: for instance, varied expression of certain molecule(s) whose presence is essential for Krev-1 protein to exert its biological effects.

To understand the molecular basis underlying the apparently opposing biological activities between Ras and Krev-1, we analysed the transforming and revertant-inducing activities of a series of chimeric H-*ras*–K*rev*-1 genes, and found that small clusters of divergent amino acid residues surrounding the putative effector binding domain are probably responsible for determining which activity the molecule exhibits (Kitayama *et al.* 1990*b*; Zhang *et al.* 1990).

Only limited information is available concerning the possible involvement of K*rev*-1 in human carcinogenesis. Expression of the K*rev*-1 gene is low in certain tumour cell lines, such as Calu 1 and Calu 6, and transfection of K*rev*-1 cDNA in these cell lines resulted in partial suppression of tumorigenicity in nude mice (Caamano *et al.* 1992). K*rev*-1 expression is also greatly reduced in sarcoma tissues of various origins (Culine et al. 1989). Krev-1 expression was found to be induced in various leukaemia cell lines, including HEL and HL-60, after treatment with phorbol ester, a differentiation inducer for these cells (Adachi et al. 1992). These findings seem to indicate that high expression of the Krev-1 gene has adverse effects on the growth of certain types of cells.

To understand the mechanism of altered expression of Krev-1 in these cells, we recently began to characterize the structure of genomic Krev-1 gene. So far, we have learned that the Krev-1 gene, which is located on human chromosome 1p13.3, is split by at least seven introns (five in the coding region), that the upstream, putative promoter region contains several known cis-regulatory elements including multiple AP-1 sites, consistent with the phorbol-responsiveness of this gene, and that at least one pseudogene exists on human chromosome 14q24.3 (Takai et al. 1993; N. Nishino et al., unpublished).

Krev-3

A 2.3 kb human cDNA named Krev-3 which can induce flat revertants in DT cells was isolated through a novel revertant-enrichment protocol using a glutamine analogue, 6-diazo-5-oxo-norleucine (DON). Krev-3 cDNA induces flat revertants at frequencies of about 4 per cent when transfected into DT cells. When Krev-3 cDNA was expressed in a human fibrosarcoma cell line, HT1080, which harbours an activated N-ras gene, suppression of growth in culture and in nude mice was observed (S. Kanazawa et al., unpublished). Structural analysis revealed that the Krev-3 cDNA represents 3′ one-third of the mRNA for an extracellular matrix protein, tenascin. Tenascin is a large multi-domain molecule (monomer >230 KDa, forms hexamer in vivo) containing EGF-like repeats, fibronectin type III repeats and a fibrinogen-like domain (Nies et al. 1991). Although tenascin shows some interesting biological activities in vitro and developmentally regulated expression patterns in vivo, the physiological function of this protein is still elusive. The Krev-3 cDNA corresponds to the region encoding three fibronectin type III repeats, the cell binding site, and the putative Ca^{2+} binding, fibrinogen-like domain (Fig. 20.5(b)). The finding may be suggestive not only of the molecular mechanisms of ras-induced cell transformation but also of the physiological function of this portion of the tenascin molecule.

rsp-1

Cutler and her colleagues (1992) have isolated a revertant-inducing 1.33 kb cDNA (rsp-1) through a similar expression cloning strategy. The source of RNA for the expression library was a chemically induced, DT-derived revertant (CHP9CJ) previously reported by Yanagihara et al. (1990). Their strategy included using ouabain to select for flat transfectants and COS fusion for the plasmid recovery. Besides its revertant-inducing activity on DT cells, rsp-1 exhibits growth-suppressive activity on non-transformed NIH3T3 cells.

Interestingly, rsp-1 encodes a 277 amino acid protein with limited homologies to the leucine-rich repeats of yeast adenylate cyclases (Fig. 20.5(c)). Since this enzyme is believed to be a direct target for the Ras protein in *Saccharomyces cerevisiae* (Kataoka *et al.* 1985) and since there is some evidence that the leucine-rich repeats are involved in this interaction (Colicelli *et al.* 1990; Field *et al.* 1990; Suzuki *et al.* 1990), the *rsp-1*-encoded protein may also interact directly with Ras through this homologous domain in mammalian cells.

Drugs suppressing transformation

Samid *et al.* reported that c-*ras*-transformed NIH3T3 cells maintained for a certain period of time in the presence of interferon gave rise to a population of persistently flat cells (persistent revertant, or PR) (Samid *et al.* 1987). These PR cells could be retransformed by treatment with 5′-azacytidine, suggesting involvement of DNA methylation in the mechanism of reversion. By differential screening, a cDNA for a novel gene named *rrg-1* which is predominantly expressed in PR cells was isolated (Contente *et al.* 1990). Expression of antisense *rrg-1* mRNA resulted in the retransformation of PR cells, suggesting an active role of the *rrg-1* gene in the interferon-induced reversion. This work represents another valuable approach to the identification of genes responsible for reversion.

An antibiotic, azatyrosine, has interesting activity in selectively suppressing the growth of transformed cells and inducing permanent reversal of the transformed phenotype after treatment (Shindo-Okada *et al.* 1989). Azatyrosine-induced flat revertants derived from the MTSV1-7 (hygro *ras*) cell line, a derivative of an immortalized human mammary epithelial cell line transformed *in vitro* by a v-Ha-*ras* gene, were commonly found to express greatly increased levels of K*rev*-1 mRNA (Kyprianou and Taylor-Papadimitriou 1992). Thus, irreversible induction of K*rev*-1 seems to be one mechanism by which azatyrosine exerts its revertant-inducing activity.

There are several antibiotics known to induce reversion in cells transformed by tyrosine kinase-type oncogenes, which include herbimycin (Uehara *et al.* 1989) and erbstatin (Umezawa *et al.* 1990). Oxanosine (Itoh *et al.* 1989) induces reversion in cells transformed by a temperature-sensitive v-Ki-*ras*. In addition to their potential clinical value, study on the mechanisms of action of these antibiotics may yield important insights into the molecular mechanism of cell transformation (Noda, 1993b).

Perspectives

Developing tissues in normal mammals eventually reach the point where they stop dividing even in the presence of enough growth factors, and the cessation of cell growth is often accompanied by cell differentiation. The mechanism of such 'active' growth suppression and differentiation may not be understood only by studying oncogenes and their products themselves. Study of transformation suppressor genes provides a promising avenue to these issues. Common obstacles in such studies include difficulties in discriminating negative regulation from non-specific toxicity and in studying growth-inhibited cells. These difficulties, however, can be circumvented in some cases (Noda 1990), and as described in this chapter, studies employing revertants have in fact led to the isolation of genes potentially important in the negative aspects of growth regulation. Although the number of genes isolated so far is still small, slight modification in experimental design (e.g. changes in the source of cDNA library, type of oncogene, selection technique, etc.) may well lead to the identification of more divergent categories of transformation suppressor genes.

Acknowledgements

I thank Nick James for his critical reading of the manuscript, Hitoshi Kitayama, Susumu Kanazawa, Setsuo Takai, and Naoki Nishino for their contribution to the study on K*rev* genes, and Tomoko Matsuzaki for technical assistance. I am also grateful to Drs Yoji Ikawa and Haruo Sugano for their support and encouragement. The work in our laboratory was supported by grants from the Japanese Foundation for Cancer Research, the Ministry of Education, Science and Culture and the Science and Technology Agency of Japanese Government, and Princess Takamatsu Cancer Research Fund.

References

Adachi, M., Ryo, R., Yoshida, A., Sugano, W., Yasunaga, M., Saigo, K., *et al.* (1992). Induction of *smg* p21/*rap*1A/K*rev*-1 p21 gene expression during phorbol ester-induced differentiation of a human megakaryocytic leukemia cell line. *Oncogene*, **7**, 323–9.

Aroian, R. V., Koga, M., Mendel, J. E., Ohshima, Y., and Sternberg, P. W. (1990). The *let-23* gene necessary for *Caenorhabditis elegans* vulva induction encodes a tyrosine kinase of the EGF receptor subfamily. *Nature*, **348**, 693–9.

Barbacid, M. (1987). *ras* genes. *Ann. Rev. Biochem.*, **56**, 779–827.

Bassin, R. H. and Noda, M. (1987). Oncogene inhibition by cellular genes. *Adv. Viral Oncol.*, **6**, 103–27.

Bassin, R. H., Noda, M., Scolnick, E. M., and Selinger, Z. (1984). Study of possible relationships among retroviral oncogenes using flat revertants isolated from Kirsten sarcoma virus-transformed cells. *Cancer Cells*, **2**, 463–71.

Bauer, M., Guhl, E., Graessmann, M., and Graessmann, A. (1987). Cellular mutation mediates T-antigen-positive revertant cells resistant to Simian virus 40 transformation but not to retransformation by polyomavirus and adenovirus type 2. *J. Virol.*, **61**, 1821–7.

Beranger, F., Goud, B., Tavitian, A., and de Gunzburg, J. (1991). Association of the Ras-antagonistic Rap1/K*rev*-1 proteins with the Golgi complex. *Proc. Natl Acad. Sci. USA*, **88**, 1606–10.

Caamano, J., DiRado, M., Iizasa, I., Momiki, S., Fernandes, E., Ashendel, C., *et al.* (1992). Partial suppression on tumorigenicity in a human lung cancer cell line transfected with K*rev*-1. *Mol. Carcinogen.*, **6**, 252–9.

Clark, S. G., Stern, M. J., and Horvitz, H. R. (1992). *C. elegans* cell-signalling gene *sem*-5 encodes a protein with SH2 and SH3 domains. *Nature*, **356**, 340–4.

Colicelli, J., Field, J., Ballester, R., Chester, N., Young, D., and Wigler, M. (1990). Mutational mapping of RAS-responsive domains of the *Saccharomyces cerevisiae* adenylyl cyclase. *Mol. Cell. Biol.*, **10**, 2539–43.

Contente, S., Kenyon, K., Rimoldi, D., and Friedman, R. M. (1990). Expression of gene *rrg* is associated with reversion of NIH3T3 transformed by LTR-c-H-*ras*. *Science*, **249**, 796–8.

Cooper, H. L., Feuerstein, N., Noda, M., and Bassin, R. H. (1985). Suppression of tropomyosin synthesis, a common biochemical feature of oncogenesis by structurally diverse retroviral oncogenes. *Mol. Cell. Biol.* **5**, 972–83.

Craig, R. and Sager, R. (1985). Suppression of tumorigenicity in hybrids of normal oncogene-transformed CHEF cells. *Proc. Natl Acad. Sci. USA*, **82**, 2062–6.

Culine, S., Olofsson, B., Gosselin, S., Honore, N., and Tavitian, A. (1989). Expression of the *ras*-related rap genes in human tumors. *Int. J. Cancer*, **44**, 990–4.

Cutler, M. L., Bassin, R. H., Zanoni, L., and Talbot, N. (1992). Isolation of *rsp*-1, a novel cDNA capable of suppressing v-*ras* transformation. *Mol. Cell. Biol.*, **12**, 3750–6.

Der, C. J. and Cox, A. D. (1991). Isoprenoid modification and plasma membrane association: critical factors for *ras* oncogenicity. *Cancer Cells*, **3**, 331–40.

Downward, J., Biehl, R., Wu, L., and Weinberg, R. A. (1990). Identification of a nucleotide exchange-promoting activity for p21*ras*. *Proc. Natl Acad. Sci. USA*, **87**, 5998–6002.

Field, J., Xu, H.-P., Michaelli, T., Ballester, R., Sass, P., Wigler, M., *et al.* (1990). Mutations of the adenylyl cyclase gene that block *ras* function in *Saccharomyces cerevisiae*. *Science*, **247**, 464–7.

Gaul, U., Mardon, G., Rubin, G. M., and Rickinson, A. B. (1992). A putative Ras GTPase activating protein acts as a negative regulator of signaling by the sevenless receptor tyrosine kinase. *Cell*, **68**, 1007–19.

Geiser, A. G., Der, C. J., Marshall, C. J., and Stanbridge, E. J. (1986). Suppression of tumorigenicity with continued expression of the c-Ha-*ras* oncogene in EJ bladder carcinoma-human fibroblasts hybrid cells. *Proc. Natl Acad. Sci. USA*, **83**, 5209–13.

Hafen, E., Basler, K., Edstroem, J. E., and Rubin, G. M. (1987). Sevenless, a cell-specific homeotic gene of *Drosophila*, encodes a putative transmembrane receptor with a tyrosine kinase domain. *Science*, **236**, 55–63.

Haynes, J. R. and Downing, J. R. (1988). A recessive cellular mutant in v-*fes*-transformed mink cells restores contact inhibition and anchorage-dependent growth. *Mol. Cell. Biol.*, **8**, 2419–27.

Hill, R. J. and Sternberg, P. W. (1992). The gene lin-3 encodes an inductive signal for vulval development in *C. elegans*. *Nature*, **358**, 470–6.

Horvitz, H. R. and Sternberg, P. W. (1991). Multiple intercellular signalling systems control the development of the *Caenorhabditis elegans* vulva. *Nature*, **351**, 535–41.

Huang, Y. K., Kung, H. F., and Kamata, T. (1990). Purification of a factor capable of stimulating the guanine nucleotide exchange reaction of *ras* proteins and its effect on *ras*-related small molecular mass G proteins. *Proc. Natl Acad. Sci. USA*, **87**, 8008–12.

Inoue, H., Yutsudo, M., and Hakura, A. (1983). Rat mutant cells showing temperature sensitivity for transformation by wild-type Moloney murine sarcoma virus. *Virology*, **125**, 242–5.

Itoh, O., Kuroiwa, S., Atsumi, S., Umezawa, K., Takeuchi, T., and Hori, M. (1989). Induction by the guanosine analogue oxanosine of reversion toward the normal phenotype of K-*ras*-transformed rat kidney cells. *Cancer Res.*, **49**, 996–1000.

Kaibuchi, K., Mizuno, T., Fujioka, H., Yamamoto, T., Kishi, K., Fukumoto, Y., *et al.* (1991). Molecular cloning of the cDNA for stimulatory GDP/GTP exchange protein for *smg* p21s (*ras* p21-like small GTP-binding proteins) and characterization of stimulatory GDP/GTP exchange protein. *Mol. Cell. Biol.*, **11**, 2873–80.

Kataoka, T., Broek, D., and Wigler, M. (1985). DNA sequence and characterization of the *S. cerevisiae* gene encoding adenylate cyclase. *Cell*, **43**, 493–505.

Kawata, M., Farnsworth, C. C., Yoshida, Y., Gelb, M. H., Glomset, J. A., and Takai, Y. (1990). Posttranslationally processed structure of the human platelet protein *smg* p21B: evidence for geranylgeranylation and carboxyl methylation of the C-terminal cysteine. *Proc. Natl Acad. Sci. USA*, **87**, 8960–4.

Kawata, M., Mastsui, Y., Kondo, J., Hishida, T., Teranishi, Y., and Takai, Y. (1988). A novel small molecular weight GTP-binding protein with the same putative effector domain as the *ras* proteins in bovine brain. *J. Biol. Chem.* **263**, 18965–71.

Kikuchi, A., Sasaki, T., Araki, S., Hata, Y., and Takai, Y. (1989). Purification and characterization from bovine cytosol of two GTPase-activating proteins specific for *smg* p21, a GTP-binding protein having the same effector domain as c-*ras* p21s. *J. Biol. Chem.*, **264**, 9133–6.

Kim, S., Mizoguchi, A., Kikuchi, A., and Takai, Y. (1990). Tissue and subcellular distributions of the *smg*-21/*rap*1/*krev*-1 proteins which are partly distinct from those of c-*ras* p21s. *Mol. Cell. Biol.*, **10**, 2645–52.

Kitayama, H., Sugimoto, Y., Matsuzaki, T., Ikawa, Y., and Noda, M. (1989)). A *ras*-related gene with transformation suppressor activity. *Cell*, **56**, 77–84.

Kitayama, H., Matsuzaki, T., Ikawa, Y., and Noda, M. (1990*a*). Genetic analysis of the Kirsten-*ras*-revertant 1 gene: protentiation of its tumor suppressor activity by specific point mutations. *Proc. Natl Acad. Sci. USA*, **87**, 4284–8.

Kitayama, H., Matsuzaki, T., Ikawa, Y., and Noda, M. (1990*b*). A domain responsible for the transformation suppressor activity in K*rev*-1 protein. *Jpn. J. Cancer Res.*, **81**, 445–8.

Krengel, U., Schlichting, L., Scherer, A., Schumann, R., Frech, M., John, J., *et al.* (1990). Three-dimensional structures of H-*ras* p21 mutants: molecular basis for their inability to function as signal switch molecules. *Cell*, **62**, 539–48.

Kuzumaki, N., Ogiso, Y., Oda, A., Fujita, H., Suzuki, H., Sato, C., *et al.* (1989). Resistance to oncogenic transformation in revertant R1 of human *ras*-transformed NIH/3T3 cells. *Mol. Cell. Biol.*, **5**, 2258–63.

Kyprianou, N. and Taylor-Papadimitriou, J. (1992). Isolation of azatyrosine-induced revertants from *ras*-transformed human mammary epithelial cells. *Oncogene*, **7**, 57–63.

McCormick, F. (1989). *ras* GTPase activating protein: signal transmitter and signal terminator. *Cell*, **56**, 5–8.

Nagata, K., Itoh, H., Katada, T., Takenaka, K., Ui, M., Kaziro, Y., *et al.* (1989). Purification, identification, and characterization of two GTP-binding proteins with molecular weights of 25,000 and 21,000 in human platelet cytosol. *J. Biol. Chem.*, **264**, 17000–5.

Nies, D. E., Hemesath, T. J., Kim J. H., Gulcher, J. R., and Stefansson, K. (1991). The complete cDNA sequence of human hexabrachion (tenascin). *J. Biol. Chem.*, **266**, 2818–23.

Noda, M. (1990). Expression cloning of tumor suppressor genes: a guide for optimists. *Mol. Carcinogen.*, **3**, 251–3.

Noda, M. (1993a). Structures and functions of the K*rev*-1 transformation suppressor gene and its relatives. *Biochim. Biophys. Acta*, **1155**, 97–109.

Noda, M. (1993b). Mechanisms of reversion. *FASEB J.*, **7**, 834–40.

Noda, M. and Ikawa, Y. (1987). Detection of genes with potential of suppressing transforming activity of the v-Ki-*ras* oncogene. In *Oncogene and cancer*, (ed. S. A. Aaronson *et al.*), pp. 261–7. Japan Sci. Soc. Press/VNU Sci. Press, Tokyo.

Noda, M., Kitayama, H., Sugimoto, Y., Okayama, H., Bassin, R. H., and Ikawa, Y. (1989). Detection of genes with a potential for suppressing the transformed phenotype associated with activated *ras* genes. *Proc. Natl Acad. Sci. USA*, **86**, 162–6.

Noda, M., Ko, M., Ogura, A., Liu, D.-g., Amano, T., Takano, T., *et al.* (1985). Sarcoma viruses carrying *ras* oncogenes induce differentiation-associated properties in a neuronal cell line. *Nature*, **318**, 73–5.

Noda, M., Selinger, Z., Scolnick, E. M., and Bassin, R. H. (1983). Flat revertants isolated from Kirsten sarcoma virus-transformed cells are resistant to the action of specific oncogenes. *Proc. Natl Acad. Sci. USA*, **80**, 5602–6.

Norton, J. D., Cook, F., Roberts, P. C., Clewley, J. P., and Avery, R. J. (1984). Expression of Kirsten murine sarcoma virus in transformed, nonproducer, and revertant NIH/3T3 cells: evidence for cell-mediated resistance to a viral oncogene in phenotypic reversion. *J. Virol.*, **50**, 439–44.

Ozer, H. L. and Jha, K. K. (1977). Malignancy and transformation: expression in somatic cell hybrids and variants. *Adv. Cancer Res.*, **25**, 53–93.

Padmanabhan, R., Howard, T. H., and Howard, B. H. (1987). Specific growth inhibitory sequences genomic DNA from quiescent human embryo fibroblasts. *Mol. Cell. Biol.*, **7**, 1894–9.

Pizon, V., Chardin, P., Lerosey, I., Olofsson, B., and Tavitian, A. (1988). Human cDNAs *rap*1 and *rap*2 homologous to the *Drosophila* gene *Dras*3 encode proteins closely related to *ras* in their 'effector' region. *Oncogene*, **3**, 201–4.

Polakis, P. G., Rubinfeld, B., Evans, T., and McCormick, F. (1991). Purification of a plasma membrane-associated GTPase-activating protein specific for *rap* 1/K*rev*-1 from HL60 cells. *Proc. Natl Acad. Sci. USA*, **88**, 239–43.

Quilliam, L. A., Mueller, H., Bohl, B. P., Prossnitz, V., Sklar, L. A., Der, C. J., *et al.* (1991). Rap1A is a substrate for cyclic AMP-dependent protein kinase in human neutrophils. *J. Immunol.*, **147**, 1628–35.

Rubin, G. M. (1989). Development of the *Drosophila* retina: inductive events studied at single cell resolution. *Cell*, **57**, 519–20.

Rubinfeld, B., Munemitsu, S., Clark, R., Conroy, L., Watt, K., Crosier, W. J., *et al.* (1991). Molecular cloning of a GTPase activating protein specific for the K*rev*-1 protein p21*rap*1. *Cell*, **65**, 1033–42.

Ryan, K. W., Christensen, J. B., Imperiale, M. J., and Brockman, W. W. (1985). Isolation of a simian virus 40 T-antigen-positive, transformation-resistant cell line by indirect selection. *Mol. Cell. Biol.*, **5**, 3577–82.

Salomon, D. S., Zwiebel, J. A., Noda, M., and Bassin, R. H. (1984). Flat revertants derived from Kirstein murine sarcoma virus-transformed cells produce transforming growth factors. *J. Cell. Physiol.*, **121**, 22–30.

Samid, D., Flessate, D. M., and Friedman, R. M. (1987). Interferon-induced revertants of *ras*-transformed cells: resistance to transformation by specific oncogenes and retransformation by 5-azacystidine. *Mol. Cell. Biol.*, **7**, 2196–200.

Schaefer, R., Iyer, J., Iten, E., and Nirkko, A. (1988). Partial reversion of the transformed phenotype in HRAS-transfected tumorgenic cells by transfer of a human genes. *Proc. Natl Acad. Sci. USA*, **85**, 1591–4.

Shindo-Okada, N., Makabe, O., Nagahara, H., and Nishimura, S. (1989). Permanent conversion of mouse and human cells transformed by activated *ras* and *raf* genes to apparently normal cells by treatment with the antibiotic azatyrosine. *Mol. Carcinogen.*, **2**, 159–67.

Simon, M. A., Bowtell, D. D., Dodson, G. S., Laverty, T. R., and Rubin, G. M. (1991). Ras 1 and a putative guanine nucleotide exchange factor perform crucial steps in signaling by the sevenless protein tyrosine kinase. *Cell*, **67**, 701–16.

Sincar, S., Rodrigues, M., and Weber, J. M. (1988). Resistance to retransformation by adenovirus but not by heterologous oncogenes in an E1-positive transformation revertant cell line may be mediated by a cellular function. *Oncogene*, **3**, 725–8.

Smith, M. R., DeGudicibus, S. J., and Stacey, D. W. (1986). Reqirement of c-*ras* proteins during viral oncogene transformation. *Nature*, **320**, 540–3.

Stephenson, J. R., Reynolds, R. K. and Aaronson, S. A. (1973). Characterization of morphologic revertants of murine and avian sarcoma virus transformed cells. *J. Virol.*, **11**, 218–22.

Suzuki, N., Choe, H.,-R., Nishida, Y., Yamawaki-Kataoka, Y., Ohnishi, S., Tamaoki, Y., *et al.* (1990). Leucine rich repeat and the carboxy terminus are required for interaction of yeast adenylate cyclase with RAS proteins. *Proc. Natl Acad. Sci. USA*, **87**, 8711–15.

Takai, S., Nishino, N., Kitayama, H., Ikawa, Y. and Noda, M. (1993). Mapping of the K*rev*1 transformation suppressor gene and its pseudogene (K*rev*1P) to human chromosome 1p13.3 and 14q24.3, respectively, by fluorescence in situ hybridization. *Cytogenet. Cell Genet.*, **63**, 59–61.

Thomas, S. M., DeMarco, M., D'Arcangelso, G., Halegoua, S., and Brugge, J. (1992). Ras is essential for nerve growth factor- and phorbol ester-induced tyrosine phosphorylation of MAP kinases. *Cell*, **68**, 1031–40.

Uehara, Y., Murakami, Y., Sugimoto, Y., and Mizuno, S. (1989). Mechanism of reversion of Rous sarcoma virus transformation by herbimycin A: reduction of total phosphotyrosine levels due to reduced kinase activity and increased turnover of p60v-src. *Cancer Res.*, **49**, 780–5.

Umezawa, K., Hori, T., Tajima, H., Imoto, M., Isshiki, K., and Takeuchi, T. (1990). Inhibition of epidermal growth factor-induced DNA synthesis by tyrosine kinase inhibitors. *FEBS Lett.*, **260**, 198–200.

Wolfman, A. and Macara, I. (1990). A cytosolic protein catalyzes the release of GDP from p21ras. *Science*, **247**, 67–9.

Wood, K. W., Sarnecki, C., Roberts, T. M., and Blenis, J. (1992). *ras* mediates nerve growth factor receptor modulation of three signal-transducing protein kinases: MAP kinase, Raf-1, and RSK. *Cell*, **68**, 1041–50.

Yamamoto, T., Kaibuchi, K., Mizuno, T., Hiroyoshi, M., Shirataki, H., and Takai, Y. (1990). Purification and characterization from bovine brain cytosol of protein that regulate the GDP/GTP exchange reaction of *smg* p21s, *ras* p21-like GTP-binding proteins. *J. Biol. Chem.*, **265**, 16626–34.

Yanagihara, K., Ciardiello, F., Talbot, N., McGeady, M. L., Cooper, H., Benede, L., *et al.* (1990). Isolation of a new class of 'flat' revertants from sras transformed NIH3T3 cells using cis-4-hydroxy-L-proline. *Oncogene*, **5**, 1179–86.

Zarbl, H., Latreille, J., and Jolicoeur, P. (1987). Revertants of v-fos-transformed fibroblasts have mutations in cellular genes essential for transformation by other oncogenes. *Cell*, **51**, 357–69.

Zhang, K., Noda, M., Vass, W. C., Papageorge, A. G., and Lowy, D. R. (1990). Identification of small clusters of divergent amino acids that mediate the opposing effects of *ras* and K*rev*-1. *Science*, **249**, 162–5.

Zhang, K., Papageorge, A. G., Martin, P., Vass, W., Olah, Z., Polakis, P., *et al.* (1991). Heterogenous amino acids in Ras and Rap1A specifying sensitivity to GAP proteins. *Science*, **254**, 1630–4.

21. Unfinished business
Henry Harris

The original experiments that revealed the existence of normal cellular genes that had the ability to suppress malignancy also revealed that this suppression was not achieved by mechanisms that inhibited the multiplication of cells *in vitro* (Harris *et al*. 1969). For the fusion of tumour cells with normal fibroblasts generated hybrids that multiplied vigorously in culture but that were unable to produce tumours; and when chromosome segregation permitted the re-emergence of the malignant phenotype, there was often little or no difference in growth rate between the non-malignant hybrids and the malignant segregants derived from them (Harris 1971). It was therefore clear that something happened when the cells were injected into the animal that did not happen when the cells were grown *in vitro*.

It is, of course, a commonplace of tumour pathology that differentiation can override progressive multiplication so that, within a malignant tumour, subpopulations of cells are generated in which multiplication is brought to a halt. Indeed, this is the normal course of events in all but the most anaplastic of solid tumours. Given that an inverse relationship between differentiation and cell multiplication underlies the whole of metazoan development, it is perhaps surprising that more than a decade elapsed between the discovery that malignancy could be suppressed in hybrid cells and the suggestion by Peehl and Stanbridge that this suppression might be effected by the imposition *in vivo* of the terminal differentiation programme of the normal cell with which the tumour cell was fused. This idea has found both supporters and opponents. It is the case that when normal fibroblasts or normal keratinocytes are fused with tumour cells, the non-malignant hybrids that are generated execute a fibroblastic or keratinocytic differentiation programme when they are injected into the animal, whereas malignant segregants derived from these hybrids have lost the ability to do this (Peehl and Stanbridge 1982; Harris 1985; Harris and Bramwell 1987). On the other hand, the observation has been made in more than one system that malignancy may be suppressed in hybrids of this kind without the appearance *in vitro* of markers characteristic of the fully differentiated state of the normal parent cell. I am less impressed by the latter evidence than by the former, for two reasons. The first is that death or *some* form of differentiation seem to me to be the only modalities so far generated by evolution that can bring cell multiplication to a halt. The second reason is more down-to-earth. No population of somatic cells that multiplies *in vitro* expresses the full panoply of differentiated traits of which the cell type is capable. Since all our experiments select for cells that continue to multiply, we always obtain *in vitro* cell populations in which the normal differentiation process for that cell type is in some measure incomplete. The absence of any particular differentiation marker under such conditions does not exclude the presence of other markers that one is not looking for, or the subsequent appearance of the marker that one is looking for, when the cells are inoculated into the animal. It is difficult to analyse over any substantial period of time the behaviour of an inoculum of non-malignant cells *in vivo*, so that there are very few cases where this last possibility has been systematically investigated.

Having thus declared my partisanship in this debate, let me now tell you what I have been doing to explore the matter further. If it is in fact true that suppression of malignancy in hybrid cells is achieved by the imposition of a normal differentiation programme, then the frustration of that programme by experimental intervention should induce the malignant phenotype to reappear. And, conversely, the restoration or imposition of a normal differentiation programme should restore a non-malignant phenotype. I have been endeavouring to test these two propositions in three different cell types, of which two are hybrid cells. Why hybrid cells? Well, it is now pretty clear that, at least in most solid tumours, several genetic events must occur before the malignant phenotype is fully established. If one began with a normal diploid cell and set about impairing its differentiation programme, it seems likely that intervention at several points would be necessary before malignancy emerged. Given the limited lifespan of most normal cells *in vitro*, the complete inhibition of an unexplored sequence of genes is hardly a practical proposition. However, in hybrid cells in which malignancy is suppressed, we know that the genome of the malignant parent cell still harbours all the genetic lesions that have contributed to the full development of the malignant phenotype, and the hybrid is non-malignant only because the consequences of all these genetic lesions are overridden by the activity of one gene or one group of genes that maps to one particular region on one particular chromosome. We know this because the elimination of that one chromosomal region from the hybrid cell permits the malignant phenotype to reappear (for reviews see Harris 1990; Stanbridge 1990). It therefore seemed a less daunting project from the technical point of view to attempt to impede a normal differentiation programme within a hybrid cell, in the knowledge that if we were successful and if it really was this differentiation programme that was suppressing the malignancy, then the complete malignant phenotype would reappear.

The first set of unfinished experiments that I wish to describe were done with hybrids between malignant cells derived from a human carcinoma and normal human keratinocytes. These are the hybrids in which Peehl and Stanbridge (1982) had shown that suppression of malignancy was associated with the ability of the hybrid cells to undergo a facsimile of normal keratinocyte differentiation when injected into the animal. Michael Bramwell and I subsequently showed that even *in vitro* the hybrids in which malignancy was suppressed continued to produce the protein involucrin, an early and, as it then appeared, critical marker of keratinocyte differentiation. But no involucrin was detectable in malignant segregants derived from these hybrids (Harris and Bramwell 1987). Elaine Griffin and I therefore decided to see whether it might be possible to inhibit the synthesis of involucrin in the non-malignant hybrid cells, and, if so, whether this might permit the malignant phenotype to reappear (Griffin and Harris 1992). Perhaps I should explain why this experiment is not as irrationally optimistic as it might at first sight appear. I have long held the view that in the execution of a differentiation programme, some of the structural proteins produced might also play a regulatory role. It is, of course, obvious that in any such programme there will be many markers whose presence or absence has no influence on the malignancy of the cell. But there must be certain nodal points at which cell differentiation and cell multiplication interact. If the differentiation programme could be interrupted at or prior to such critical junctures, then it was not unreasonable to hope that the programme might be blocked at that point with the consequence that cell multiplication could then continue without restraint. Since involucrin was a very early (although not the earliest) marker of keratinocyte differentiation, and a marker of obvious physiological importance, it seemed as good a place to start as any (Watt 1989).

But there was an additional, essentially technical, reason for choosing involucrin as our *ballon d'essai*. The human involucrin gene, which codes for a 68 kDa polypeptide, contains within its more recently evolved region a series of 39 repeats of a sequence of 30 strongly homologous nucleotides (Eckert and Green 1986). The gene and its transcript thus present a particularly favourable target for the application of antisense methodology, with which Diana Steel and I had previously had some modest success (Steel and Harris 1989). For, in the case of involucrin, an antisense sequence of nucleotides would have multiple sites within the gene and its transcript with which it could interact. Because the great majority of experiments in which antisense RNA has been used to inhibit the expression of a particular gene have been only partially successful, if at all, and because the precise mode of action of antisense RNA has been elucidated only in much simplified systems, we devised a novel strategy that was not sensitive to the mode of action of the antisense RNA and that relied on *selective* procedures to obtain clones of cells having the properties we were looking for. Since antisense RNA may act in a number of different ways, among others by complexing with the gene, by interfering with splicing or transport of the gene product, or by inhibition of its translation, we made antisense constructs that straddled as many of these potential targets as possible. I shall not burden you with our preliminary efforts but simply say that we settled on a 1.835 kb fragment of the involucrin gene that included part of the intron, the 3' splice site, the translation initiation codon, and 1433 base pairs of coding sequence. This was ligated into the pKG4 expression vector in both the sense and the antisense orientation with respect to the promoter. Both sense and antisense constructs were then transfected by standard procedures into the carcinoma × normal keratinocyte hybrids in which malignancy was suppressed. Fifty-six clones stably transfected with the sense construct and 50 clones stably transfected with the antisense construct were isolated by the usual selective procedures. None of the clones carrying the sense construct showed any significant reduction in the amount of involucrin produced by the cells, but of the 50 clones carrying the antisense construct 12 showed a greater than 50 per cent reduction in involucrin production, and one of these produced only 3 per cent of the amount of involucrin made by the untransfected cells. This variability in the efficacy of the antisense procedure is, of course, in agreement with common experience. It was not, in our case, due to variation in the number of antisense constructs stably incorporated into the genome of the recipient cell. All the clones that showed reduction in involucrin synthesis contained only one integrated copy of the antisense plasmid. The trouble with partial suppression of a particular marker is that you do not know how much or how little of the molecule a cell actually needs to achieve the end result you are interested in. Now it was at this point that I had an idea of Einsteinian proportions. Why not give the cells in which the transfection of the antisense construct had produced the greatest reduction in involucrin synthesis a second helping? So we made a second involucrin antisense construct in a plasmid containing a different selectable marker, pBabe Hygro. Into this we ligated a 2.356 kb involucrin gene fragment containing a part of the intron, the 3' splice site, the translation initiation codon, and the entire coding region. Fifty-seven stably transfected clones were isolated, and of these involucrin could not be detected at all in 17, either by analysis of Western blots or by immunoperoxidase staining of fixed cells. We had therefore, in two steps, achieved total inhibition of the synthesis of the marker we had targeted. I see no reason why, in culture, this selective approach to the problem of inactivating specific genes could not be generally applicable. And if, for any particular gene, complete inactivation is not achieved by two helpings of antisense DNA, there are enough selectable markers available to permit the cell to be given an Edwardian banquet of antisense RNAs.

I should like to have been able to tell you that the total inhibition of involucrin synthesis did indeed frustrate the execution of the terminal differentiation programme of the keratinocyte, and that, in the hybrid cells in which malignancy was initially suppressed, the malignant phenotype reappeared. Regrettably I bring the sad news that the non-malignant cells remained non-malignant and that the differentiation programme of the keratinocyte was not frustrated. The cells in which involucrin synthesis had been abolished still made at least some of the keratins that appear late in the differentiation process, and cells making as little as 3 per cent of the normal level of involucrin still produced highly differentiated static keratin nests when injected into the animal. And that is the first of my three items of unfinished business. We now propose to apply our antisense strategy to the keratins of the cells in which involucrin synthesis has been abolished. What will these keratinocyte hybrids do in the animal when they can no longer keratinize?

My second item concerns the formation of the extracellular matrix in cells of fibroblastic lineage. My experimental interest in this question stems from an observation I made some years ago on the behaviour *in vivo* of hybrids between a variety of different tumour cells and normal fibroblasts (Harris 1985). I found that when hybrids in which malignancy was suppressed were injected into the animal, they formed a copious collagenous extracellular matrix, assumed the elongated morphology characteristic of mature fibrocytes, and stopped multiplying. Malignant segregants derived from these hybrids failed to produce this collagenous matrix, showed no sign of fibrocytic morphological change, and continued to multiply. Michael Copeman and I made a systematic study of the main extracellular matrix components produced by these hybrids *in vitro* (Copeman and Harris 1988). We found that when malignant segregants were derived from hybrids in which malignancy was initially suppressed, there was a profound reduction in the amount of Type I procollagen produced by the cells. We found no change in the total amount of fibronectin produced, but there was a reduction in the amount retained in the extracellular matrix. There was, however, ample evidence in other systems that a loose correlation did exist between some parameters of fibronectin synthesis and 'transformed' morphology *in vitro*. In a few cases there was also circumstantial evidence that fibronectin might play a role in determining malignant behaviour *in vivo*. With Helen Mardon and Jayne Devlin I have been looking at fibronectin more closely in some hybrids that I thought might be especially informative. The HT1080 human fibrosarcoma cell line has been an object of study in several laboratories. It makes very little fibronectin mRNA and very little fibronectin protein compared with the normal fibroblast which is its presumptive ancestor; and what fibronectin it does make it fails to incorporate into a normal extracellular matrix. This failure is not due to the synthesis of aberrant fibronectins but is determined by some defect in the surface of the cell. When HT1080 cells are grown in intimate contact with normal fibroblasts, which provide an ample supply of normal fibronectins, the tumour cells still fail to incorporate them into an extracellular matrix even though the fibroblasts with which they are in contact do (J. Devlin and H. Harris, in preparation). We fused the HT1080 cells with normal human fibroblasts and, in the usual way, derived malignant segregants from the hybrids in which malignancy was initially suppressed. In this panel of cells we examined in detail the synthesis and organization of fibronectin as a whole and of two of the isoforms generated by alternative splicing of the fibronectin gene transcript (Mardon *et al.* 1992). EDIIIA and EDIIIB are fibronectin splice variants that each contain a sequence of repeating amino acids that can be excluded from the completed fibronectin molecule by alternative splicing or included within it. These variants are of particular interest within the present context because it has been reported that fibronectins containing EDIIIB are present in carcinomatous and sarcomatous tumours but not in normal adult tissues.

In agreement with our earlier observations on other hybrids, we found that the total amount of fibronectin mRNA made by the malignant segregants was not significantly different from that made by the hybrids in which malignancy was suppressed: the hybrids made less fibronectin mRNA than the parental fibroblasts, but at least 10 times more than the HT1080 cells. We then established that the two parental cells and both the malignant and the non-malignant hybrids processed the EDIIIA and EDIIIB exons accurately. The levels of EDIIIA+ and EDIIIB+ mRNA were found to be similar in all four cell lines, and all four made much less EDIIIB+ mRNA than EDIIIB- mRNA. We are therefore loath to believe that the presence of the EDIIIB+ variant described in certain tumours in closely linked to the malignant phenotype itself. Indeed, we found only marginal differences in splicing pattern between the hybrids in which malignancy was suppressed and the malignant segregants derived from them.

However, when we came to look at the protein components in the extracellular matrix of the four cell lines, we did find profound differences between the malignant and the non-malignant hybrids. The defect in HT1080 cells that prevents them from incorporating fibronectin into an extracellular matrix was corrected in the non-malignant hybrids, but reappeared in the malignant segregants. Normal fibroblasts incorporate more than 70 per cent of the fibronectin that they make into the extracellular matrix, but fibronectin is barely detectable in the extracellular matrix of HT1080 cells. The hybrids in which malignancy was suppressed incorporated more than 80 per cent of their fibronectin into the extracellular matrix, whereas the malignant segregants incorporated less than 10 per cent. Moreover, there were striking differences in the incorporation of the fibronectin splice variants. Approximately 90 per cent of the EDIIIA+ variant made by the non-malignant hybrids was incorporated into the extracellular matrix; the malignant segregants incorporated less than 10 per cent. In the case of the EDIIIB+ variant, the difference was even more extreme: the non-malignant hybrids incorporated about 60 per cent of the EDIIIB+ variant that they made into the extracellular matrix, whereas for the malignant segregants this figure was less than 2 per cent. It is thus clear that in the malignant cells the EDIIIB+ variant is virtually excluded from the extracellular matrix. I conclude that malignancy is not linked to any abnormality in the fibronectin splicing pattern, nor is it linked to the exclusive or preferential synthesis of any particular fibronectin isoform. It is, however, linked to a profound impairment of the cell's ability to incorporate the fibronectin that it does make into a normal extracellular matrix, and this impediment affects some fibronectin isoforms much more severely than others.

What might this impediment be? And if we could repair it, would the malignant segregants become non-malignant once again? Inspired by some work done a decade or so ago, we decided to examined the effect of corticosteroids on the ability of the defective cells to incorporate fibronectin into an extracellular matrix. We found, to our considerable satisfaction, that after 48 h in 100 nM dexamethasone, the malignant segregants had incorporated fibronectin into an extracellular matrix that was indistinguishable by immunocytochemical methods from that of the hybrid cells in which malignancy was suppressed (J. Devlin and H. Harris, in preparation). We are now exploring this effect in detail. In particular, we want to know, of course, whether a respectable extracellular matrix can be produced *in vivo* by the malignant segregants under the influence of corticosteroids, and, if so, whether that has any effect on their ability to grow progressively in the animal. It takes no great insight to see where this investigation might go if we are lucky.

My third item of unfinished business concerns melanomas in which I have long had an abiding interest. Many years ago, I did a very primitive experiment with an experimental mouse melanoma which, when injected subcutaneously into the animal, conveniently generates pigmented metastases in the lungs. I excised some of these black metastases and explanted them in the hope that I might thus obtain sublines of cells that were heavily pigmented. To my surprise, I found that all the clones formed *in vitro* from the melanotic metastases were initially amelanotic or almost so, and they did not develop pigment until the centre of the clone became so crowded that multiplication there was brought to a halt. It was in these crowded centres that pigment first began to accumulate in the cells and then only in some of the clones that were generated. The black cells in the metastases did not generate black clones; and when I came to examine the fate of these black cells more closely I found that they were not viable. It was clear that the growth of the black metastases was not due to the multiplication of black cells, but was driven by amelanotic cells or cells with low levels of pigmentation. Within the explanted cells accumulation of pigment beyond a certain point brings multiplication to a stop. Melanomas are thus no different from the great majority of malignant tumours in generating subpopulations of cells in which the process of differentiation (in this case pigmentation) overrides cell multiplication. From the point of view of the patient with melanoma, black is beautiful.

If you hope to stay in experimental science of any length of time you must remain an optimist. That explains why I began to look for compounds that would stimulate the formation of pigment in melanomas. I screened about sixty plausible compounds and found several that readily turned a culture of melanoma cells black and destroyed it. But when tested against melanomas growing *in vivo*, these compounds were either ineffective or, when pushed to the limit, destroyed the animal. So I eventually stopped playing roulette and decided that the time had come to undertake a serious examination of the mechanisms by which pigmentation in melanoma cells is regulated. In this I had the great good fortune to be joined by a D. Phil. student called Alison Taylor (now Winder) who soon took the matter out of my hands. It became apparent almost at once that the standard assay for the enzyme tyrosinase (EC1.14.18.1) which catalyses the first and critical step in the synthesis of melanin was a poor thing. It was non-linear, it relied on a coupled reaction between tyrosine hydroxylase activity and dopa oxidase activity without distinguishing between the two, and it was relatively insensitive. So the first thing that Alison Winder did was to develop new and separate assays for the tyrosine hydroxylase and the dopa oxidase activities of the enzyme. These new assays are specific, linear, and much more sensitive than the procedure currently in use and should become standard methods (Winder and Harris 1991). With these new methods we set about determining what tyrosinase actually did. To begin with, we decided to measure tyrosine hydroxylase and dopa oxidase activity independently in the one cell type under a variety of conditions. The cell line we chose was a human melanoma that produced pigment *in vitro*, in increasing amounts as the cells became confluent and crowded. Although the agreed doctrine was that tyrosine hydroxylation and dopa oxidation were tightly coupled and catalysed by the one enzyme, it soon became apparent that the ratio of these two activities in the cell cultures was not constant, but changed with time. Moreover, we were able to show that the increase in tyrosinase activity that occurred as the cells became confluent could not be accounted for simply by an increase in tyrosinase mRNA (Winder and Harris 1992). Some regulation of the enzyme activity was achieved by post-transcriptional mechanisms. We were obviously faced with a problem of some complexity.

In order to reduce this complexity and because cDNA from the tyrosinase gene (the *c* locus in the mouse) was available, we decided to see what the tyrosinase gene would do when it was transfected into a cell not normally involved in any way in the synthesis of melanin, the Swiss 3T3 mouse fibroblast (Winder 1991). The gene, in an appropriate expression vector, was transfected into the cells by standard procedures together with a selectable marker. Of 63 clones isolated in selective medium, four were pigmented and found on further cultivation to contain tyrosinase activity. Subcloning eventually generated a cell line in which tyrosinase was stably retained and which continued to form pigment. As in the human melanoma cell line we had worked with, pigmentation and tyrosinase activity increased as the cells became confluent. The transfected cell line contained a single integrated copy of the tyrosinase gene, but produced five different tyrosinase mRNAs, perhaps by alternative splicing. The significance of these different mRNAs is at present obscure. The tyrosinase specified by the interpolated cDNA was processed normally. Proteolytic cleavage generated both the membrane-bound and the soluble forms of the enzyme, the latter having the expected molecular weight. The enzyme protein appeared to be appropriately glycosylated. Neither catalase nor dopachrome tautomerase activity could be detected in the transfected cells, thus indicating that the protein encoded in the *c* locus of the mouse is indeed a specific tyrosinase. Analysis of the products of this tyrosinase activity in the transfected cells showed that all the normal intermediates of pigment biosynthesis were generated, and the pigment formed appeared to be normal phaeomelanin (Winder *et al.* 1992*a*). The pigment was sequestered into vesicles that, by morphological criteria, resembled melanosomes. It thus appeared that the expression of the mouse tyrosinase *c* locus is all that is required to produce normal pigmentation of the cell.

In order to see whether it was possible to increase the degree of pigmentation by increasing the dosage of the tyrosinase gene, the transfected, stably pigmented, cell line was given a second dose of the tyrosinase cDNA, this time coupled with a different selectable marker, just as we had done with the involucrin antisense cDNA. Although highly pigmented clones could be isolated, the level of tyrosinase activity and of pigmentation in the cell lines generated from these clones was not significantly higher than that found in the cell line before the second transfection, despite the fact that the cells could be shown to have incorporated a second copy of the tyrosinase cDNA. This result supports our general thesis that pigmentation beyond a certain modest level is incompatible with cell multiplication.

The *b* (*brown*) locus in the mouse in thought to increase the synthesis of eumelanin, which is black, relative to phaeomelanin, which is brown. The *b* gene is also supposed to affect the size and shape of the melanosomes. In order to explore its function further a complete copy of the cDNA was transfected into the Swiss 3T3 fibroblasts, as I have described for the *c* locus (Winder *et al.* 1992*b*). Again, clones containing a single integrated copy of the gene were isolated and expanded into cell lines. The transfected cells synthesized a protein that was indistinguishable in its distribution and immunological reactivity from the normal product of the *b* locus. Among the guesses that have made about the function of this product are the suggestions that it might be another form of tyrosinase, a catalase, a peroxidase or dopachrome tautomerase. Assays of the transfected cells, however, showed that it was none of these things. Nor were the transfected cells pigmented. However, when the *b* locus cDNA was transfected into the Swiss 3T3 cells that had already received the *c* locus, cell lines could be isolated which expressed the products of both genes and in which the tyrosinase activity was 5 to 10-fold higher than the levels achieved by the transfection of the *c* locus alone. The doubly transfected cells were also more heavily pigmented. The *b* locus thus acts in some way to protect the cells from levels of tyrosinase activity that would otherwise be incompatible with their multiplication or simply lethal. This interpretation is supported by the fact that two mutations at the *b* locus are known, one of them a simple amino acid substitution, that cause premature death of melanocytes. It seems to me to be a matter of some importance that the function of the b protein be elucidated in molecular terms, for it is not difficult to see that a molecular definition of its mode of action might well suggest new ways in which the death of melanoma cells could be encompassed.

Those are my three items of unfinished business. We all know, of course, that in science no business is ever finished. But it is my hope, nonetheless, that I shall be able to pursue these problems a little further; and that should see me out.

References

Copeman, M. C. and Harris, H. (1988). The extracellular matrix of hybrids between melanoma cells and normal fibroblasts. *J. Cell Sci.* **91**, 281–6.
Devlin, J. and Harris, H. (In preparation.)
Eckert, R. L. and Green, H. (1986). Structure and function of the human involucrin gene. *Cell*, **46**, 583–9.
Griffin, E. F. and Harris, H. (1992). Total inhibition of involucrin synthesis by a novel two-step antisense procedure. Further examination of the relationship between differentiation and malignancy in hybrid cells. *J. Cell Sci.* **102**, 799–805.
Harris, H., Miller, O. J., Klein, G., Worst, P., and Tachibana, T. (1969). Suppression of malignancy by cell fusion. *Nature*, **223**, 363–8.
Harris, H. (1971). Cell fusion and the analysis of malignancy. The Croonian Lecture. *Proc. Roy. Soc. Land. B*, **179**, 1–20.
Harris, H. (1985). Suppression of malignancy in hybrid cells: the mechanism. *J. Cell Sci.*, **79**, 83–94.
Harris, H. and Bramwell, M. E. (1987). Suppression of malignancy by terminal differentiation: evidence from hybrids between tumour cells and keratinocytes. *J. Cell Sci.* **87**, 383–8.
Harris, H. (1990). Tumor suppressor genes: studies with mouse hybrid cells. In *Tumor suppressor genes*, (ed. G. Klein), pp. 1–13. Marcel Dekker, New York.
Mardon, H. J., Grant, R. P., Grant, K. E., and Harris, H. (1992). Fibronectin splice variants are differentially incorporated into the extracellular matrix of tumorigenic and non-tumorigenic hybrids between normal fibroblasts and sarcoma cells. *J. Cell Sci.* **104**, 783–92.
Peehl, D. M. and Stanbridge, E. J. (1982). The role of differentiation in the suppression of tumorigenicity in human hybrid cells. *Int. J. Cancer*, **30**, 113–20.

Stanbridge, E. J. (1990). Human tumor suppressor genes. *Annu. Rev. Genet.* **24**, 615–57.
Steel, D. M. and Harris, H. (1989). The effect of antisense RNA to fibronectin on the malignancy of hybrids between melanoma cells and normal fibroblasts. *J. Cell Sci.* **93**, 515–24.
Watt, F. M. (1989). Terminal differentiation of epidermal keratinocytes. *Curr. Opin. Cell Biol.*, **1**, 1107–15.
Winder, A. J. (1991). Expression of a mouse tyrosinase cDNA in 3T3 Swiss mouse fibroblasts. *Biochem. Biophys. Res. Commun.*, **178**, 739–45.
Winder, A. J. and Harris, H. (1991). New assays for the tyrosinase hydroxylase and dopa oxidase activities of tyrosinase. *Eur. J. Biochem.*, **198**, 317–26.
Winder, A. J. and Harris, H. (1992). Induction of tyrosinase in human melanoma cells by l-tyrosine phosphate and cytochalasin *D*. *Exp. Cell Res.*, **199**, 248–54.
Winder, A. J., Wittbjer, A., Rosengren, E., and Rorsman, H. (1993*a*). Fibroblasts expressing mouse *c* locus tyrosinase produce an authentic enzyme and synthesise phaeomelanin. *J. Cell Sci.*, **104**, 467–75.
Winder, A. J., Wittbjer, A., Rosengren, E., and Rorsman, H. (1993*b*). The mouse brown *b* locus protein has dopachrome tautomerase activity and is located in lysosomes in transfected fibroblasts. *J. Cell Sci.* **106**, 153–166.

Index

'A rule'
  DNA polymerase 54
  xeroderma variants 54
agarose microbeads for cell encapsulation 88
allelic exclusion and immortalization,
    background 159–60
allelotyping, first use 172
APC (familial adenomatous polyposis), and colon
    cancer 250–4
APC gene, somatic cell approaches 251–2
  characteristics 254–5
  chromosome deletions 251–2
  hybrid mapping panel 251–2
  irradiation hybrids 253
  other physical mapping methods 253–4
arginine synthesis 20
astrocytomas, malignant progression 223–5
ataxia telangiectasia, inhibition of replication
    defect 52
ATPase, cell membrane, mobility/immobility
    105
azatyrosine, selective suppression of transformed
    cell growth 280

bacteria, DNA polymerase bypass, in DNA
    damage 54
bacteriophage T7 RNA polymerase,
    immobilization 96–7
breast carcinoma, *neu* oncogene 266
Brownian motion, tracking 110–12
Burkitt's lymphoma, cell fusion 161–2
  tetraploid cells 161–2

c-Has-*ras* gene 174
Caenorhabditis elegans, sex determination 8
caffeine
  accumulation of DSBs 57
  effect on post-replication repair 62–3
cancer
  predisposition, genetics 216–22
    accumulation of genetic alterations
      216–17
    alteration at 'tumour locus' 217–18
    chromosome loci 218–19
    osteosarcomas 220–1
    retinoblastomas 218–20
    RFLP analysis 218
  prevailing theories 169, 170
    tumorigenesis theory 199–200
  progression, genetics 222–5
    low/high grade tumours 225
  suppression by retinoblastoma gene 227–40
  syndromes, inherited 249–50
cancer cells
  colon, DCC tumour suppressor gene 143
  discovery of oncogenes and tumour suppressor
    genes 169–73
  genetic analysis 169–82
  genetic defects 172
  hit-kinetics analysis 141
  increase in information flow 75–8
  karyotype stability, children 142
  loss-of-function mutations 215–26
carbamylphosphate synthetase (CPS-I) assay
    23–7
cell fusion, milestones 153–6
  see also heterokaryons; hybridoma;
    rodent–human cells
cell membrane, fluid mosaic and lateral mobility
    of proteins 101–14
cell proliferation, in *Drosophila*, genes in control
    183–98
cellular genes
  role in cancer 169–73
  see also oncogenes; tumour suppressor genes
chick erythrocyte heterokaryons 31–6
  competition phenomena 36–8
  inhibitory factors released 38–9
  loss of nuclear components 34
  nuclear reactivation 33–4
  nucleic acid hybridization 40–1
  nucleocytoplasmic protein exchange,
    reprogramming 40–1
  nucleolar proteins, formation 35–6
  see also heterokaryons
chloramphenicol transferase assay 72
chromatin
  activation 31–2
  dispersion 33
  in transcription
    aggregation 88
    removal 89–91
chromosomes in cancer
  deletions for APC gene 251–2
  instability 171–2
  predisposition to cancer 218–22

chromosomes in cancer (*cont.*)
  progression of cancer 222–5
  translocation, *myc* activation 263
cis-acting DNA sequence elements 9
colon cancer
  and familial adenomatous polyposis 250–2
  neoplastic progression 177
  tumour suppressor genes 177
colon cancer genes, somatic cell genetics 249–60
  APC genes 250–4
  DCC gene 256
  DCC tumour suppressor gene 143
  p53 254–5
  see also APC; colorectal carcinoma; familial adenomatous polyposis
corticosteroids, effect on fibronectin 291
cyclobutane pyrimidine dimer (CB)
  excision, selective/non-selective 63
  UV-induced DNA damage 51–2

damaged DNA *see* DNA repair, in mammalian cells
differentiation
  basis of dedifferentiation 19–22
  cell growth control 22–3
  Faofl-C2 mutant 21
  genetic analysis 19
  in heterokaryons
    continuous regulation 3–16
    trans-acting regulators 5–8
  somatic cell genetics 17–30
  see also hepatomas
DNA binding, retinoblastoma gene 231–2
DNA isolation from cancer cells 169–70
DNA polymerase, bypass of lesions in DNA damaged bacteria 54
DNA repair
  in mammalian cells 50–67
    DNA damage, normal recognition pattern 63–5
    double-strand breaks (DSBs) 54–5
    'post-replication repair (PRR) 53–4
    replication of damaged template 52–4
    responses to UV-induced DNA damage 51–4
    selective damage during post-replication repair period 54–6
DNA rotation during transcriptional elongation 94–5
DNA transfection, genetic complementation 10
dopa oxidase, new assays 292
double-strand breaks (DSBs), replication-associated 54–5
Drosophila
  cell proliferation 184

cloned tumour suppressor genes 191, 191–2
genes in control of cell proliferation and tumorigenesis 183–98
helix–loop–helix motifs 8
homeotic selector gene products, autoregulation 8
identifying overgrowth mutations 184–6
  three different approaches 185–6
lethal (2) giant larvae gene 192–5
neoplastic and hyperplastic mutants 188
tissue overgrowth 190
tumorigenesis
  genes in control 183–98
  as model 183–4
tumour phenotype 186–7
  classification 187
  hyperplastic growth 187
  neoplastic growth 186–7
tumour suppressor genes 187–92
X-chromosome mutations 190

EBO fragment (Epstein–Barr virus replicon plus hygromycin resistance gene), pcD vector 11–12
Ehrlich's tumour cells, first fusion with HVJ virus 155
encapsulation, agarose microbeads 88
endoplasmic reticulum
  form, and transition vesicles 115
  transport of secretory proteins to Golgi apparatus 115–30
epidermal growth factor receptor, first descriptions 171
eumelanin 293
excision repair enzymes, flag theory 63–5
extinction
  causes and definition 17, 18
  dissection by use of microcell hybrids 18–19
  tse-1 on mouse chromosome 11 18–19

familial adenomatous polyposis
  APC gene mapping 251–4
  and colon cancer, genetics 250
  somatic cell genetic approaches 251–2
Faofl-C2 mutant, 'differentiation' 21
fibroblast growth factor (bFGF)
  gliomas 77
    NORs of endothelial cells 76
    NORs of glioma cells 76
  and tumour angiogenesis 77–8
fibroblasts
  as heterokaryons 286–7
  lineage, extracellular matrix 289–90
fibronectin
  and corticosteroids 291
  EDIIIA and EDIIIB 290–1

HT1080 human fibrosarcoma cell line 290–1
  role in malignant behaviour 290–1
fibrosarcoma cell line 174, 175
  HT1080 human 290–1
fluorescence photobleaching and recovery (FPR) 102–7
  and spatial heterogeneity of membranes 107–9
fos oncogene, suppression of differentiation 22
fruit-fly *see Drosophila*

gene mapping
  radiation hybrid mapping 133–9
  RHMAP program 137
  selectable markers 133–4
  tumour suppressor genes 140–9
gene regulation
  identification of regulatory genes 9–10
  regulatory feedback loops 3–12
genetic complementation
  DNA transfection 10
  mutant myogenic cell lines 10–11
  recovery of complementing DNA 11–12
glioblastomas, malignant progression 223–5
gliomas
  and fibroblast growth factor (bFGF) 76–7
  loss of heterozygosity 223–4
  malignant progression, genetics 222–5
gluconeogenesis
  dedifferentiation vs back-mutation 21–2
  dexamethassone stimulation 21
glycosylation
  effects on lateral mobility, membrane proteins 106
  tunicamycin treatment 106–7
Golgi apparatus, transport of secretory proteins from endoplasmic reticulum 115–30
growth factor receptors, first descriptions 171

HaCaT cell line 174
hamster UV-5 (PRR) mutant 54–5
  SVM counterpart 57
c-Has-*ras* gene 174
HAT system
  and Ag-resistant cell line 156
  definition 156
HeLa cells
  first fusion with Ehrlich's tumour cells 155–6
  hepG2–HeLa fusions 117–22
helix–loop–helix motifs, complexes in regulatory proteins 6
hepatomas
  dedifferentiation variants 19–22
  HTC TG3{S}neo{s} hybrids 23–7
  instability of differentiation 17–30

liver-specific trf/ 18
  somatic cell genetics 17
  tse-1 on mouse chromosome 11 18–19
  urea cycle enzyme loss, and mitogen independence 23–7
heterokaryons
  characterization, stability 4
  chick erythrocyte heterokaryons 31–6
  continuous regulation of differentiation 3–16
    activation of silent genes 4–5
  cross-species 4
  hepG2–HeLa fusions 117–22
  human–mouse myoblasts 43
  instability of differentiation 17–30
  intracellular transport of secretory proteins 115–22, 122–30, 125–8
  merging of Golgi apparatuses 116
  nuclear protein sorting 31–49
  nuclear ratio 7
  X-inactivation in female mammalian cells 85–6
  see also chick erythrocyte heterokaryons
histiocytoma cells, *myc* expression 262
histone
  hyperacetylation 39
  loss from CE nuclei 38–40
  modification reactions 39
HIV, and nucleolus 70–2
HIV Tat protein, *trans*-acting transcriptional activator 74
homokaryons
  nuclear protein sorting 41–5
  tHeLa–HeLa fusions 122–3
HT1080 cell line 174, 175
HTC TG3^{neo}, carbamylphosphate synthetase (CPS-I) assay 23–7
human chromosomal fragments
  isolation method 134–5
  X-irradiation 135
human chromosome 9, position of tumour suppressor gene for melanoma 147
human chromosome 11, stability in J1 hybrid series 143–5
Human Genome Mapping Center project 138
human markers in rodent–human hybrid cells, statistical analysis 134
HVJ virus, Ehrlich's tumour cells 155
hybrid cells
  suppression of malignancy 287
  see also heterokaryons; rodent–human hybrid cells
hybrid mapping panel for APC gene 251–2
hybridoma, first descriptions 160, 161
hyperplasia, increase in information flow 75–8
hyperplastic mutants, *Drosophila* 189–92
hypoxanthine aminopterin, thymidine *see* HAT system

immunofluorescence staining, N-CAMs 6
immunoglobulins
    /*myc* translocation 265
    fusion, early isoelectric focusing 159–60
    Ig-producing cells, fusion 158–9
        first experiments 161
Indian muntjac fibroblast (SVM) line (defective in PRR) 56–65
    caffeine effects 62–3
    dominant mutant 61–2
    flag theory 64–5
    superinduction of sister chromatid exchange 57–8
information flow, nucleolus 75–8
interference contrast microscopy 110
interleukin-2 receptor, Rex extending half-life, cytoplasmic RNA 69
involucrin
    antisense methodology 288–9
    synthesis inhibition 288
irradiation hybrids and FAP chromosome deletions 253

J1 hybrid series 143–7
    chromosome 11 stability 143–5
    schema 144

keratinocytes
    /carcinoma hybrids
        involucrin 288–9
        suppression of malignancy 287–8
    fused with tumour cells 286–7
Ki-*ras* oncogene 177
kidney cells, fusion by infection 154–5
k*rev*, *see also* Rev protein
K*rev*-1 transformation suppressor gene 276–9
    biochemical activities 276–7
    comparison with Ras 276, 277, 278
    regulatory proteins for 277–8
    structural features of proteins encoded by 276, 277
K*rev*-3 transformation suppressor gene 279

LDL receptor, fluorescence visualization 110–11
lethal giant larvae gene
    Drosophila 192–6
    expression 194–5
    function 195–6
    genetics and development biology 192–3
    isolation 193
    protein 193–4
leukaemia cells, *myc* expression 262, 263, 265
liver-specific enzymes, arginine synthesis 20

liver-specific transcription factors
    differentiation of hepatomas 18
    see also hepatomas
loss of heterozygosity
    colon cancer 249–56
        tumour suppressor genes 249
    in FAP, and genetic linkage 251
    gliomas 223–4
    osteosarcomas 220–1
    in other human tumours 221–2
    retinoblastomas 172, 219–20
    RFLP analyses 172
loss-of-function mutations in human cancer 215–26
lung carcinoma, *myc* amplification 265–6
lymphoma, *myc* gene 264
    translocations 264–5
lysolecithin, first used 156

malignancy, suppression 287–8
melanin, eumelanin 293
melanomas
    black metastases 291–2
    flag theory 64
    malignant melanoma, cutaneous
        human chromosome 9 tumour suppressor gene 140
        karyotype instability 142
        position of tumour suppressor gene 142, 147
    pigmentation 293–4
    Xiphophorus, genetic control 200–3
membrane proteins
    behaviour of small numbers of molecules in a fluid bilayer 109–12
    lateral mobility 101–14
        diffusion coefficient D 102
        fluorecence photobleaching and recovery (FPR) 102–7
        glycosylation effects 106
        presence of impermeable domains 108–9
        spatial heterogeneity, and FPR 107–9
MHC Class II molecules, diffusion coefficient changes 106
mice *see* rodent(s)
microtubules, hepG2 cells 123–5
mitogen independence, and urea cycle enzyme loss in hepatomas 23–7
monoclonal antibodies
    early work 153–61
    history 156–62
    technique, introduction 161
mRNA, and nucleolus 69
muscle cell differentiation (human) 4–12
    mutant myogenic cell lines 10–11
    N-CAMs 6, 7
    novel regulators and feedback mechanisms 8–10

myc oncogene   174–5
  biological background   261–2
  in oncogene complementation   262–3
  roles   261–9
    activation by chromosomal translocation   263–5
    amplification   265–6
myeloma cells
  cloning   158, 162–3
    MOPC 21 line   162–3
  fusion   156, 157
    mouse spleen/mouse muscle cells   162
    isoelectric focusing   162–3
  milestones of biology   156–62
myoblasts, homo and heterokaryons   41–5
MyoD, DNA-binding   6–8
myogenesis, schema   42
myogenic cell lines, as test cells for genetic complementation   10–11
myotubes, as homokaryons   43–5

N-CAMs
  immunofluorescence staining   6
  muscle cell differentiation   6, 7
neu oncogene in breast carcinoma   266
neuroblastoma cells, *myc* expression   262
nuclear transplantation, experiments   3–4
nucleocytoplasmic protein transport   31–41
  role of nucleolus   68–83
  schema   36
  transport and sorting mechanisms   39–40
  see also chick erythrocyte heterokaryons
nucleoid cages
  collapse and DNA release   86–7
  site of transcription   88
nucleolus
  cellular mRNA   69
  and HIV   70–2
  information flow   75–8
  nucleolar organizer regions   78
  phosphorylation   72–3
  position in nucleus   74
  retroviral mRNA   68–9
  targeting signals (NOS)   39
    function   69
    and Rex   69
  transcriptional activators   74–5
  transfer of information   75–8
nucleoskeleton
  associated active RNA polymerases   89–92
  attachment of RNA polymerases   92
  concept   84
  role during transcription   84–100

ocular tumours, hereditary, *Xiphophorus*   203–4
oligonucleotide probes, design   9

oncogenes
  classification of other genes   262
  complementation, role of *myc*   262–3
  definition   169
  discovery   169–73
  dominantly acting   171, 173–4
    myc and *ras*   174–5, 261–9
  functional evidence   173–7
  functional significance in human cancer   174, 178–9
  identification   170–1
  temperature-sensitive, suppression of differentiation   22
  see also specific genes; tumour suppressor genes
ornithine, arginine synthesis   20

p53 gene and protein
  deletions and mutations   175–6
  expression in human tumours   243
  functional evidence   176
  functional regulation   245
  levels, regulation   243–5
  low/high grade tumours   225
  pathway and model   244–5
  properties   241–2
  tumour suppressing qualities   175, 177
pcD vector, incorporation of EBO fragment (Epstein–Barr virus replicon plus hygromycin resistance gene)   11–12
phaeomelanin   293
phage lambda repressor, autoregulation   8
phosphorylation
  nucleolus   72–3
  retinoblastoma protein   230–1
pigmentation, and melanoma   293–4
platelet-derived growth factor and receptor, first descriptions   171
polyethylene glycol, fusion of protoplasts   156
'post-replication repair' (PRR)
  caffeine effects   62–3
  flag theory   63–5
  Indian muntjac fibroblast (SVM) line (defective in PRR)   56–65
  PRR correcting gene, isolation   58–60
  transfectants, resistance to 6-thioguanine   61
protein kinase c
  activator TPA   73
  inhibitor H-7, phosphorylation   72
pulsed field gel electrophoresis, APC gene   253
pyrimidine products, UV-induced DNA damage   51–2

R gene *see Xiphophorus*
radiation hybrid mapping   133–9

ras genes, mutated 270–85
  drugs suppressing transformation 280
  GTP-binding and hydrolysing activity 270
  perspectives 280–1
  revertants
    chemical mutagens 272–4
    DNA transfection 274–6
    'flat' revertants 271–2
  structures 270
  transformation suppressor genes 276–80
  see also oncogenes
ras oncogene, suppression of differentiation 22
regulatory proteins, complexes, helix–loop–helix motifs 6
replication
  damaged DNA, mammalian cells 50–67
  'post-replication repair' (PRR) 53–4
    BND-cellulose chromatographic analysis 56
    Indian muntjac fibroblast (SVM) line (defective in PRR) 56–65
retinoblastoma
  functional evidence 175
  inherited genes 249–50
  loss of heterozygosity 172–3, 219–20
  protein, biochemical properties 231–4
  rate-limiting events, loss of tumour suppressor gene 141
retinoblastoma gene
  binding to SV40 T antigen 231–2
  cancer suppression by 227–40
    biochemical properties of RB protein 231–4
    methods used 227–8
    mouse model 229
    RB protein and cell cycle regulation 229–31
    tumour suppressor genes 175
  control of cell cycle 175
  DNA binding 231–2
  kinase 231
  RFLP analyses 172–3
  tumorigenesis 218
retinoic acid, mitogen 23–7
retroviral mRNA, and nucleolus 68–9
Rev protein
  compatibility with Rex 70
  nonfunctional mutant pH2drev 71
  nucleolar targeting signal (NOS) 71
    interference of dRev 71
  phosphorylation 72
  properties 71
  see also k*rev*-1
Rex protein
  deficient phosphorylation 72–3
  nucleolar location 75
  nucleolar targeting signal (NOS) function 69
  regulation of mRNA splicing 69
RNA polymerases
  attachment to nucleoskeleton 92

immobilization, rotation of DNA 94–5
'Miller spreads' 86
nucleoid cages 86–8
resisting electroelution 89–92
solubility 86
'textbook' model of transcription 86
topology in transcription 92–6
transcription by 84–100
rodent(s)
  transgenic mice 174
  'knock-out' 179
rodent cells
  hybrids, mouse spleen/mouse muscle cells 162
  lethal-albino mutants, liver differentiation failure 17
  mouse chromosome 4, position of tumour suppressor gene for melanoma 147
  transformed, flag theory 64
rodent–human hybrid cells
  first description of cloning 159–60
  in gene mapping 133–9
  myeloma cells 159–60
rsp-1, transformation suppressor gene 279–80

secretory proteins, intracellular transport within homo and heterokaryons 115–30
signal transducers, c-Jun 8
sister chromatid exchange (SCE), superinduction 57–8
small cell lung carcinoma, amplification of *myc* 265–6
snRNP antigens, expression in CE nuclei 37
somatic cell genetics
  approaches to APC gene 251–4
  background 162–3
  search for colon cancer genes 249–60
spleen cells, mouse, fusion with mouse myeloma cells 162
statistical analysis, human markers in rodent–human hybrids 134
SV40 T antigen, binding to retinoblastoma gene 231–2
SVM (Indian muntjac fibroblast line, defective in PRR 56–65

Tat protein, *trans*-acting transcriptional activator 74–5
teratocarcinoma cells, *myc* expression 262
12-*O*-tetradecanoylphorbol-13-acetate (TPA), activator of PKC 73
6-thioguanine, resistance, post-replication repair transfectants 61
thyroidal tumours, hereditary, *Xiphophorus* 203–4
TMR, 'tumour' locus 217–18
trans-acting regulators 5–8
transcription
  autoregulation by transcription factors 8

by immobile RNA polymerases 84–100
fluorescence microscopy 92
model, generalized 98
'textbook' model 86
topology 92–6
transcriptional activators, nucleolus 74–5
transcriptional elongation, models 93–6
transcriptional regulators
   novel regulators and feedback mechanisms 8–10
   synergism 8
transformation suppressor genes 276–80
transition vesicles, ER 115
translocations, *myc* activation 263–5
Trebonema vulgare, transition vesicle formation 128
tse-1 locus, mouse chromosome 11 18–19
Tu gene *see Xiphophorus*
tumorigenesis
   alteration at 'tumour locus' 217–18
   cancer-causing genes 199–200
   Drosophila, genes in control 183–98
   early research 171–2
   genetics 216–22
   retinoblastoma 218
   Xiphophorus 199–214
tumour angiogenesis, and fibroblast growth factor (bFGF) 77–8
tumour suppressor genes
   candidate human 172–3
   cloning 140–9
   colorectal carcinoma 177
   defined 140, 169
   discovery 169–73
   Drosophila 187–92
      cloned 191–2
   and functional evidence for oncogenes 173–7
   functional significance in human cancer 178–9
   identification 171
      regulatory genes 10
   location and isolation 141–3
   promotion of differentiation 23
   recessiveness of malignancy 4
   RFLP/LOH analysis 173
   transcription factors 173
tunicamycin, glycosylation 106–7
tyrosinase
   in 3T3 Swiss mouse fibroblasts 292–4
   activity 292–3
tyrosine hydroxylase, new assays 292

urea cycle
   carbamylphosphate synthetase 23–7
   enzyme loss in hepatoma 23–7
UV-induced DNA damage 50–6
   pyrimidine products 51–2

Wilms' tumour
   cell line J1-11 145
   children 142–3
   deletions 146–7
   homozygous deletion 143
   rate-limiting events, loss of tumour suppressor gene 141
   suppressor gene 140–7
   WT33 (WT1) gene, zinc finger amino acid sequence 146

X-chromosome mutations, *Drosophila* 190
X-inactivation in female mammalian cells 85–6
X-irradiation
   determination of dose 137
   rodent–human hybrids 135
Xenopus oocytes, Tat protein 75
xeroderma
   defective excision repair of DNA damage 50, 51
   excision repair-defective cells 54
   flag theory 64
   hamster counterpart UV-5 mutant 54–5
   post-replication repair-defective mutants 53
   variants, 'A rule' 54
Xiphophorus tumours
   genetics 199–214
      molecular biology 204–11
      other oncogenes possibly involved in neoplastic phenotype 209
      reverse genetic approaches towards isolation of *Tu* and *R* 204–5
      theory 199–200
   hereditary ocular and thyroidal tumours 203–4
   perspective 210–11
   spontaneous melanoma 200–3
      X*mrk* oncogene 210–11
X*mrk* oncogene
   cloning and characterization 205–6
   duplication/recombination model for generation 210
   new upstream region acquired by 209–10
   proto-oncogene and oncogenic copies 206–7
   transcriptional activation 207
X*mrk* protein
   autophosphorylation 207–8
   cellular substrates 208
   transforming ability 208–9

yeast, gene product MIG1, zinc finger amino acid sequence of WT1 146
yeast artificial chromosomes 254